Cyclometalation Reactions

Iwao Omae

Cyclometalation Reactions

Five-Membered Ring Products as Universal Reagents

 Springer

Iwao Omae
Omae Research Laboratories
Sayama, Japan

ISBN 978-4-431-56146-0   ISBN 978-4-431-54604-7 (eBook)
DOI 10.1007/978-4-431-54604-7
Springer Tokyo Heidelberg New York Dordrecht London

Printed on acid-free paper

Springer is part of Springer Science+Business Media (www.springer.com)

Preface

Our laboratory at Osaka University began studies on catalysis for direct reactions of alkyl halides with tinfoil in 1959. The studies revealed that alcohols and amines show high catalytic activity in these reactions. This catalytic activity is considered to be caused by the coordination of lone electron pair of hetero atoms such as oxygen and nitrogen to a tin atom. That is, metal activation of the tin atom is caused by these compounds. Two or more reaction products, R_nSnX_{4-n} (R: alkyl, X: halogen atom; $n = 0$, 1, 2, 3, or 4), are always obtained.

In 1965, we began studies on direct reactions of compounds containing functional groups such as haloesters, haloketones, and haloamides with tinfoil in the presence of the above catalysts. Reactions with compounds containing functional groups such as carbonyl and amino groups proceed surprisingly more quickly than those with the above alkyl halides, and a single five-membered ring product is easily isolated. In these reactions, so-called cyclometalation reactions proceed easily and form organometallic intramolecular-coordination five-membered ring compounds.

Two factors are thought to explain why these organometallic intramolecular-coordination five-membered ring compounds are very easily synthesized through cyclometalation reactions: The first is metal activation by the coordination of the lone electron pair to the metal atom, and the second is a chelate effect caused by the formation of a five-membered ring due to bonding between the metal atom and a carbon atom at the γ-position to the hetero atom.

In cyclometalation reactions with metal compounds instead of the metal atoms discussed above, ligands such as hetero atom groups (e.g., bipyridines, phosphines, and carboxylates), unsaturated groups (e.g., aryl, allyl, Cp and Cp*), carbonyl groups, halogen atoms (F, Cl, Br, Cl), etc., also act as metal activators.

These activities are understood to reflect the catalytic actions of metal phosphines, e.g., Wilkinson hydrogenation catalyst, $(PPh_3)_3RhCl$, metal carbonyls , e.g., hydrogenation catalysts, $Ni(CO)_5$), and cyclopetadienyl metal compounds, e.g., olefin polymerization catalysts, Cp_2ZrCl_2.

Cyclometalation reactions yield two types of compounds: σ-coordination compounds and π-coordination compounds. Organometallic intramolecular-coordination five-membered ring compounds belong to the former, very stable σ-coordination compound type and are very easily prepared regioselectively. Hence, an immense number of articles on organometallic intramolecular-coordination five-membered ring compounds have been published since the 1960s. These compounds are utilized not only for synthetic applications, but also in other fields, such as catalysts, organic electronic devices, pharmaceuticals, dye-sensitized solar cells, carbon dioxide utilizations, and sensors. Catalysts for chiral reactions, metathesis reactions and cross-coupling reactions, organic light-emitting diodes for TV 4K screens, solar cells, carbon dioxide utilizations, and sensors are expected to generate especially great demand.

Sayama, Japan   Iwao Omae

Acknowledgments

The author wishes to express his sincere appreciation to Dr. Sumio Chubachi for reading the full manuscript, enhancing its accuracy and clarity, and providing much valuable constructive criticism. He also expresses his sincere gratitude to Dr. Michael A. Smith for making substantial editorial corrections and improvements.

Contents

Chapter 1
Introduction

Abstract Cyclometalation reactions are the most actively investigated reactions in the world today among the many synthetic organic reactions, and the studies have resulted in the publication of a large number of articles, since organometallic intramolecular-coordination five-membered ring compounds are produced regioselectively and extremely easily. These five-membered ring products have applications in many fields, such as catalysts for a wide range of reactions, including chiral reactions, metathesis reactions, and cross-coupling reactions, organic electronic devices, pharmaceuticals, dye-sensitized solar cells, carbon dioxide utilizations, and sensors.

Keywords Catalyst • Chelate effect • Cyclometalation • Five-membered ring • Intramolecular coordination

Cyclometalation reactions [1–56] are the most actively investigated reactions among the synthetic organic reactions in the world today, and studies concerning them have resulted in the publication of a large number of articles, since organometallic intramolecular-coordination five-membered ring compounds are produced regioselectively and extremely easily.

For purposes of comparison, the numbers of articles related to the most representative recent synthetic organic reactions, i.e., chiral reactions [24, 28, 31, 57–67], metathesis reactions [68–79], and cross-coupling reactions [24, 29, 30, 80–90], are shown. Information retrieval conducted with the Chemical Abstract Database SciFinder System on January 26, 2012 (total data 31,409,203) produced the following results for the three types of reaction:

Chiral reaction	38,494
Metathesis	28,759
Cross-coupling	34,410

I. Omae, *Cyclometalation Reactions: Five-Membered Ring Products as Universal Reagents*, DOI 10.1007/978-4-431-54604-7_1, © Springer Japan 2014

Similar search results for cyclometalation, including orthometalation and cyclopalladation and orthopalladation (i.e., keywords: cyclometalation or ortho-metalation or cyclopalladation or orthopalladation), showed 240,422 studies. These results indicate that the number of articles related to cyclometalation reactions is more than six times that concerning any of the three recent synthetic organic reactions shown above.

Just as this Nobel Prize winning research may be considered to provide a basis for synthetic organic reactions, cyclometalation reactions may also be considered as a basis for synthetic organic reactions. Many organometallic intramolecular-coordination five-membered ring compounds are used, moreover, as catalysts for synthetic organic reactions.

Among the chiral reactions reported, for example, BINAP, shown below, is an important chiral substrate of the chiral reactions, and BINAP is also prepared in the presence of organometallic intramolecular-coordination five-membered ring compounds **1.1** as a catalyst, as shown in Scheme 1.1 [57].

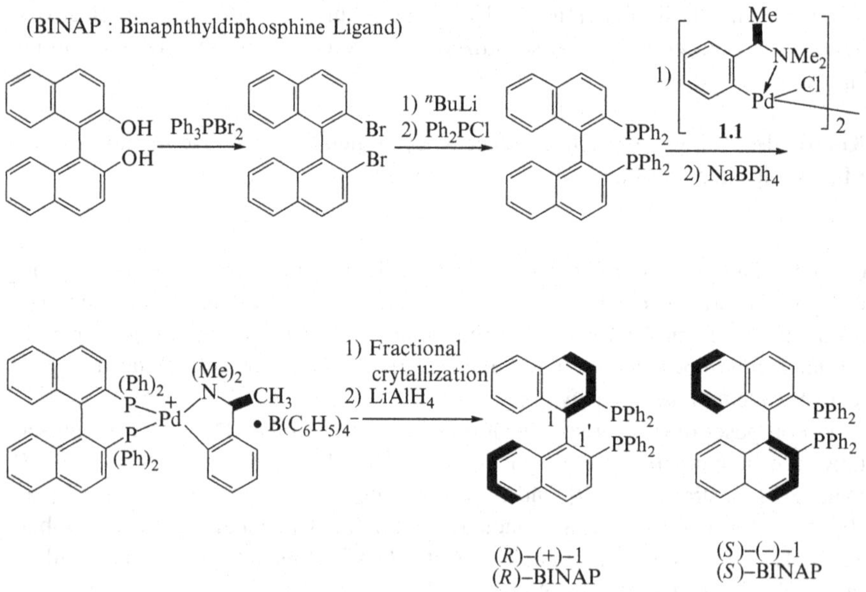

Scheme 1.1 Preparation of BINAP [57]

Fig. 1.1 Structures of
catalysts for metathesis

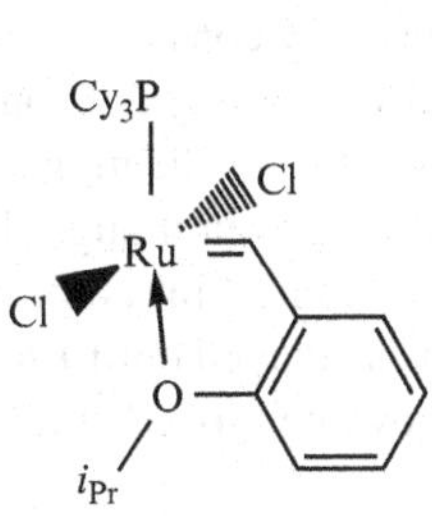

Hoveyda-Grubbs
1st generation

1.2

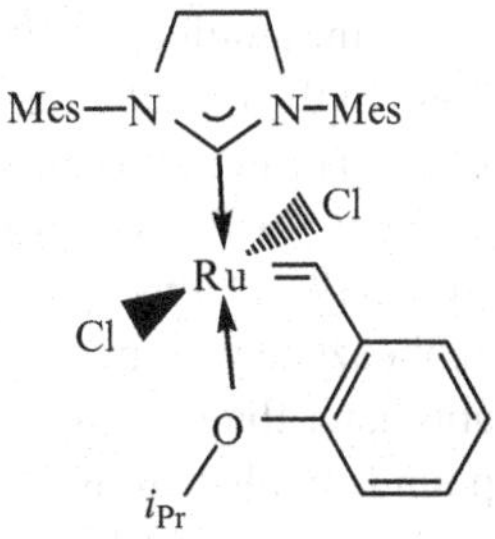

Hoveyda-Grubbs
2nd generation

1.3

Fig. 1.2 Structure of a
catalyst for cross-coupling
reaction

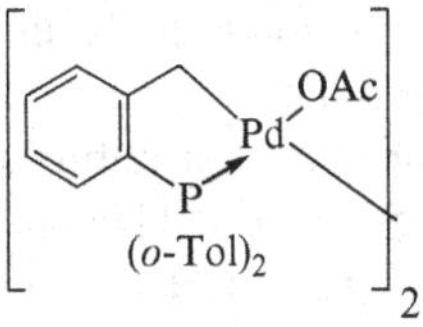

1.4

In the articles on metathesis reactions, ruthenium compounds of organometallic intramolecular-coordination five-membered ring compounds are used for what are generally referred to as Hoveyda–Grubbs first-generation **1.2** and second-generation **1.3** catalysts (Fig. 1.1) [68, 69].

In the articles on cross-coupling reactions, the palladacycles **1.4** of tri-*o*-tolylphosphines are used as catalysts for various kinds of cross-coupling reactions (Fig. 1.2) [84–86, 90, 91].

The representative catalysts for the three most recent synthetic organic reactions are shown as compounds **1.1–1.4**, and further examples of such catalysts are also shown in Figs. 8.1–8.3, 8.5 and 8.6.

This monograph discusses the six items enumerated below:

The first item concerns the first article verifying the intramolecular-coordination bond in cyclometalation reactions [92]. Published in 1966, this article reported on cyclometalation reactions with main group metal compounds, which are organotin compounds. The intramolecular-coordination bond is verified with IR spectrum studies. Unfortunately, this study was not widely known, because it was published in a Japanese academic journal.

Most of the reports on cyclometalation reactions have focused on transition metal compounds. The first widely known article was thus a 1963 article on the reaction of an azobenzene with a nickel compound [93]. However, the verification of the intramolecular bond was incorrectly considered to be an N=N π-coordination bond to the nickel atom. For this reason, in an article on the reaction of azobenzene with Pd or Pt atoms published in 1965, the verification of nitrogen lone electron pair to the Pd or Pt atom was not mentioned, even though the chemical equations were shown [94]. Consequently, the verification of intramolecular-coordination bonds of azobenzene with transition metals was delayed until 1969 [95], which was considerably later than our report on our abovementioned study of main group metal compounds published in 1966.

The second item for discussion concerns organometallic intramolecular-coordination compounds. These compounds have two types of bonds: a π-coordination bond in which the intramolecular-coordination bond is an unsaturated π-coordination bond and a σ-coordination bond formed by lone electron pair of a hetero atom to a metal atom. The former compounds are π-coordination compounds, and the latter are σ-coordination compounds. In these two types of compounds, an overwhelming majority of the latter organometallic intramolecular-coordination five-membered ring compounds are synthesized by cyclometalation reactions, because the latter products are very easily prepared regioselectively. The author has compiled many articles and published many reviews [41–55, 91, 96–99] and three monographs [56, 100, 101] concerning them since 1972.

The third item concerns the characteristics of cyclometalation reactions with respect to organometallic intramolecular-coordination five-membered ring compounds. These reactions show very high reactivities, and the compounds in question are extremely easily synthesized. To date, 69 kinds of metal atoms and many substrates have been studied.

The fourth item concerns the reasons why organometallic intramolecular-coordination five-membered ring compounds are extremely easily synthesized through cyclometalation reactions. There are three reasons: The first is metal activation by the coordination of lone electron pair of a hetero atom such as N, P, O, or S to the metal atom; the second is the chelate effect resulting from the formation of a five-membered ring; and the third is also metal activation of the ligands bonding with the metal atom, such as hetero atom groups (e.g., bipyridines, phosphines, and carboxylates), unsaturated groups (e.g., aryl, allyl, Cp, and Cp*), carbonyl groups, and halogen atoms (F, Cl, Br,Cl).

The fifth item concerns agostic interactions, C–H activation, C–X activation, C–H functionalization, chelation-assisted reactions, cross-coupling reactions, etc., which are indicated as titles. The reactions indicated by these titles are mostly related to cyclometalation reactions. The reaction mechanisms of these reactions include metal activation by the coordination of a hetero atom to the central metal atom and the chelate effect of the formation of a five-membered ring.

The sixth item concerns applications for both cyclometalation reactions and their products, which are organometallic intramolecular-coordination five-membered ring compounds. The applications for cyclometalation reactions are syntheses of organometallic intramolecular-coordination five-membered ring compounds, and these products and their reaction intermediates are used to obtain their derivatives. Pincer products are also utilized for synthesis of their derivatives.

Applications other than synthetic applications of organometallic intramolecular-coordination five-membered ring compounds include catalysts, organic electronic devices, pharmaceuticals, dye-sensitized solar cells, carbon dioxide utilizations, and sensors.

In these studies on cyclometalation reactions for producing organometallic intramolecular-coordination five-membered ring compounds, the author et al. first reported on the verification of intramolecular coordination bonds in organotin carbonyl compounds (see Eq. (2.4)) in 1966, as mentioned above. He then published a review entitled "Organometallic Intramolecular-Coordination Compounds Containing Carbonyl Groups" in 1972 concerning direct reactions of halo-carbonyl compounds with tinfoil as the first in a series of reviews on silicon, germanium, tin, and lead compounds [41], which covered carbonyl oxygen atoms in oxygen-coordinating atoms and which was also the first review in the "Organometallic Intramolecular-Coordination Compounds" series. The second review concerned nitrogen-coordinating atoms. Entitled "Organometallic Intramolecular-coordination Compounds Containing a Nitrogen Donor Ligand," it was published in *Chemical Reviews* in 1979 [42].

The author has also published many reports on coordinating atoms, such as nitrogen [38, 44], phosphorus [45], arsine [49], oxygen [44, 50], and sulfur [43] atoms, and on coordinating groups, such as cyclopentadienyl [46], π-olefin [47], diolefin [51], and π-aryl [52], as donor groups.

In 1986, he compiled these reports and published a monograph entitled "Organometallic Intramolecular-coordination Compounds" with Elsevier Publishers [56]. In 1989, he published another monograph entitled "Organotin Chemistry [100]," which included organotin intramolecular-coordination five-membered ring compounds, also with Elsevier Publishers. In 1998, moreover, he published another monograph entitled "Applications of Organometallic Compounds [101]" regarding the 20 representative kinds of organometallic compounds, including many kinds of organometallic intramolecular-coordination five-membered ring compounds, with John and Wiley Publishers.

Because organometallic intramolecular-coordination five-membered ring compounds are now extremely easily synthesized by cyclometalation reactions, cyclometalation reactions are used as one of the most easily applicable organometallic synthetic methods. Organometallic intramolecular-coordination five-membered ring compounds, in turn, are used to produce many fine chemicals for use as catalysts, etc. The author has published two reports on the applications of organometallic intramolecular-coordination five-membered ring compounds. These include "Intramolecular Five-membered Ring Compounds and Their Applications [91]" and "Three Types of Reaction with Intramolecular Five-membered Ring Compounds in Organic Synthesis [97]," published in 2004 and 2007, respectively.

In 2010, the author published a report entitled "Carbonyl Group-containing Organometallic Intramolecular-coordination Five-membered Ring Compounds [98]," and in 2011, he published another report entitled "Agostic Bonds in Cyclometalation" clarifying the reaction mechanism of cyclometalation reactions [99].

At the same time, as described above, an enormous number of articles on cyclo-metalation reactions have been published. Many reviews [1–20, 36–40] and three books [21–35] by researchers other than the author have also been published. The first review having cyclometalation as the title word, "Some Studies of the Cyclopalladation Reaction [2]," was published by Trofimenko in 1973. This was followed by a review entitled "Cyclometallated Compounds" by Dehand and Pfeffer [3] in 1976 and by a review entitled "Cyclometallation Reactions" by Bruce [4] in 1977.

This research field is generally referred to as "Cyclometalation Reactions," although the author had already reported "Organometallic Intramolecular-Coordination Compounds Containing Carbonyl Groups" in 1972 [41].

These cyclometalation reactions are regarded as involving almost all transition metal compounds, especially palladium compounds [2, 6, 8, 10, 16–18, 20, 26–34, 60, 85, 86, 89, 90, 102–111]. Many reviews on pincer compounds have also reported on transition metal compounds [21–25, 65, 102, 112–116]. Few reviews have reported on both transition metal and main group metal compounds [9, 13, 15] or on main group metal compounds alone [14], except the reviews published by the author.

This monograph reports on these six items, including recent articles.

References

1. Parshall GW (1970) Acc Chem Res 3:139
2. Trofimenko S (1973) Inorg Chem 12:1215
3. Dehand J, Pfeffer M (1976) Coord Chem Rev 18:327
4. Bruce MI (1977) Angew Chem Int Ed 16:73
5. Constable EC (1984) Polyhedron 3:1037
6. Ryabov AD (1985) Synthesis 223
7. Newkome GR, Puckett WE, Gupta VK, Kiefer GI (1986) Chem Rev 86:451
8. Dunina VV, Potatov VM (1988) Russ Chem Rev 57:250
9. van Koten G (1989) Pure Appl Chem 61:1681
10. Pfeffer M (1990) Recl Trav Chim Pay B 109:567
11. Ryabov AD (1990) Chem Rev 90:403
12. Lyndsay M, Nicholson BK (1994) Adv Met Org Chem 3:1
13. Kiplinger JL, Richmond TG, Osterberg CE (1994) Chem Rev 94:373
14. Gruter G-JM, van Klink GPM, Akkerman OS, Bickelhaupt F (1995) Chem Rev 95:2405
15. Rietveld MHP, Grove DM, van Koten G (1997) New J Chem 21:751
16. Dunina VV, Gorunova ON (2004) Russ Chem Rev 73:309
17. Dunina VV, Gorunova ON (2005) Russ Chem Rev 74:871
18. Dupont J, Consorti CS, Spencer J (2005) Chem Rev 105:2527
19. Mohra F, Privér SH, Bhargava SK, Bennett MA (2006) Coord Chem Rev 250:1851
20. Vicente J, Saura-Llamas I (2007) Comment Inorg Chem 28:39
21. Morales-Morales D, Jensen CM (eds) (2007) The chemistry of pincer compounds. Elsevier, Amsterdam
22. Richards CJ, Fossey JS (2007) Chiral pincer complexes and their application to asymmetric synthesis. In: Morales-Morales D, Jensen CM (eds) The chemistry of pincer compounds. Elsevier, Amsterdam, p 45

23. Panzner MJ, Tessier CA, Youngs WJ (2007) Pincer complexes of N-heterocyclic carbenes. Potential uses as pharmaceuticals. In: Morales-Morales D, Jensen CM (eds) The chemistry of pincer compounds. Elsevier, Amsterdam, p 139

24. Sommer WJ, Jones CW, Weck M (2007) Stability of supported pincer complex-based catalysts in Heck catalysis. In: Morales-Morales D, Jensen CM (eds) The chemistry of pincer compounds. Elsevier, Amsterdam, p 385

25. Chase PA, van Koten G (2007) Dendrimers incorporating metallopincer functionalities: synthesis and applications. In: Morales-Morales D, Jensen CM (eds) The chemistry of pincer compounds. Elsevier, Amsterdam, p 399

26. Dupont J, Pfeffer M (eds) (2008) Palladacycles. Synthesis, characterization and applications. Wiley-VCH, Weinheim

27. Albrecht M (2008) C–H bond activation. In: Dupont J, Pfeffer M (eds) Palladacycles. Synthesis, characterization and applications. Wiley-VCH, Weinheim, p 13

28. Djukic J-P (2008) Cyclopalladated compounds As resolving agents for racemic mixtures. In: Dupont J, Pfeffer M (eds) Palladacycles. Synthesis, characterization and applications. Wiley-VCH, Weinheim, p 123

29. Nájera C, Alonso DA (2008) Application of cyclopalladated compounds As catalysts for Heck and Sonogashira reactions. In: Dupont J, Pfeffer M (eds) Palladacycles. Synthesis, characterization and applications. Wiley-VCH, Weinheim, p 155

30. Bedford RB (2008) Palladacyclic pre-catalysts for Suzuki coupling, Buchwald-Hartwig amination and related reactions. In: Dupont J, Pfeffer M (eds) Palladacycles. Synthesis, characterization and applications. Wiley-VCH, Weinheim, p 209

31. Spencer J (2008) Other uses of palladacycles in synthesis. In: Dupont J, Pfeffer M (eds) Palladacycles. Synthesis, characterization and applications. Wiley-VCH, Weinheim, p 227

32. Ryabov AD (2008) Cyclopalladated compounds as enzyme prototypes and anticancer drugs. In: Dupont J, Pfeffer M (eds) Palladacycles. Synthesis, characterization and applications. Wiley-VCH, Weinheim, p 307

33. Pijnenburg NJM, Korstanje TJ, van Koten G, Gebbink RJMK (2008) Palladcycles on dendrimers and star-shaped molecules. In: Dupont J, Pfeffer M (eds) Palladacycles. Synthesis, characterization and applications. Wiley-VCH, Weinheim, p 361

34. Donnio B, Bruce DW (2008) Liquid crystalline ortho-palladated complexes. In: Dupont J, Pfeffer M (eds) Palladacycles. Synthesis, characterization and applications. Wiley-VCH, Weinheim, p 239

35. Chatani N (ed) (2007) Topics in organometallic chemistry, directed metallation, vol 24. Springer, Heidelberg

36. Djukic J-P, Sortais J-B, Barloy L, Pfeffer M (2009) Eur J Inorg Chem 817

37. Albrecht M (2010) Chem Rev 110:576

38. Selander N, Szabó KJ (2011) Chem Rev 111:2048

39. Lyons TW, Sanford MS (2010) Chem Rev 110:1147

40. Aeockiam PB, Bruneau C, Dixneuf PH (2012) Chem Rev 112:5879

41. Omae I (1972) Rev Silicon Germanium Tin Lead 1:59

42. Omae I (1979) Chem Rev 79:287

43. Omae I (1979) Coord Chem Rev 28:97

44. Omae I (1979) Kagaku No Ryoiki 33:767

45. Omae I (1980) Coord Chem Rev 32:235

46. Omae I (1982) Coord Chem Rev 42:31

47. Omae I (1982) Angew Chem Int Ed 21:889; [Angew Chem (1982) 94:902]

48. Omae I (1982) Yuki Gosei Kagaku Kyokaishi 40:147

49. Omae I (1982) Coord Chem Rev 42:245

50. Omae I (1982) Kagaku Kogyo 33:989

51. Omae I (1983) Coord Chem Rev 51:1

52. Omae I (1984) Coord Chem Rev 53:261

53. Omae I (1988) Coord Chem Rev 83:137

54. Omae I (1998) Kagaku Kogyo 49:303

55. Omae I, Aoki A, Horiguchi K (1998) Kagaku Kogyo 49:469
56. Omae I (1986) Organometallic intramolecular-coordination compounds, J Organomet Chem
    Library 18. Elsevier, Amsterdam
57. Noyori R, Takaya H (2001) Kagaku 56:32
58. Cheow YL, Pullarkat SA, Li Y, Leung P-H (2012) J Organomet Chem 696:4215
59. Dunina VV (2011) Curr Org Chem 15:3415
60. Dunina VV, Gorunova ON, Zykov PA, Kochetkov KA (2011) Rus Chem Rev 80:51
61. Ohshima T, Kawabata T, Takeuchi Y, Kakinuma T, Iwasaki T, Yonezawa T, Murakami H,
    Nishiyama H, Mashima K (2011) Angew Chem Int Ed 50:6296
62. Nishiyama H, Ito J (2010) Chem Commun 46:203
63. Fischer DF, Barakat A, Xin Z, Weiss ME, Peters R (2009) Chem Eur J 15:8722
64. Djukic J-P, Hijazi A, Flack HD, Bernardinelli G (2008) Chem Soc Rev 37:406
65. Nishiyama H (2007) Chem Soc Rev 36:1133
66. Yu J-Q, Giri R, Chen X (2006) Org Biomol Chem 4:4041
67. Albert J, Granell J, Muller G (2006) J Organomet Chem 691:2101
68. Volugioukalakis GC, Grubbs RH (2010) Chem Rev 110:1746
69. Samoiłowicz C, Bieniek M, Grela K (2009) Chem Rev 109:3708
70. Schrodi Y, Pederson RL (2007) Adv Synth Catal 349:1
71. Grubbs RH (2004) Tetrahedron 60:7117
72. Grubbs RH (ed) (2003) Handbook of metathesis, vol 1–3. Wiley-VCH, Weinheim
73. Hryniewicka A, Kozłowska A, Witkowski S (2012) J Organomet Chem 701:87
74. Hérisson J-I, Chauvin Y (1971) Macromol Chem 141:161
75. Stewart IC, Ung T, Pletnev AA, Berlin JM, Grubbs RH, Schrodi Y (2007) Org Lett 9:1589
76. Murelli RP, Snapper ML (2007) Org Lett 9:1749
77. Ritter T, Hejl A, Wenzel AG, Funk TW, Grubbs RH (2006) Organometallics 25:5740
78. Benitez D, Goddard WA III (2005) J Am Chem Soc 127:12218
79. Garber SB, Kingsbury JS, Gray BL, Hoveyda AH (2000) J Am Chem Soc 122:8168
80. Ohshima K (2010) Kagaku 65:Dec 12
81. Kagaku's editors (2010) Kagaku 65:Dec 18
82. Yamaguchi S, Tamao K (2002) Kagaku To Kogyo 55:550
83. Su Y, Jiao N (2011) Curr Org Chem 15:3362
84. Alonso DA, Nájera C (2010) Chem Soc Rev 39:2891
85. Zapf A, Beller M (2005) Chem Commun 431
86. Beletskaya IP, Cheprakov AV (2004) J Organomet Chem 689:4055
87. Bedford RB (2003) Chem Commun 1787
88. Singleton JT (2003) Tetrahedron 59:1837
89. Littke AF, Fu GC (2002) Angew Chem Int Ed 41:4176
90. Herrmann WA, Böhm VPW, Reisenger C-P (1999) J Organomet Chem 576:23
91. Omae I (2004) Coord Chem Rev 248:995
92. Matsuda S, Kikkawa S, Omae I (1966) Kogyo Kagaku Zasshi 69:646
93. Kleiman JP, Dubeck M (1963) J Am Chem Soc 85:1544
94. Cope AC, Siekman RW (1965) J Am Chem Soc 87:3272
95. Balch AL, Petridis D (1969) Inorg Chem 8:2247
96. Omae I (2004) Phosphorus Sulfur 179:891
97. Omae I (2007) J Organomet Chem 692:2608
98. Omae I (2010) Appl Organomet Chem 24:347
99. Omae I (2011) J Organomet Chem 696:1128
100. Omae I (1989) Organotin chemistry, J Organomet Chem Library 21. Elsevier, Amsterdam
101. Omae I (1998) Applications of organometallic compounds. Wiley, New York
102. Moreno I, SanMartin R, Inés B, Churruca F, Domínguez E (2010) Inorg Chim Acta 363:1903
103. Sun C-L, Li B-J, Shi Z-J (2010) Chem Commun 46:677
104. Xu L-M, Li B-J, Yang Z, Shi Z-J (2010) Chem Soc Rev 39:712
105. Muñiz K (2009) Angew Chem Int Ed 48:9412

106. Deprez NR, Sanford MSD (2007) Inorg Chem 46:1924
107. Ghedini M, Aiello I, Crispini A, Golemme A, Deda ML, Pucci D (2006) Coord Chem Rev 250:1373
108. Dupont J, Pfeffer M (2001) J Spencer Eur J Inorg Chem 1917
109. Wild SB (1997) Coord Chem Rev 166:291
110. Chen X, Engle KM, Wang D-H, Yu J-Q (2009) Angew Chem Int Ed 48:5094
111. Baranoff E, Yum J-H, Graetzel M, Nazeeruddin MK (2009) J Organomet Chem 694:2661
112. Niu J-L, Hao X-Q, Gong J-F, Song M-P (2011) Dalton Trans 40:5135
113. Selander N, Szabó KJ (2009) Dalton Trans 6267
114. Leis W, Mayera HA, Kaska WC (2008) Coord Chem Rev 252:1787
115. van der Boom ME, Milstein D (2003) Chem Rev 103:1759
116. Albrecht M, van Koten G (2001) Angew Chem Int Ed 40:3750

Chapter 2
The First Discovery of Intramolecular-Coordination Bonds in Cyclometalation Reactions

Abstract The author et al. first reported on the verification of intramolecular-coordination bonds in products of cyclometalation reactions of halo-dicarboxylic acid esters with tinfoil based on IR spectra data in 1966. The stereoisomers of these reaction products were resolved into two kinds of stereoisomers by the use of mixed solvents, and each of the two structures of these stereoisomers was determined by X-ray diffraction studies in 1968 and 1969, respectively.

Keywords Cyclometalation • Dicarboxylic acid ester • Intramolecular coordination • IR spectrum • Organotin

In 1959, our laboratories began investigating direct reactions of alkyl halides with tinfoil, as shown in Eq. (2.1), and various halo-alkyltin halides were obtained as products of reactions conducted under many reaction conditions in the presence of catalysts [1–4].

$$2RX \quad + \quad Sn \quad \xrightarrow{\text{Cat}} \quad R_4Sn \quad + \quad R_3SnX \quad + \quad R_2SnX_2 \quad + \quad RSnX_3 \tag{2.1}$$

R = Alkyl group
X = Halogen atom
cat = Mg, ^{n}BuOH, I, etc.

I. Omae, *Cyclometalation Reactions: Five-Membered Ring Products as Universal Reagents*, DOI 10.1007/978-4-431-54604-7_2, © Springer Japan 2014

Then in 1965, we initiated studies on the reactions of halo-carbonyl compounds with tinfoil in the presence of the catalysts we had used in the former reactions (Eq. (2.1)) [5–23].

$$2 \ X\text{-}C \underset{R^2}{\overset{R^1}{\underset{|}{\overset{|}{C}}}} \left(\underset{R^4}{\overset{R^3}{\underset{|}{\overset{|}{C}}}} \right)_n \overset{O}{\underset{}{\overset{\parallel}{C}}}\text{-}Z \quad + \quad Sn \quad \xrightarrow{\ \text{Cat}\ } \quad X_2Sn \left[\underset{R^2}{\overset{R^1}{\underset{|}{\overset{|}{C}}}} \left(\underset{R^4}{\overset{R^3}{\underset{|}{\overset{|}{C}}}} \right)_n \overset{O}{\overset{\parallel}{C}}\text{-}Z \right]_2 \quad (2.2)$$

R^1 - R^4 = H, alky, phenyl, $COOR^5$, CH_2COOR^6

R^5, R^6 = H, alkyl

Z = R^7, OR^8, NR^9R^{10}, $NHCH_2COOEt$ **2.1**

R^7- R^{10} = H, alkyl

X = Halogen atom

n = 0, 1, 2

In these reactions, the reaction products **2.1** shown in Eq. (2.2) were obtained only from halo-β-carbonyl compounds ($n=1$). In the reactions with the other halo-carbonyl compounds ($n=0, 2, 3$), the starting halo-carbonyl compounds were decomposed during the reaction process, or the reaction products were decomposed during isolation of the products.

The halo-carbonyl compounds used as starting materials are shown in Fig. 2.1.

In the reactions of halo-α-carbonyl compounds ($n=0$), e.g., $BrCH_2COOEt$, with tinfoil, the reaction product could not be isolated from the reaction mixture. The reaction of $BrCH(Et)COOEt$ with tinfoil, on the other hand, gave a product $[Br_2Sn(CH(Et)COOEt)]_2$ as the condensation product of two molecules.

In the reactions of halo-γ-carboxylic acid esters, e.g., $BrCH_2CH_2CH_2COOEt$, decomposed products of the raw material were mainly obtained, as shown in Eq. (2.3) [19].

$$BrCH_2CH_2CH_2COOEt \quad + \quad Sn \quad \xrightarrow{\ \text{Cat}\ } \qquad (2.3)$$

$$Br_2SnEt_2 \quad + \quad CH_2CH_2CH_2C{=}O \quad + \quad Br_2EtSnCH_2CH_2CH_2C\text{-}OEt$$

These results show very high selectivity in the formation of five-membered rings in cyclometalation reactions, and this high selectivity is understandable based on data retrieved from the Cambridge Structural Database, which are shown in Table 5.3 (selectivity: 95 %).

Halo-monocarboxylic acid esters [15-18]

XCH_2CH_2COOR $X = Br, I$

                            $R = Me, Et, {}^nPr, {}^nBu, {}^nOct, CH_2Ph$

$XCHRCOOR$ $X = Br, I$; $R = H. Et$

$XCH_2CHRCOOR$ $X = Br, I$; $R = Me, Et, {}^nPr, {}^iPr, {}^iBu, {}^tBu$

$XCHRCH_2COOR$ $X = Br, I$; $R = Me, Et, {}^nPr$

$XCH_2CH_2CH_2COOR$ $X = Br, I$; $R = Me, Et$

β-Halo-ketones [22]

$RCOCH_2CH_2X$ $X = Cl, Br, I$; $R = Me, Et, {}^nPr, Ph$

$CH_3COCH_2C(CH_3)_2X$ $X = Cl, Br, I$

Halo-amido compounds [19-21]

XCH_2CH_2CONRR' $X = Br, I$

                            $R = H, Me, Et$

                            $R' = H, Me, Et, Ph, C_6H_4\text{-}CH_3\text{-}p$

$XCH_2CHRCONR'R''$ $X = Br, I$

                            $R = Me$

                            $R',R'' = H, Me, Ph,   C_6H_4\text{-}CH_3\text{-}p$

$XCHRCH_2CONRR'$ $X = Br, I$

                            $R = H, Me, Et$

                            $R',R'' = H, Me, Ph,   C_6H_4\text{-}CH_3\text{-}p$

$XCHRCHR'CONHCH_2COOEt$ $X = Br, I$

                            $R, R' = H, Me$

Halo-dicarboxylic acid esters [6-10]

$XCHCOOR$ $X = Br, I$
$\mid$ $R = Me, Et, {}^nPr, {}^nBu, {}^nOct$
CH_2COOR

$XCH_2CHCOOR$ $X = Br, I$
    $\mid$ $R = Me$
    CH_2COOR

Fig. 2.1 Halo-carbonyl compounds

The author was responsible for the investigation of the reactions of halo-dicarboxylic acid esters, such as halo-succinic acid esters and halo-methylsuccinic acid esters [5].

A mixture of ethyl bromosuccinate and tinfoil in the presence of catalysts was stirred for 4 h, for example, and the tinfoil soon disappeared from the reaction mixture in the process. A product was easily isolated as a solid form from the reaction mixture, and two crystalline diastereomeric species were obtained by fractional crystallization of the above product. Based on the IR spectra of the products, the author first proposed a five-membered ring structure for organotin compounds in which lone electron pair of carbonyl oxygen coordinates to a tin atom, as shown in Eq. (2.4).

In the IR spectra of the starting materials, only a strong C=O absorption at 1,740 cm^{-1} was observed, but for the two crystalline species of the products **2.2**, two bands appeared at 1,710 and at about 1,660 cm^{-1} in the solid state as well as in organic solvents. As the results of molecular weight measurements of the products suggested, the products **2.2** had both monomeric forms in the organic solvents, and the strong absorption at 1,660 cm^{-1} must be due to a strongly shifted carbonyl stretching vibration, as would be expected from the coordination of the oxygen atom of C=O in the γ-position with respect to the tin atom. The strong absorption found at 1,710 cm^{-1} was assigned to the C=O stretching mode of the ester group in the β-position.

$$\begin{array}{ccccc}
& & & & 110\ ^\circ C \\
2\ \text{BrCHCOEt} & + & \text{Sn} & \xrightarrow{\hspace{1.5cm}} & \\
& & & 4\ h &
\end{array}$$

Yield 63.6%

Diasteromers **2.2**
mp 114 - 115 °C
mp 122 - 123 °C

(2.4)

Infrared spectra data

ν_βC=O : 1740 cm^{-1} ν_βC=O : 1710 cm^{-1} (shift = 30 cm^{-1})

ν_γC=O : 1740 cm^{-1} ν_γC=O : 1660 cm^{-1} (shift = 80 cm^{-1})

Many stereoisomers on these six-coordinated products **2.2** were expected, because each of the starting materials had a chiral carbon, but only two kinds of m.p. isomers were, in fact, isolated by fractional crystallization from the mixed solvents of ethanol and ethyl ether. It was found, however, that these compounds were optically inactive. The six-coordination structure has been confirmed by two X-ray diffraction studies performed on the products **2.2**, for the low-m.p. isomer [9] in 1968 and the high-m.p. isomer [10, 11] in 1969, as shown in Figs. 2.2 and 2.3, respectively.

Fig. 2.2 Molecular structure
of the low-m.p. isomer of
bis(1,2-diethoxycarbonylethyl)
tin dibromide [9]

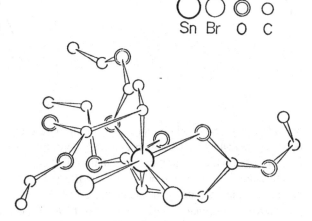

Fig. 2.3 Molecular structure
of the high-m.p. isomer of
bis(1,2-diethoxycarbonylethyl)
tin dibromide [10]

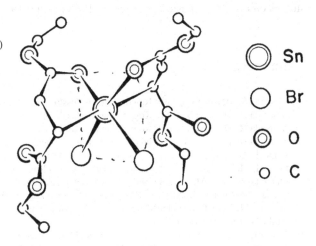

Both crystals have almost identical structures. The two bromine atoms attached
to the tin atom are in the *cis* position. The two ligands, both 1,2-bis(ethoxycarbonyl)
ethyl groups, are bound to the tin atom through carbon and oxygen atoms, forming
somewhat puckered five-membered rings, with the groups around the tin atoms in a
six-coordinated, distorted octahedron. The Sn–C distance is 2.24 Å (high-m.p.
isomer) or 2.26 Å (low-m.p. isomer); these values are slightly longer than the sum
of the covalent radii (2.17 Å). The Sn–O bond distance, 2.49 Å or 2.45 Å, is considerably longer than the sum of the covalent radii (2.06 Å). These data suggest
that the Sn–O bond is only weakly coordinated. Each of the two ligands, both
1,2-bis(ethoxycarbonyl)ethyl groups, has an asymmetric carbon atom. In the low-
m.p. isomer, one of the two asymmetric carbons is in the R-form and the other is in
the S-form. In the high-m.p. isomer, however, since the molecule has C2/c symme-
try, both are in the R-form or both in the S-form. The molecule must then display

Fig. 2.4 Two optically active antipodes [10, 11]

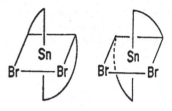

optical activity. Since glide planes are present in the unit cell (space group: C2/c), however, there are equal numbers of molecules of the two optical antipodes (Fig. 2.4) in the crystal, which exhibit no particular optical activity [10, 11].

Thus, the author first reported the verification of intramolecular-coordination bonds in organometallic intramolecular-coordination compounds in the Japanese journal *Kogyo Kagaku Zasshi,* Vol. 69, April 1966 issue [5], including the transition metal compounds, as described in Chap. 3, and the verification with X-ray diffraction studies on these two diastereoisomers in 1968 [9] and 1969 [10], respectively.

References

1. Matsuda S, Matsuda H (1960) Kogyo Kagaku Zasshi 63:114
2. Matsuda H, Matsuda S (1960) Kogyo Kagaku Zasshi 63:1958
3. Matsuda H, Matsuda S (1960) Kogyo Kagaku Zasshi 63:1965
4. Matsuda H, Taniguchi H, Matsuda S (1961) Kogyo Kagaku Zasshi 64:541
5. Omae I, Kikkawa S, Matsuda S (1966) Kogyo Kagaku Zasshi 69:646
6. Omae I, Matsuda S, Kikkawa S, Sato R (1967) Kogyo Kagaku Zasshi 70:705
7. Omae I, Onishi S, Matsuda S (1967) Kogyo Kagaku Zasshi 70:1755
8. Omae I, Matsuda S, Kikkawa S (1967) Kogyo Kagaku Zasshi 70:1759
9. Yoshida M, Ueki T, Yasuoka N, Kasai N, Kakudo M, Omae I, Kikkawa S, Matsuda S (1968) Bull Chem Soc Jap 41:1113
10. Kimura T, Ueki T, Yasuoka N, Kasai N, Kakudo M, Omae I, Kikkawa S, Matsuda S (1968) 21th Annual Meeting of the Chemical Society of Japan, Osaka, Vol 1, p 176
11. Kimura T, Ueki T, Yasuoka N, Kasai N, Kakudo M (1969) Bull Chem Soc Jpn 42:2479
12. Matsuda S, Kikkawa S, Omae I (1969) J Orgonometal Chem 18:95
13. Omae I, Onishi S, Matsuda S (1970) J Orgonometal Chem 22:623
14. Omae I, Yamaguchi K, Matsuda S (1970) J Orgonometal Chem 24:663
15. Omae I (1968) Studies on organotin compounds containing Halo-dicarboxylic acid esters, Thesis
16. Matsuda S, Kikkawa S, Nomura M (1966) Kogyo Kagaku Zasshi 69:649
17. Nomura M, Matsuda S, Kikkawa S (1967) Kogyo Kagaku Zasshi 70:710
18. Nomura M, Kashiwagi S, Kikkawa S (1968) Kogyo Kagaku Zasshi 71:1021
19. Nomura M, Ando S, Matsuda S (1968) Kogyo Kagaku Zasshi 71:394
20. Hayashi T, Uchimura J, Matsuda S, Kikkawa S (1967) Kogyo Kagaku Zasshi 70:714
21. Hayashi T, Kikkawa S, Matsuda S (1967) Kogyo Kagaku Zasshi 70:1389
22. Hayashi T, Kikkawa S, Matsuda S, Fujita K (1967) Kogyo Kagaku Zasshi 70:2298
23. Matsuda S, Kikkawa S, Kashiwa S (1966) Kogyo Kagaku Zasshi 69:1036

Chapter 3
Verification of the Formation of Intramolecular-Coordination Bonds in Cyclometalation Reactions with Transition Metal Compounds

Abstract Cyclometalation reactions have been carried out with most transition metal compounds. The first article on these reactions is generally thought to be a 1963 article on the reaction of an azobenzene with a nickel compound. However, the verification of the intramolecular bond was incorrectly assigned in that article as the N=N π-coordination bond to the nickel atom. For this reason, an article on the reaction of azobenzene with a Pd or Pt atom published in 1965 did not include verification of the coordination of nitrogen lone electron pair with a Pd or Pt atom, although the chemical equations were shown. Consequently, the verification of intramolecular-coordination bonds of azobenzene with the transition metals was delayed until 1969, which was later than the abovementioned report on our research on main group metal compounds in 1966.

Keywords Azobenzene • Cyclometalation • Intramolecular coordination • Lone electron pair • Verification

The first set of articles on the main group of metal organometallic intramolecular-coordination compounds is considered to be the series of articles on organoaluminum compounds published by Bähr and Müller [1–3]. Since the IR and NMR spectra were not yet known in those days, however, no verification of intramolecular-coordination bonds in the product compounds was conducted utilizing these spectral data. They therefore showed data for the elementary analysis, molecular weight, and decomposition compounds of the products and two reaction routes for synthesis of the final product; hence, they only guessed at the structures of intramolecular products described above.

As discussed in the previous chapter, with respect to main group metal elements, the verification of the intramolecular-coordination bonds had already been published with IR spectra data in a Japanese academic journal in 1966.

I. Omae, *Cyclometalation Reactions: Five-Membered Ring Products as Universal Reagents*, DOI 10.1007/978-4-431-54604-7_3, © Springer Japan 2014

With regard to transition metal compounds, on the other hand, a report on the reaction of azobenzene with a dipentadienyl nickel compound by Kleiman and Dubeck [4] in 1963 is generally considered to be the first on organometallic intramolecular-coordination compounds, as shown in Eq. (3.1).

$$\text{(3.1)}$$

3.1

Their report showed that intramolecular-coordination bonds provided the coordination of N=N π-electrons with nickel metal. They reported that "the loss of absorption in the 318 mμ region, attributed to the conjugation of the unsaturated nitrogen with the phenyl rings, strongly suggests that the nitrogen system in the complex is bonded to nickel in the manner shown in the structure **3.1**."

It can be considered that this conclusion was obtained for the following two reasons:

1. Coordination bonds of π-electrons to a metal atom in organometallic compounds were familiar from the following three compounds:

   In 1827, Zeise's salt, ethylene C=C π-electrons bond to platinum atom ($K^+ [C_2H_4PtCl_3]^-$) [5].

   In 1919, Hein complex, benzene π-electrons bond to chromium atom (($C_6H_6)_2Cr$) [6].

   In 1951, ferrocene, cyclopentadienyl π-electrons bond to iron atom (Cp_2Fe) [7, 8].

   The intramolecular bonds of lone electron pair in a hetero atom in organometallic compounds, although it is well established in inorganic chelate complexes, were not known, however.

2. Although use of the IR, UV, and NMR spectra had begun by this time, the relatively limited volume of accumulated data made it difficult to verify the intramolecular bonds from these data yet.

In 1967, the same intramolecular-coordination bond of N=N π-electrons to nickel atoms was also reported in a patent [9]. Consequently, the intramolecular-coordination bond of the cyclopentadienyl azobenzene nickel compound was correctly elucidated by X-ray diffraction studies and published in 1989, after 26 years had passed since 1963, as shown in compound **3.2** in Fig. 3.1 [10].

In 1965, on the other hand, Cope and Siekman [11] reported the intramolecular-coordination bonds of nitrogen lone electron pair to a metal atom **3.3**, as shown in

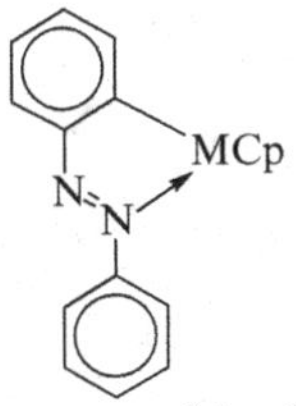

Fig. 3.1 Structures of
cyclopentadienyl
diazobenzene metal
compounds **3.2** correctly
elucidated by X-ray diffraction
studies in 1989 [10]

M = Ni, Pd, Pt

3.2

Cp = cyclopentadienyl

Eq. (3.2). They made their report without assigning credit for data on the IR, UV, mass, and NMR spectra. Cope and Siekman were unable to demonstrate the correct structure with their data, because they considered that Kleiman and Dubeck had also published correct intramolecular-coordination data regarding the coordination of N=N π-electrons to the Ni atom. This can be understood from the following description: they pointed out that their data are similar to the data of the Kleiman and Dubeck studies, except that the authors represented coordinated bonding to nickel as occurring through the π-electrons of the nitrogen–nitrogen bond [11].

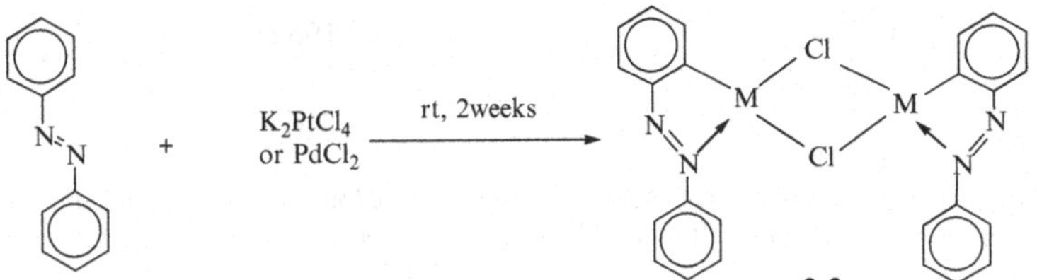

$$(3.2)$$

3.3

Pd, mp 279-281 °C
Pt, dc 270 °C

In 1969, Balch and Petridis [12] showed that IR and NMR spectral observations of *trans* $(RN=NR)_2PdX_2$ **3.4** indicate that each of the azo groups is coordinated to palladium through a nitrogen lone electron pair. They also showed X-ray diffraction data for two naphthylazo inorganic chelate complexes, complex **3.5** in 1961 [13] and complex **3.6** in 1967 [14] as shown in Fig. 3.2.

An investigation conducted by Bach and Petridis in 1969 verified that the results obtained by Cope and Siekman for the intramolecular coordination of diazobenzene metal complexes were correct.

As for the coordination bonds of N=N π-electrons to metal, which Kleiman and Dubeck [4] insisted were diazobenzene metal complexes, on the other hand, no similar intramolecular-coordination bonds of N=N π-electrons to metal were found, according to a report by the Cambridge Structural Database on July 21, 2012.

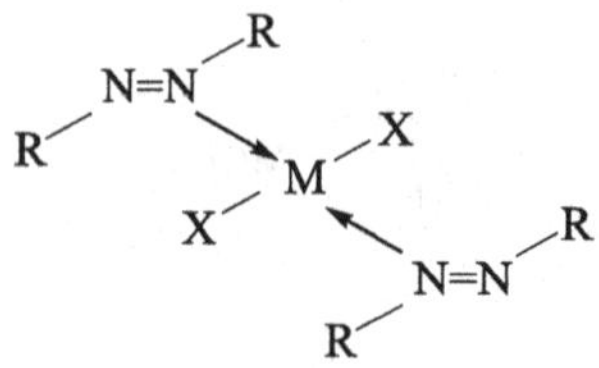

R = Me, Ph
M = Pd
X = Cl, Br

3.4

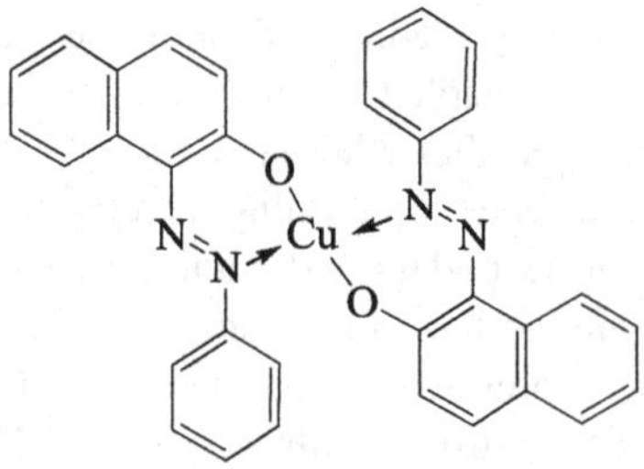

1961

3.5

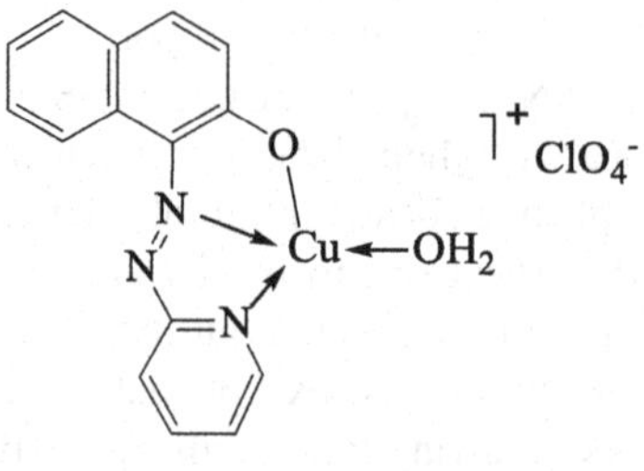

1967

3.6

Fig. 3.2 Verification of an intramolecular-coordination bond of the lone electron pair of a nitrogen atom in azo group to a metal atom by IR and NMR data of compounds **3.4** and X-ray diffraction data of compounds **3.5** and **3.6** by Balch and Petridis [12] in 1969

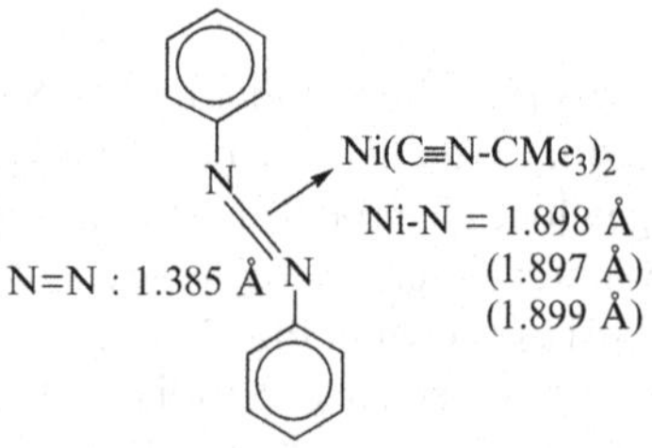

3.7

Fig. 3.3 Example of a π-bonded azobenzene nickel complex **3.7** (bis (*t*-butyl isocyanide) (azobenzene)nickel) [15]

The intermolecular bonds of N=N π-electrons to metal were, however, reported as a π-bonded azo-transition complex **3.7** (bis(*tert*-butyl isocyanide)(azobenzene)nickel) by Dickson and Ibers in 1972 as shown in Fig. 3.3 [15].

References

1. Bähr G, Müller GE (1955) Chem Ber 88:251
2. Bähr G, Müller GE (1955) Chem Ber 88:1765
3. Bähr G, Thiele K-H (1957) Chem Ber 90:1578
4. Kleiman JP, Dubeck M (1963) J Am Chem Soc 85:1544
5. Zeise WC (1827) Pogg Ann 9:632; (1931) 21:497
6. Hein F (1919) Chem Ber 52:195
7. Kealy TJ, Pauson PL (1951) Nature 168:1039
8. Miller SA, Tebboth JA, Tremaine JF (1952) J Chem Soc 632
9. Kleiman JP, Park O, Dubeck M (1967) US Patent 3,300,472
10. Anderson GK, Cross RJ, Muir KW, Manojlović-Muir L (1989) J Organomet Chem 362:225
11. Cope AC, Siekman RW (1965) J Am Chem Soc 87:3272
12. Balch AL, Petridis D (1969) Inorg Chem 8:2247
13. Jarvis JAJ (1961) Acta Cryst 14:961
14. Ooi S, Carter D, Fernando Q (1967) J Chem Soc Chem Commun 1301
15. Dickson RS, Ibers JA (1972) J Am Chem Soc 94:2988

Chapter 4
Organometallic Intramolecular-Coordination Compounds

Abstract Organometallic intramolecular-coordination compounds are classified into two types, i.e., σ-coordination compounds and π-coordination compounds. Most studies have been on σ-coordination compounds, however, because the products are five-membered ring compounds that are extremely easily regioselectively synthesized by cyclometalation reactions. Since the 1970s, the author has compiled a great many articles on them and has published many reviews and a monograph, mainly on coordinating atoms such as N, P, As, O, and S and some coordinating groups, as described in the preface and Chap. 1.

Keywords π-Coordination compound • σ-Coordination compound • Cyclometalation • Intramolecular coordination • *N*-Heterocyclic carbene

Organometallic intramolecular-coordination compounds are classified into two types, i.e., π-coordination compounds and σ-coordination compounds. The former, π-coordination compounds, are those having an unsaturated π-coordination bond to the metal atom, e.g., compound **4.1** (though this is an imaginary compound); the latter, σ-coordination compounds, are those having a σ-coordination bond of lone electron pair of a hetero atom to a metal atom, e.g., compounds **4.2** and **4.3** as shown in Fig. 4.1. The former were reported in 1963 as the coordinating group and the latter in 1965 as the coordinating atom, respectively.

Organometallic intramolecular-coordination compounds are defined schematically as shown in compound **4.4** in Fig. 4.2 [1].

Examples of organometallic intramolecular-coordination compounds containing a coordinating atom, such as N [2], P [3–6], As [7], O [8], S [9], F [10], Cl [11], Br [12], I [13], and Se [14], and of some organometallic intramolecular-coordination compounds containing a coordinating group, such as π-vinyl [15], π-ally [16], cyclopentadienyl [17], and aryl [18], are shown in Figs. 4.3 and 4.4, respectively.

I. Omae, *Cyclometalation Reactions: Five-Membered Ring Products as Universal Reagents*, DOI 10.1007/978-4-431-54604-7_4, © Springer Japan 2014

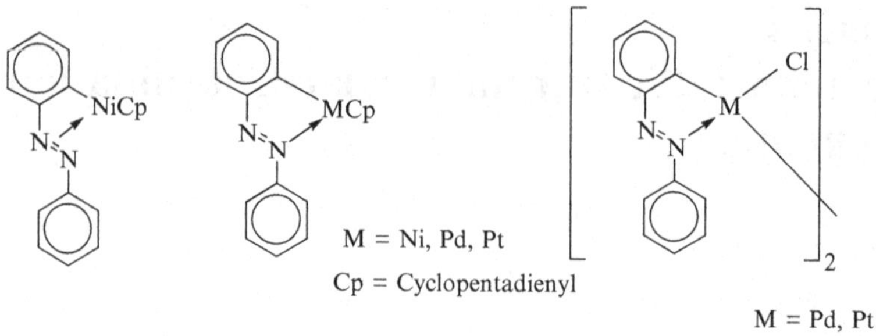

4.1 **4.2** **4.3**

Fig. 4.1 Examples of intramolecular σ-bond (**4.1**) and π-bond (**4.2, 4.3**) coordination metal compounds

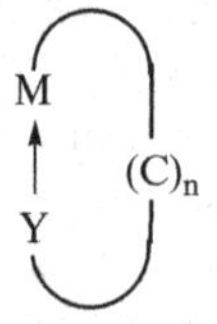

M : Metal atom
Y : Coordinating atom: N, P, O, S, etc.
Coordinating group: π-allyl, π-aryl, π-alkenyl, Cp, etc.
n : 1, 2, 3, etc.

Cp = Cyclopentadienyl group

4.4

Fig. 4.2 Definition of organometallic intramolecular-coordination compounds [1]

No π-coordination compounds with a coordinate bond from an aryl group to the metal of the type shown in compound **4.19** that are as stable as those in which the ligand atom is nitrogen, phosphorus, arsine, oxygen, or sulfur have yet been reported [19, 20]; the synthesis of these **4.19** compounds is difficult, because β-elimination readily occurs in these alkyl compounds and the Group 6 metals chromium, molybdenum, and tungsten as shown in Fig. 4.5.

However, coordination compounds of the Group 6 metals which have a π-bond from a carbon–carbon double bond, rather than a single σ-coordinate metal–carbon bond, may be obtained readily by the reaction shown in Eq. (4.1) [21–24].

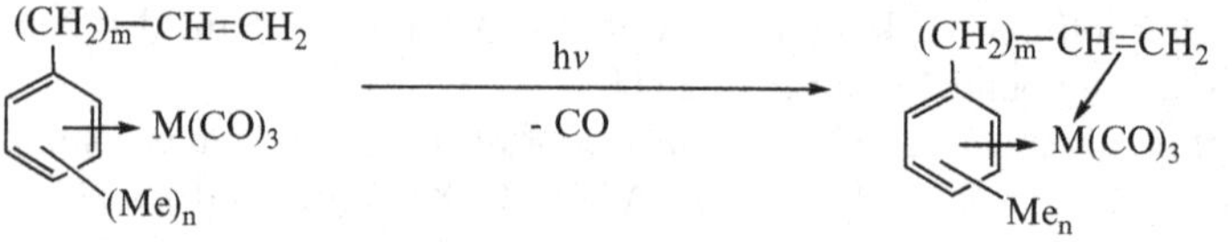

(4.1)

m = 2 - 4
n = 2,3
M = Cr, Mo, W

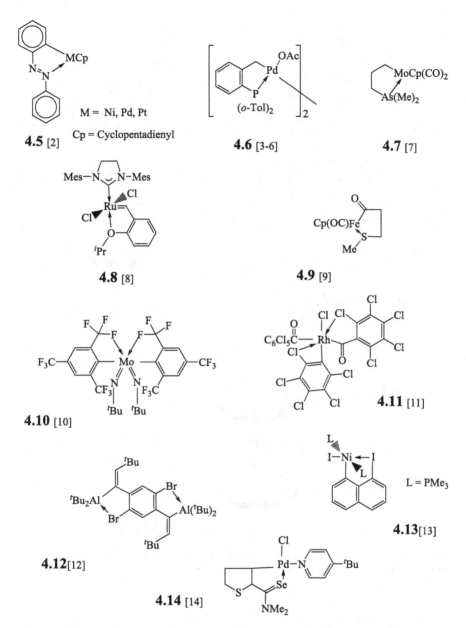

Fig. 4.3 Examples of organometallic intramolecular-coordination compounds containing coordinating atom, such as N, P, As, O, S, F, Cl, Br, I, or Se

The classifications of the organometallic intramolecular-coordination compounds shown in Fig. 4.6 have been reported in a monograph published in 1986 [1].

The number of articles on *N*-heterocyclic carbene compounds, such as on their use as metathesis catalysts (see compound **1.3** in Chap. 1), has increased recently.

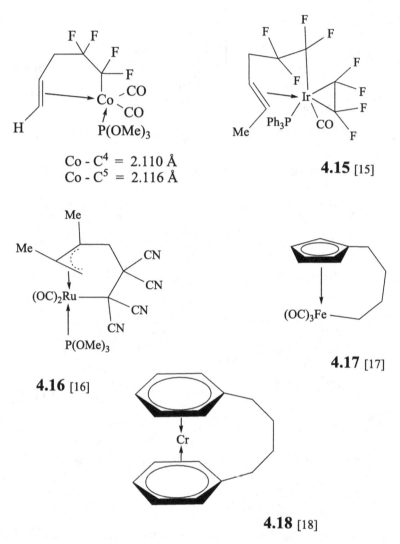

$$Co - C^4 = 2.110 \text{ Å}$$
$$Co - C^5 = 2.116 \text{ Å}$$

4.15 [15]

4.16 [16]

4.17 [17]

4.18 [18]

Fig. 4.4 Examples of organometallic intramolecular-coordination compounds containing a coordinating group, such as a π-vinyl, π-ally, cyclopentadienyl, or the aryl group

Fig. 4.5 π-Coordination compound of a coordinate bond from an aryl group to the metal as shown in compound **4.19**

4.19

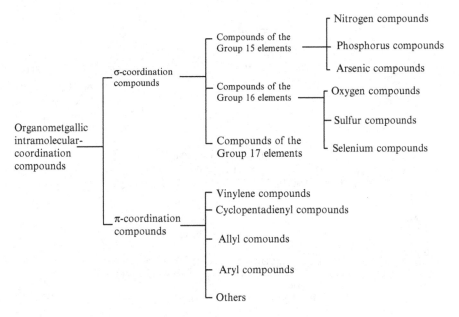

Fig. 4.6 Classifications of organometallic intramolecular-coordination compounds [1]

Examples are shown in Fig. 4.7 [25–30]. These reactions proceed through the coordination of two electrons in a carbene atom to a metal atom, in the same way as cyclometalation reactions with a hetero atom such as N, P, O, and S proceed, as shown in Eq. (6.7) .

As a consequence, these carbene atoms may be considered as coordinating atoms. Hence, the carbene compounds of Group 14 elements may be added to the coordinating atoms in the classification of organometallic intramolecular-coordination compounds in Fig. 4.8.

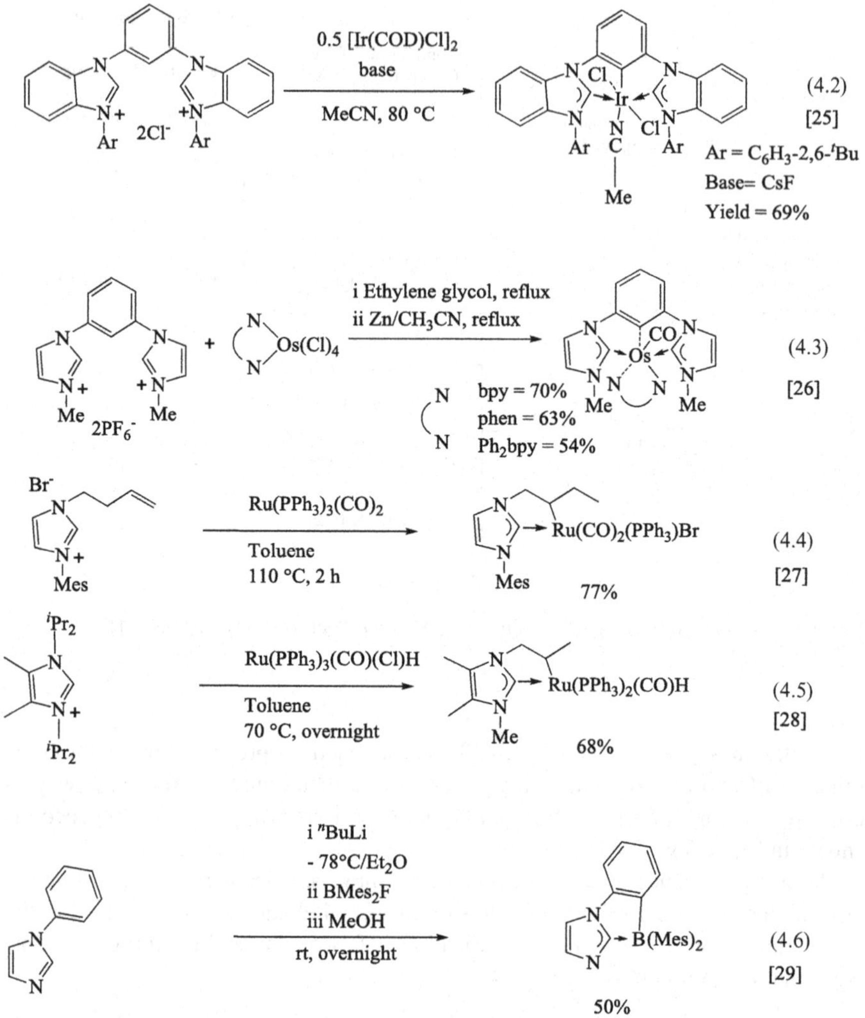

2Cl⁻
0.5 [Ir(COD)Cl]$_2$
base
MeCN, 80 °C
(4.2)
[25]
Ar = C$_6$H$_3$-2,6-^{t}Bu
Base= CsF
Yield = 69%
2PF$_6^-$
Os(Cl)$_4$
i Ethylene glycol, reflux
ii Zn/CH$_3$CN, reflux
bpy = 70%
phen = 63%
Ph$_2$bpy = 54%
(4.3)
[26]
Ru(PPh$_3$)$_3$(CO)$_2$
Toluene
110 °C, 2 h
Ru(CO)$_2$(PPh$_3$)Br
77%
(4.4)
[27]
Ru(PPh$_3$)$_3$(CO)(Cl)H
Toluene
70 °C, overnight
Ru(PPh$_3$)$_2$(CO)H
68%
(4.5)
[28]
i ^{n}BuLi
- 78°C/Et$_2$O
ii BMes$_2$F
iii MeOH
rt, overnight
B(Mes)$_2$
50%
(4.6)
[29]

Fig. 4.7 Cyclometalation reactions with *N*-heterocarbene as a coordinating atom

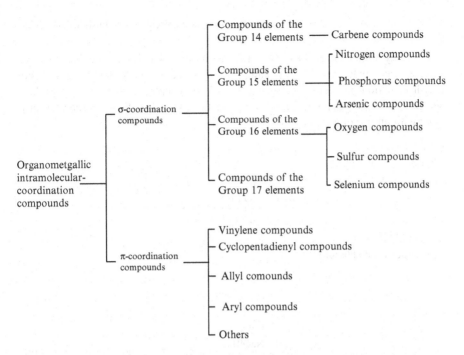

Fig. 4.8 Revised classifications of organometallic intramolecular-coordination compounds

In the reaction scheme shown in Eq. (4.7), the cyclometalation reaction provides six-membered ring products. However, more stable five-membered ring compounds are formed by warming the obtained six-membered ring products at 60 °C in toluene through rearrangement of a phosphine moiety [30].

$$ (4.7) $$

Two six-membered rings

R = H, 77%
R = Me, 55%

Six-membered ring

Five-membered ring

These cyclometalation products also are used as catalysts. The borations of arenes proceed in high yields, for example, as shown in Eq. (4.8). On the other hand, dehydrogenation proceeds without the use of an acceptor such as *t*-butylethylene (see Eqs. 8.33, 8.38, and 8.39), as shown in Eq. (4.9). The TON 68 reaction proceeds without use of an acceptor [25].

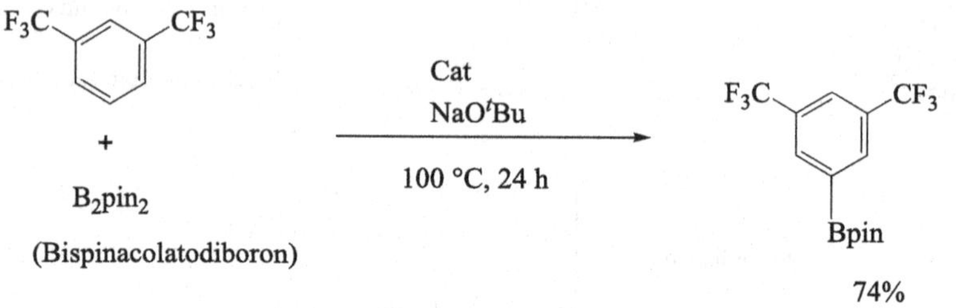

$$\begin{array}{c} \text{F}_3\text{C} \diagdown \diagup \text{CF}_3 \\ + \\ \text{B}_2\text{pin}_2 \\ \text{(Bispinacolatodiboron)} \end{array} \xrightarrow[\text{100 °C, 24 h}]{\begin{array}{c}\text{Cat}\\ \text{NaO}^t\text{Bu}\end{array}} \begin{array}{c}\text{F}_3\text{C}\diagdown\diagup\text{CF}_3\\ |\\ \text{Bpin}\\ 74\%\end{array}$$

(4.8)

$$\begin{array}{c}\bigcirc\\ \text{bp 151 °C}\end{array} \xrightarrow[\begin{array}{c}\text{Reflux, 20 h}\\ \text{Ar stream}\end{array}]{\begin{array}{c}\text{Cat}\\ \text{NaO}^t\text{Bu}\end{array}} \begin{array}{c}\bigcirc\\ \text{TON 68}\end{array}$$

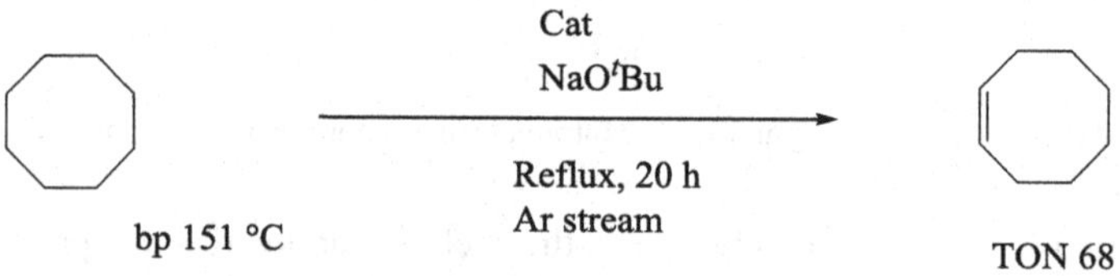

(4.9)

Cat =

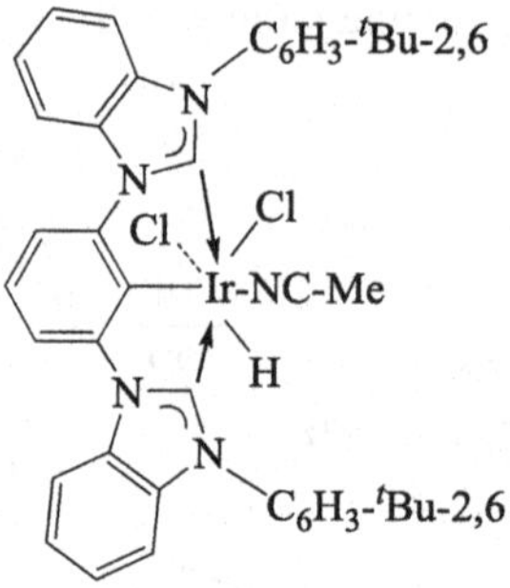

References

1. Omae I (1986) Organometallic intramolecular-coordination compounds, J Organomet Chem Library 18. Elsevier, Amsterdam
2. Anderson GK, Cross RJ, Muir KW, Manojlović-Muir L (1989) J Organomet Chem 362:225
3. Alonso DA, Nájera C (2010) Chem Soc Rev 39:2891
4. Zapf A, Beller M (2005) Chem Commun 431

5. Beletskaya IP, Cheprakov AV (2004) J Organomet Chem 689:4055
6. Herrmann WA, Böhm VPW, Reisenger C-P (1999) J Organomet Chem 576:23
7. Mickiewicz M, Wainwright KP, Wild SB (1976) J Chem Soc Dalton Trans 262
8. Volugioukalakis GC, Grubbs RH (2010) Chem Rev 110:1746
9. King RB, Bisnette MB (1964) J Am Chem Soc 86:1267
10. Dillon KB, Gibson VC, Howard JAK, Redshaw C (1997) J Organomet Chem 528:179
11. García MP, Jiménez MV, Cuesta A, Siurana C, Oro LA, Lahoz FJ, López JA, Catalán MP, Tripicchio A, Lanfranchi M (1997) Organometallics 16:1026
12. Uhl W, Claesener M, Hepp A, Jasper B, Vinogradov A, van Wüllen L, Köster TK-J (2009) Dalton Trans 10550
13. Brown JS, Sharp PR (2003) Organometallics 22:3604
14. Mizuno H, Kita M, Fujita J, Nonoyama M (1992) Inorg Chim Acta 202:183
15. Green M, Taylor SH (1975) J Chem Soc Dalton Trans 1128
16. Amos KL, Connelly NG (1980) J Organomet Chem 194:C57
17. Eilbracht P (1976) Chem Ber 109:1429
18. Nesmeyanov AN, Zaitseva NN, Domrachev GA, Zinov'ev VD, Yur'eva LP, Tverdokhlebova II (1976) J Organomet Chem 121:C52
19. Glockling F, Sneeden RPA, Zeiss H, Bonfinlioli R (1964) J Organomet Chem 2:109
20. Zeiss H, Sneeden RPA (1967) Angew Chem Int Ed 6:435
21. Trahanovsky WS, Hall RA (1975) J Organomet Chem 96:71
22. Nesmeyanov AN, Krivykh VV, Petrovskii PV, Kaganovich VS, Rybinskaya MI (1978) J Organomet Chem 162:323
23. Nesmeyanov AN, Rybinskaya MI, Krivykh VV, Kaganovich VS (1975) J Organomet Chem 93:C8
24. Sruchkov YT, Andrianov VG, Nesmeyanov AN, Krivykh VV, Kaganovich VS, Rybinskaya MI (1976) J Organomet Chem 117:C81
25. Chianese AR, Mo A, Lampland NL, Swartz RL, Bremer PT (2010) Organometallics 29:3019
26. Chung L-H, Chan S-C, Lee W-C, Wong C-Y (2012) Inorg Chem 51:8693
27. Benhamou L, Wolf J, César V, Labande A, Poli R, Lugan N, Lavigne G (2009) Organometallics 28:6981
28. Häller LJL, Page MJ, Erhardt S, Macgregor SA, Mahon MF, Naser MA, Vélez A, Whittlesey MK (2010) J Am Chem Soc 132:18408
29. Rao Y-L, Chen LD, Mosey NJ, Wang S (2012) J Am Chem Soc 134:11026
30. Shaw BK, Patrick BO, Fryzuk MD (2012) Organometallics 31:783

Chapter 5
Characteristics of Cyclometalation Reactions for Organometallic Intramolecular-Coordination Five-Membered Ring Compounds

Abstract One of the characteristics of cyclometalation reactions for the production of organometallic intramolecular-coordination five-membered ring compounds is high reaction rates. Many reactions consequently proceed under mild reaction conditions, and their reaction products are very easily isolated from the reaction mixtures, since five-membered rings are generally stable. Many kinds of metal elements, a total of 69 kinds of both transition metal and main group metal compounds, are utilized for these reactions. Many substrates are also used.

Keywords Cyclometalation • Five-membered ring • Intramolecular coordination • Substrate

5.1 Introduction

Cyclometalation reactions produce three-, four-, six-, and seven-membered ring compounds in addition to five-membered ring compounds. Cyclometalation reactions are mainly used, however, for the preparation of five-membered ring compounds. Since the preparation of five-membered ring compounds is easily performed, many of these reactions were actually tried using many kinds of metal compounds, and many reaction products are used for convenient applications in various fields. Cyclometalation reactions are therefore widely utilized as reaction methods in recent synthetic organic chemistry.

I. Omae, *Cyclometalation Reactions: Five-Membered Ring Products as Universal Reagents*, DOI 10.1007/978-4-431-54604-7_5, © Springer Japan 2014

5.2 High Reactivity and Extremely Easy Synthesis

Cyclometalation reactions for the synthesis of organometallic intramolecular-coordination five-membered ring compounds proceed very promptly, because five-membered ring compounds are more stable as compared with other ring compounds, such as four- and six-membered ring compounds. The cyclometalation reactions generally proceed regioselectively as a result.

The selectivity of five-membered ring compounds is recognizable in all cyclometalation reactions, as shown in Table 5.1 [1]. The mean selectivity ratio shown in the tables is approximately 80 %. For organotin compounds, however, the ratio of five-membered ring compounds increases to approximately 85 %, as shown in Table 5.2 [2]. Furthermore, the selectivity of reactions of halo-carbonyl compounds with tin (see Eq. (2.2)) is 95 %, as shown in Table 5.3 [1].

The high reactivity is also understandable from the mild reaction conditions under which many cyclometalation reactions proceed with many kinds of metal elements, such as Ti, Fe, Ru, Os, Rh, Ir, Ni, Pd, Pt, B, and Cu, and with many kinds of substrates, such as amines, imines, 2-phenylpyridines, benzo[h]quinones, quinolines, diarylpyridines, naphthyridines, azobenzenes, oxazolines, aldehydes, ketones, amides, phosphines, phosphites, and alkyl sulfides, as shown in Figs. 5.1, 5.2, 5.3, 5.4, 5.5, 5.6, 5.7, 5.8, and 5.9.

Table 5.1 Number of articles published in the Cambridge Structural Database[a] on organometallic intramolecular-coordination compounds [1]

Ring configuration	Number of articles (%)
Five-membered ring	5,284 (78.0)
Four-membered ring	656 (9.7)
Six-membered ring	838 (12.3)

[a]X-ray and neutron diffraction analyses of carbon-containing molecules having up to 1,000 atms
The 2008 version of the CSD contained 456,637 entries

Table 5.2 Numbers of articles reporting on compounds with four-, five-, and six-membered rings in the Cambridge Structural Database [2]

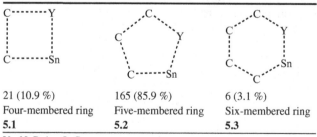

21 (10.9 %)	165 (85.9 %)	6 (3.1 %)
Four-membered ring	Five-membered ring	Six-membered ring
5.1	**5.2**	**5.3**

Y = N, P, As, O, S

Table 5.3 Number of articles published in the Cambridge Structural Database[a] on organometallic intramolecular-coordination compounds having a carbonyl group [1]

	Number of articles (%)	
Ring configuration	All kinds of metals	Sn
Five-membered ring	623 (87.2)	86 (94.5)
Four-membered ring	24 (3.3)	0
Six-membered ring	68 (9.5)	5 (5.5)

[a]X-ray and neutron diffraction analyses of carbon-containing molecules having up to 1,000 atms
The 2008 version of the CSD contained 456,637 entries
Subsequently added articles numbered 12,801 in May 2009

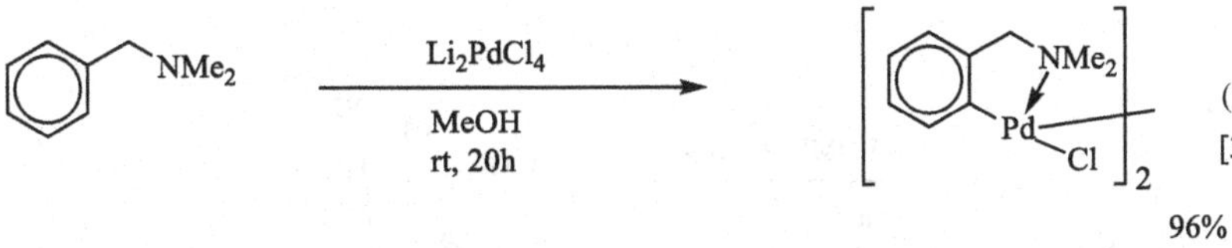

(5.1) [3]

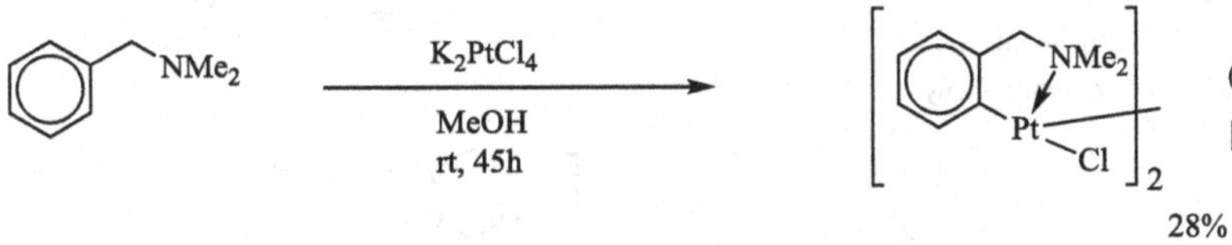

(5.2) [3]

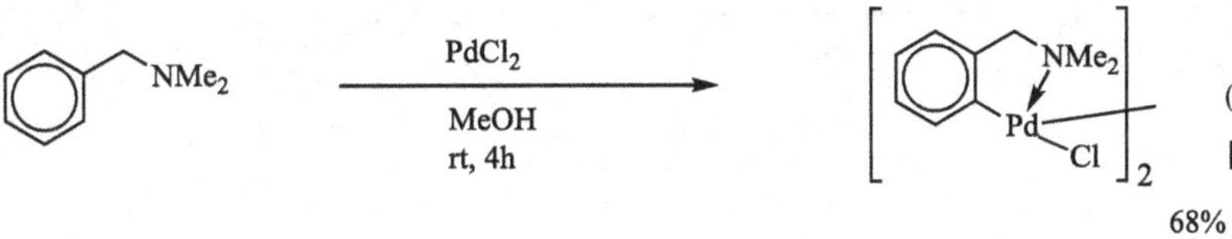

(5.3) [3]

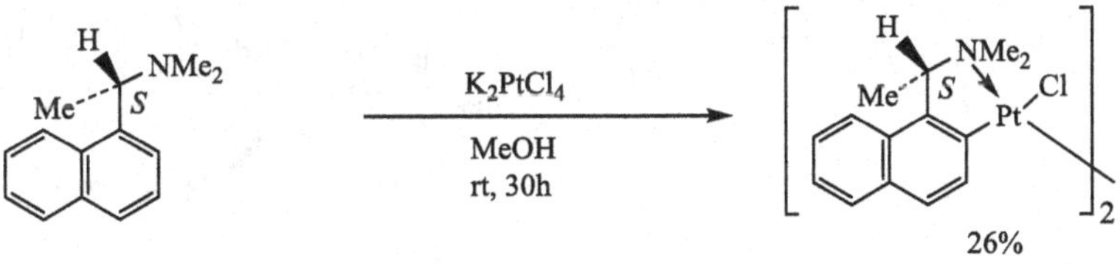

(5.4) [4]

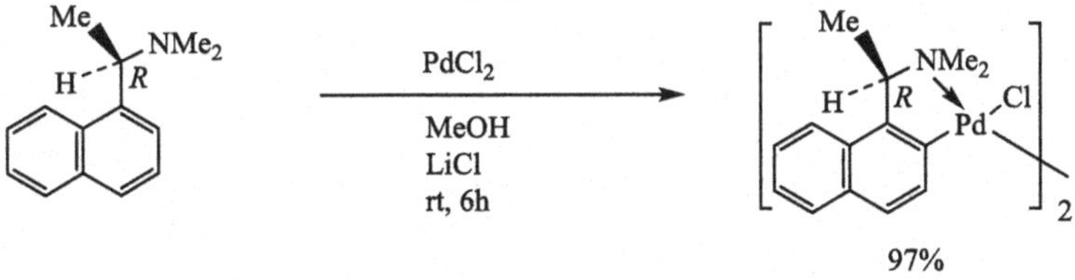

(5.5) [5]

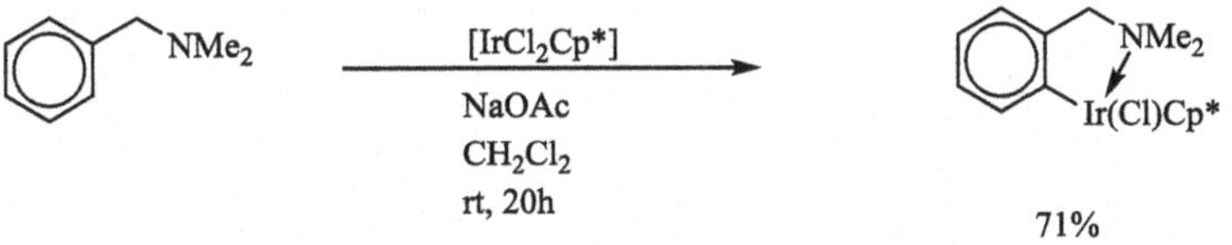

(5.6) [6]

Fig. 5.1 Cyclometalation reactions with amines at room temperature

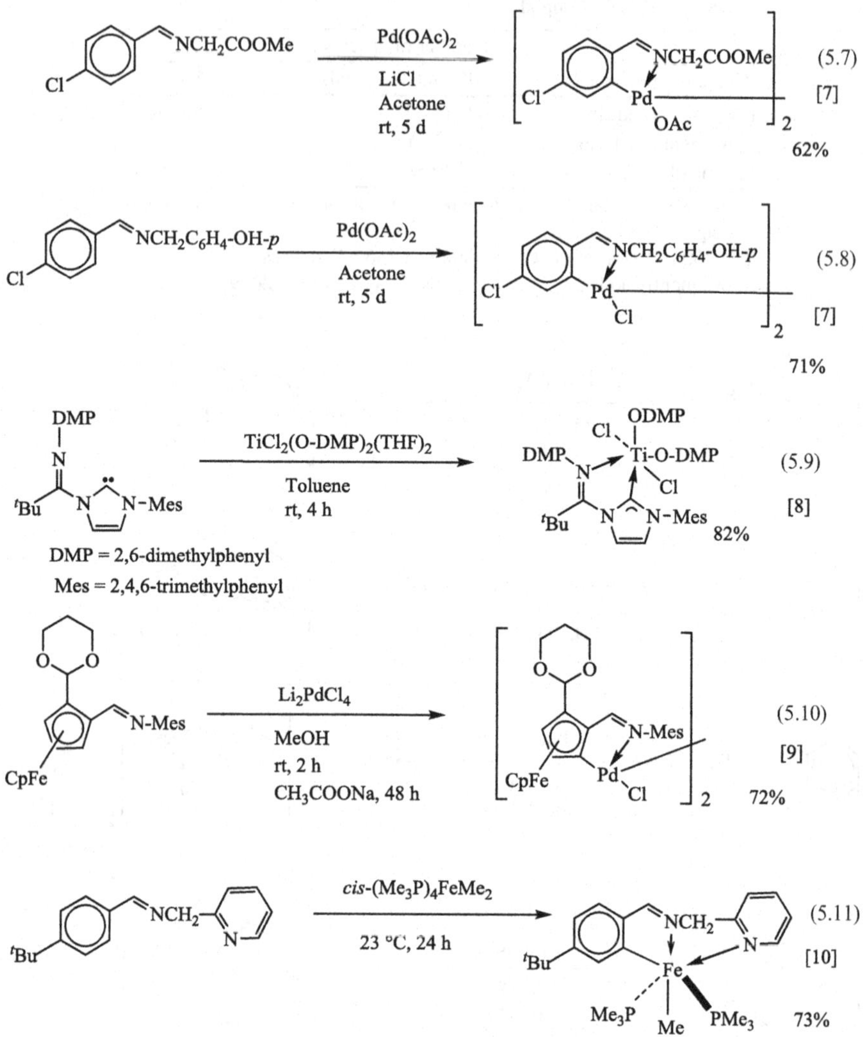

(5.7)

[7]

62%

(5.8)

[7]

71%

(5.9)

[8]

DMP = 2,6-dimethylphenyl

Mes = 2,4,6-trimethylphenyl

82%

(5.10)

[9]

72%

(5.11)

[10]

73%

Fig. 5.2 Cyclometalation reactions with imines at room temperature

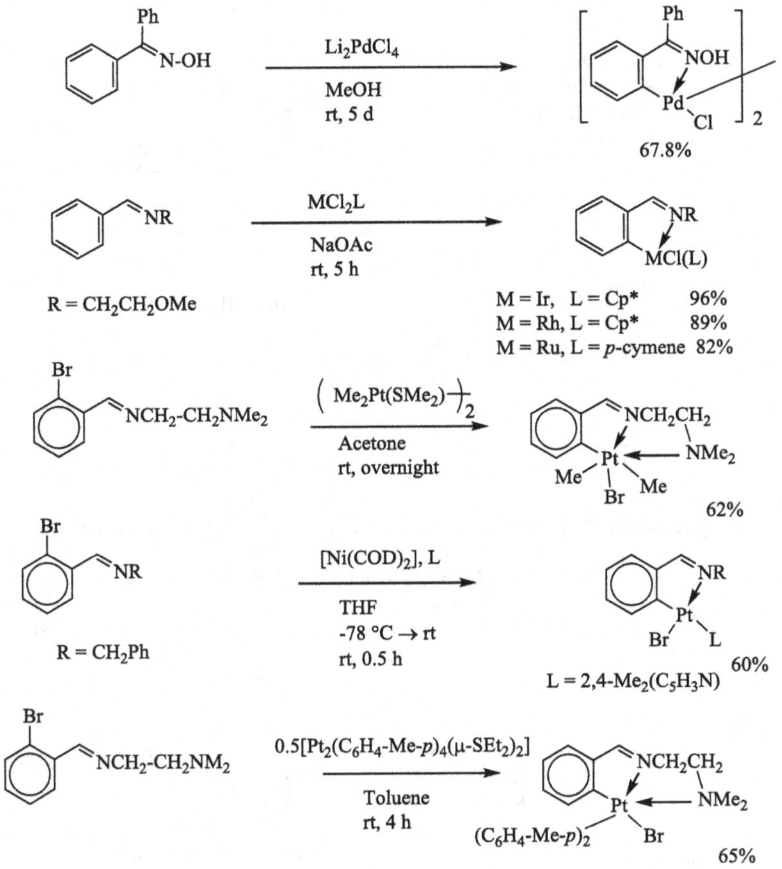
Ph
N-OH
Li2PdCl4
MeOH
rt, 5 d
Ph
NOH
Pd
Cl
2
67.8%
NR
MCl2L
NaOAc
rt, 5 h
NR
MCl(L)
R = CH2CH2OMe
M = Ir, L = Cp* 96%
M = Rh, L = Cp* 89%
M = Ru, L = p-cymene 82%
Br
NCH2-CH2NMe2
(Me2Pt(SMe2))2
Acetone
rt, overnight
NCH2CH2
Pt
NMe2
Me
Me
Br
62%
Br
NR
[Ni(COD)2], L
THF
-78 °C → rt
rt, 0.5 h
NR
Pt
Br
L
R = CH2Ph
60%
L = 2,4-Me2(C5H3N)
Br
NCH2-CH2NM2
0.5[Pt2(C6H4-Me-p)4(μ-SEt2)2]
Toluene
rt, 4 h
NCH2CH2
Pt
NMe2
(C6H4-Me-p)2
Br
65%

(5.12) [11]

(5.13) [6]

(5.14) [12]

(5.15) [13]

(5.16) [14]

Fig 5.2 (continued)

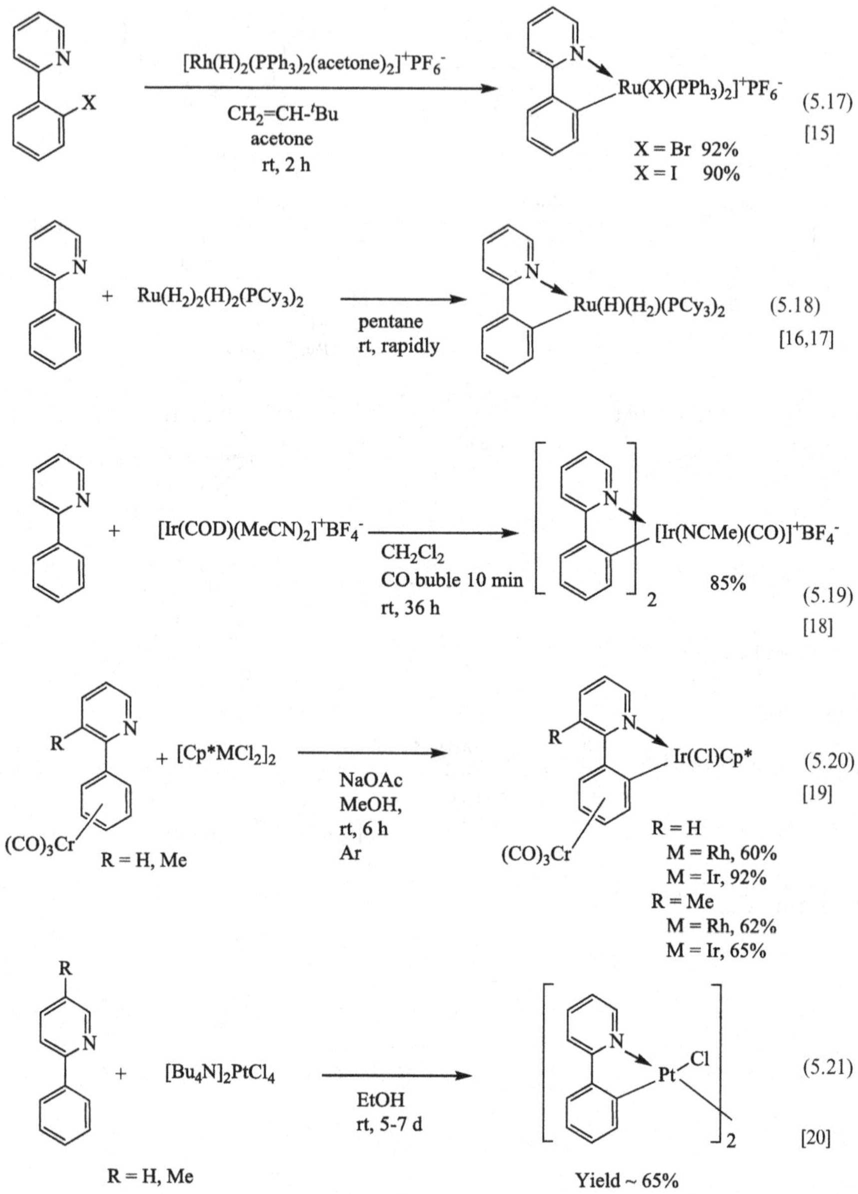

$$[Rh(H)_2(PPh_3)_2(acetone)_2]^+PF_6^-$$

CH$_2$=CH-^{t}Bu
acetone
rt, 2 h

Ru(X)(PPh$_3$)$_2$]$^+$PF$_6^-$ (5.17)

X = Br 92%   [15]
X = I   90%

+ Ru(H$_2$)$_2$(H)$_2$(PCy$_3$)$_2$

pentane
rt, rapidly

Ru(H)(H$_2$)(PCy$_3$)$_2$ (5.18)

[16,17]

+ [Ir(COD)(MeCN)$_2$]$^+$BF$_4^-$

CH$_2$Cl$_2$
CO buble 10 min
rt, 36 h

[Ir(NCMe)(CO)]$^+$BF$_4^-$

85%

(5.19)

[18]

+ [Cp*MCl$_2$]$_2$

NaOAc
MeOH,
rt, 6 h
Ar

Ir(Cl)Cp*   (5.20)

[19]

R = H
   M = Rh, 60%
   M = Ir, 92%
R = Me
   M = Rh, 62%
   M = Ir, 65%

(CO)$_3$Cr   R = H, Me

(CO)$_3$Cr

+ [Bu$_4$N]$_2$PtCl$_4$

EtOH
rt, 5-7 d

Pt^{Cl} (5.21)

[20]

R = H, Me   Yield ~ 65%

Fig. 5.3 Cyclometalation reactions with 2-phenylpyridines at room temperature

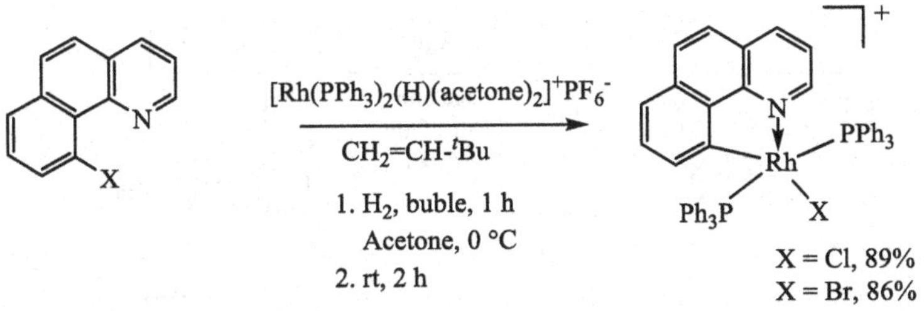

(5.22)

[15]

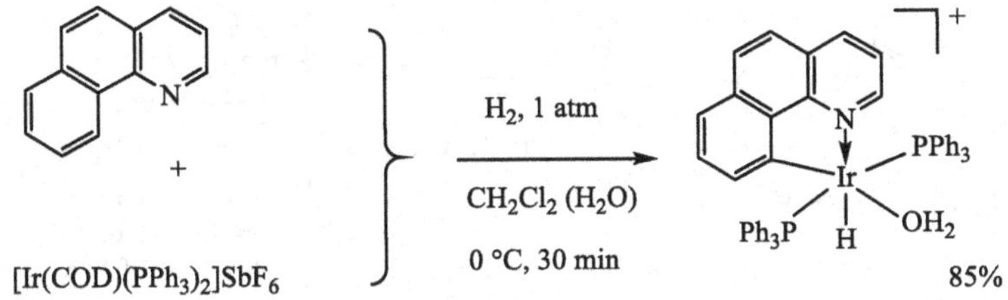

(5.23)

[21]

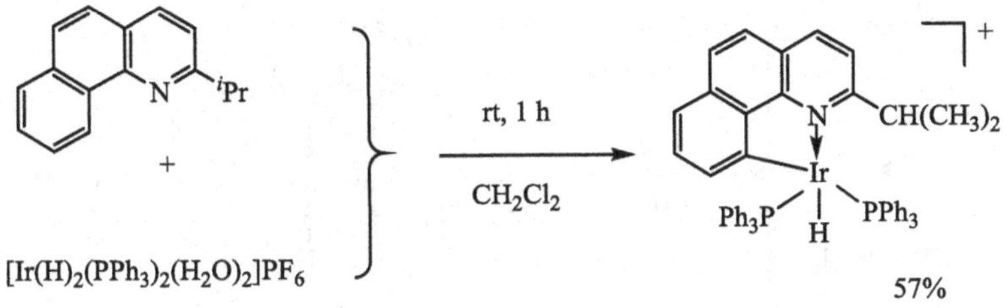

(5.24)

[22]

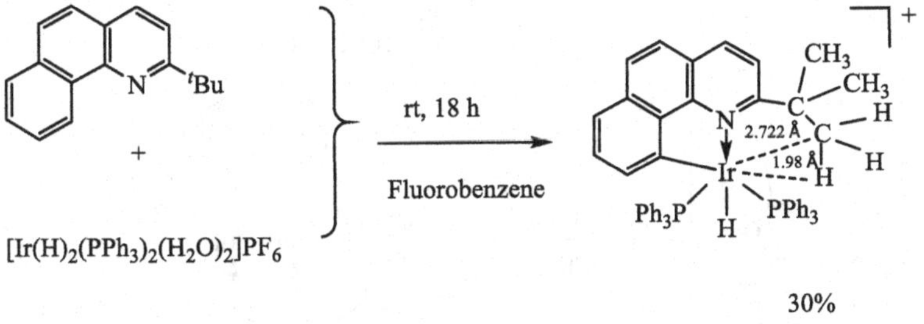

(5.25)

[22]

Fig. 5.4 Cyclometalation reactions with benzo[*h*]quinones at room temperature

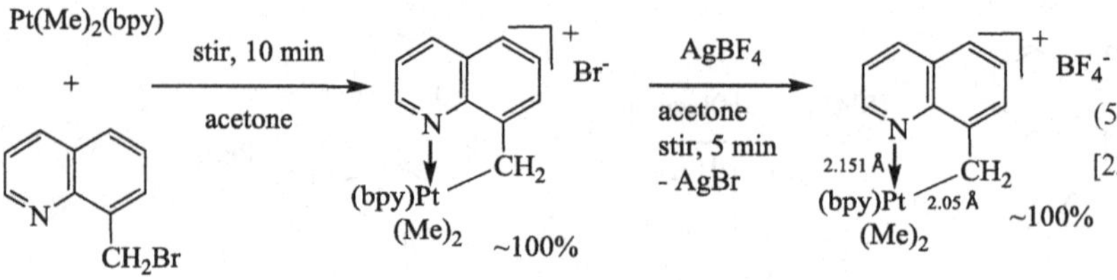

(5.26)
[23]

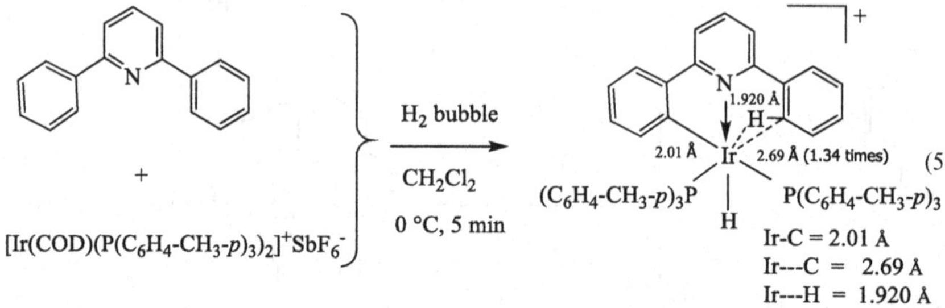

(5.27)
[24]

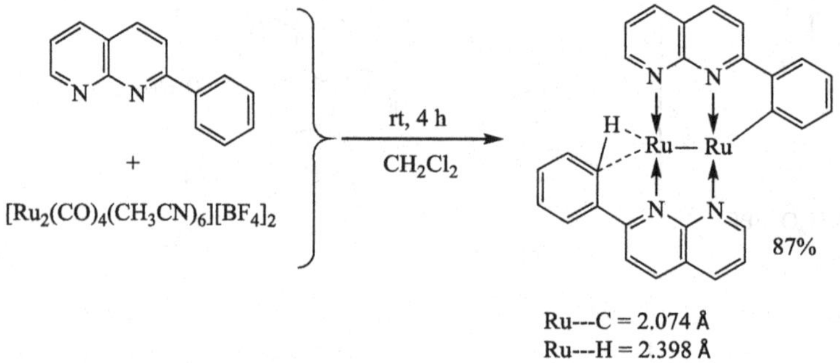

(5.28)
[25]

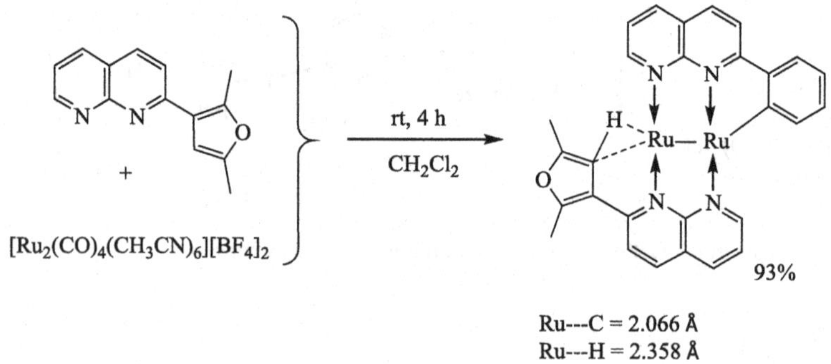

(5.29)
[25]

Fig. 5.5 Cyclometalation reactions with other nitrogen compounds at room temperature

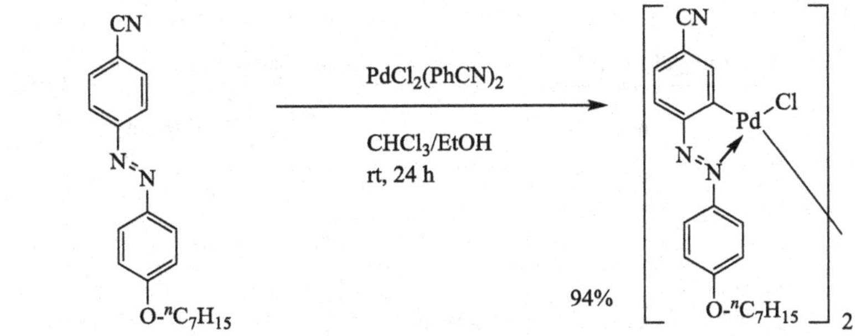

(5.29)

[26]

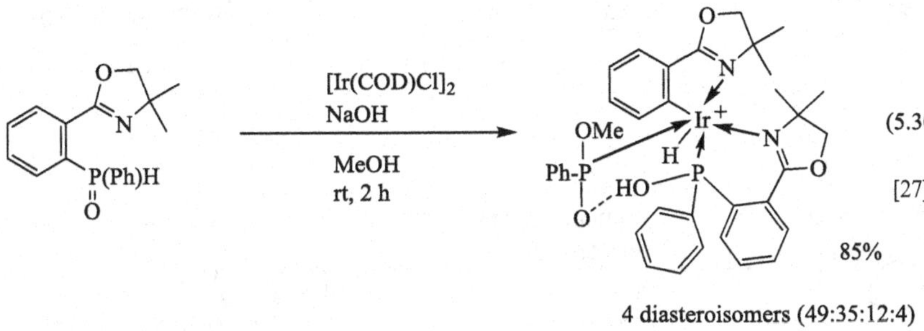

(5.30)

[27]

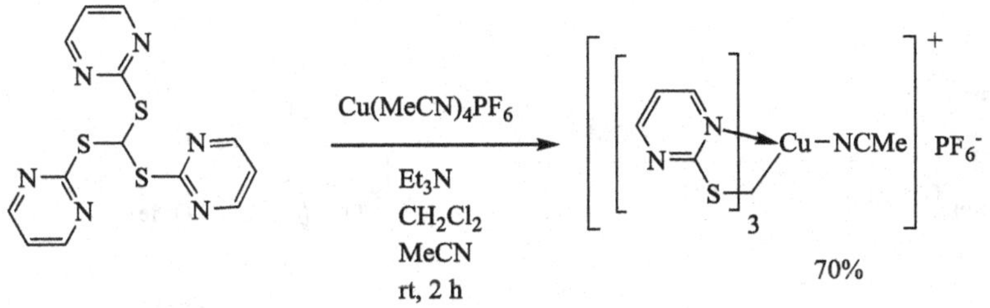

4 diasteroisomers (49:35:12:4)

(5.31)

[28]

Fig 5.5 (continued)

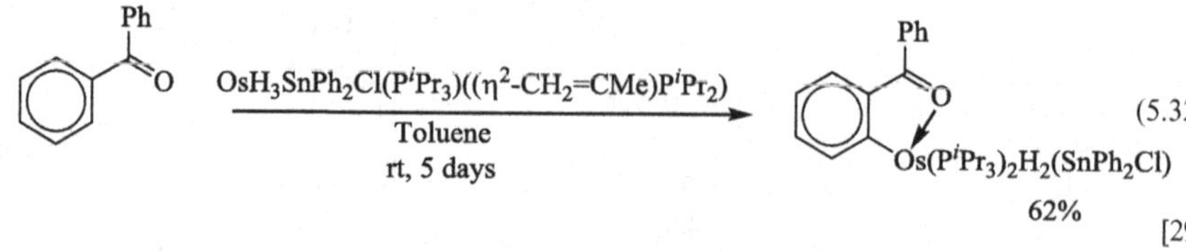
Ph
O
OsH3SnPh2Cl(PiPr3)((η2-CH2=CMe)PiPr2)
Toluene
rt, 5 days
Ph
O
Os(PiPr3)2H2(SnPh2Cl)
62%

(5.32)

[29]

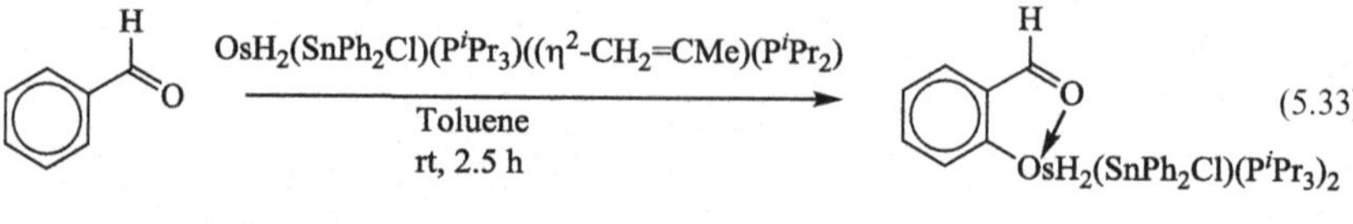
H
O
OsH2(SnPh2Cl)(PiPr3)((η2-CH2=CMe)(PiPr2)
Toluene
rt, 2.5 h
H
O
OsH2(SnPh2Cl)(PiPr3)2

(5.33)

48.4 % [30]

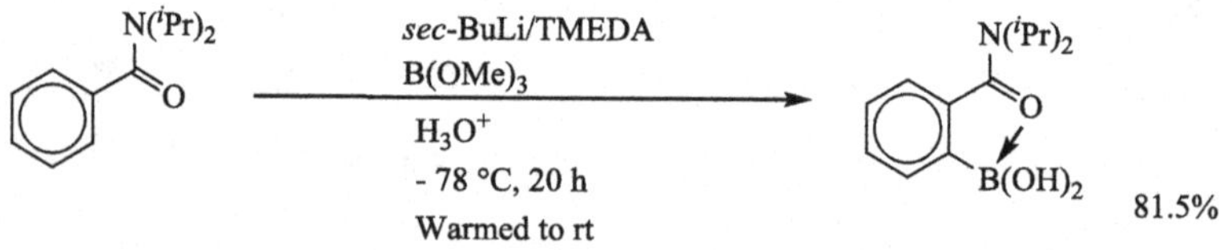
N(iPr)2
O
sec-BuLi/TMEDA
B(OMe)3
H3O+
- 78 °C, 20 h
Warmed to rt
N(iPr)2
O
B(OH)2

(5.34)

81.5% [31]

TMEDA = Tetramethylethylenediamine

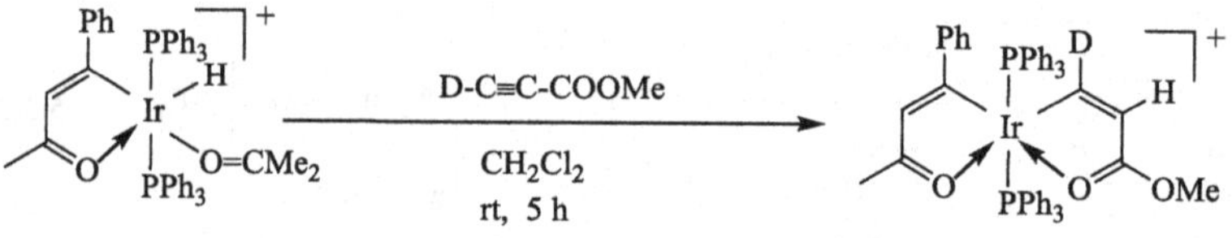
Ph
PPh3
H
Ir
O
O=CMe2
PPh3
+
D-C≡C-COOMe
CH2Cl2
rt, 5 h
Ph
PPh3
D
H
Ir
O
O
OMe
PPh3
+

(5.35)

[32]

83 %

Fig. 5.6 Cyclometalation reactions with oxygen-containing compounds at room temperature

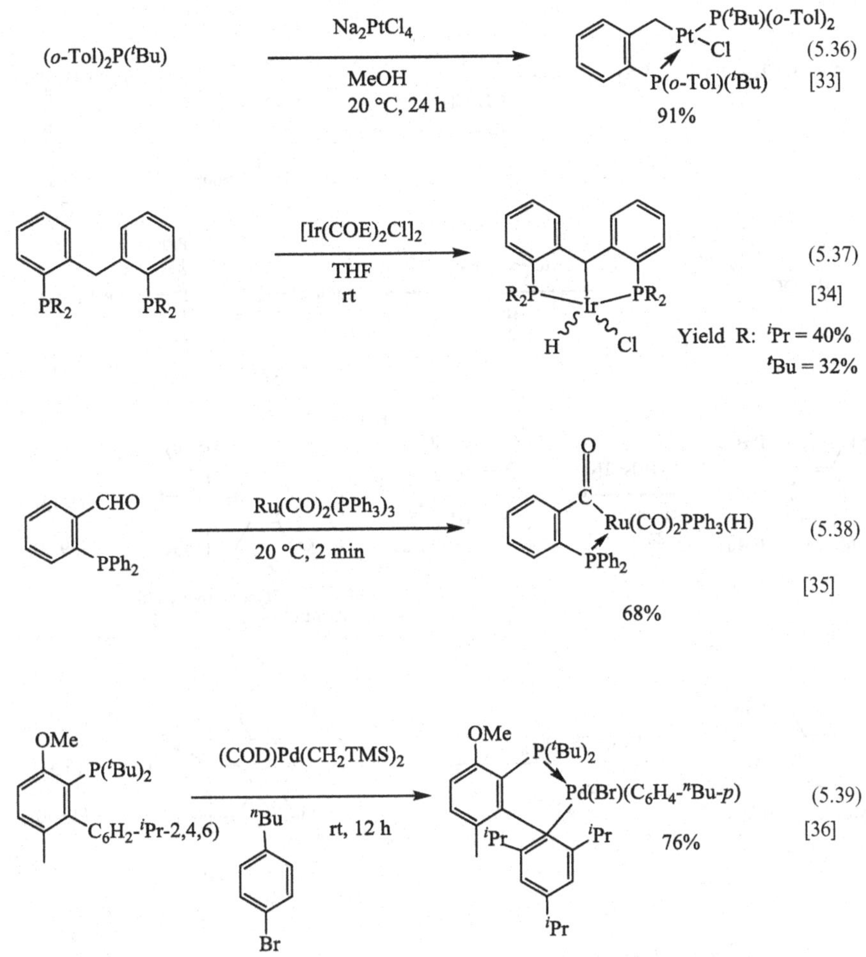

$(o\text{-Tol})_2P({}^tBu)$ $\xrightarrow[\begin{array}{c}\text{MeOH}\\20\ °C,\ 24\ h\end{array}]{Na_2PtCl_4}$ [structure: Pt with $P({}^tBu)(o\text{-Tol})_2$, Cl, $P(o\text{-Tol})({}^tBu)$] (5.36)

91% [33]

[structure with PR_2 and PR_2] $\xrightarrow[\begin{array}{c}\text{THF}\\rt\end{array}]{[Ir(COE)_2Cl]_2}$ [structure: R_2P—Ir—PR_2, H, Cl] (5.37)

[34]

Yield R: ${}^iPr = 40\%$

${}^tBu = 32\%$

[structure with CHO and PPh_2] $\xrightarrow[20\ °C,\ 2\ min]{Ru(CO)_2(PPh_3)_3}$ [structure: $Ru(CO)_2PPh_3(H)$, PPh_2] (5.38)

[35]

68%

[structure: OMe, $P({}^tBu)_2$, $C_6H_2\text{-}{}^iPr\text{-}2,4,6$] $\xrightarrow[\begin{array}{c}{}^nBu\text{—}C_6H_4\text{—}Br\\rt,\ 12\ h\end{array}]{(COD)Pd(CH_2TMS)_2}$ [structure: OMe, $P({}^tBu)_2$, $Pd(Br)(C_6H_4\text{-}{}^nBu\text{-}p)$, ${}^iPr$, ${}^iPr$, ${}^iPr$] (5.39)

76% [36]

Fig. 5.7 Cyclometalation reactions with phosphorus compounds at room temperature

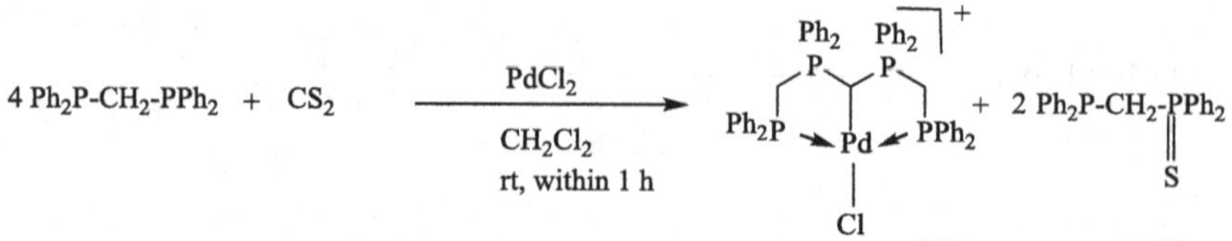
4 Ph2P-CH2-PPh2 + CS2
PdCl2
CH2Cl2
rt, within 1 h
+ 2 Ph2P-CH2-PPh2

>60%

(5.40)
[37]

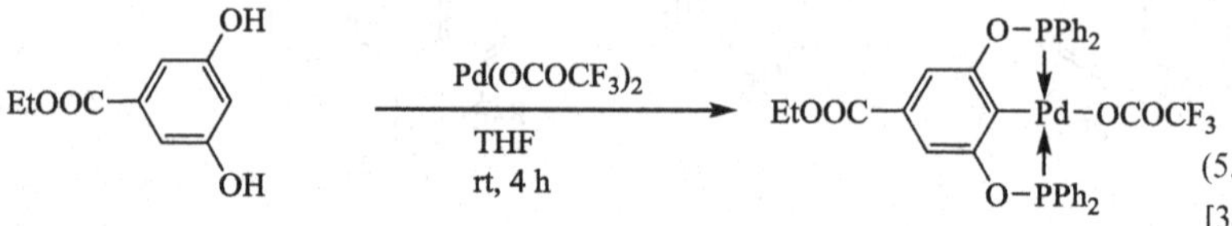
EtOOC
Pd(OCOCF3)2
THF
rt, 4 h

56%

(5.41)
[38]

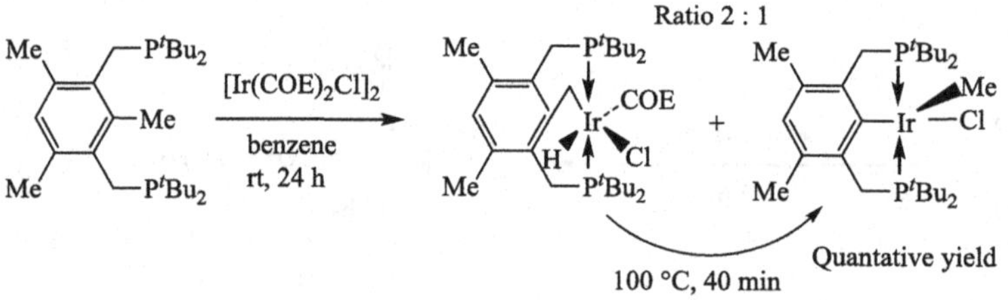
Ratio 2 : 1
[Ir(COE)2Cl]2
benzene
rt, 24 h
100 °C, 40 min
Quantative yield

(5.42)
[39,40]

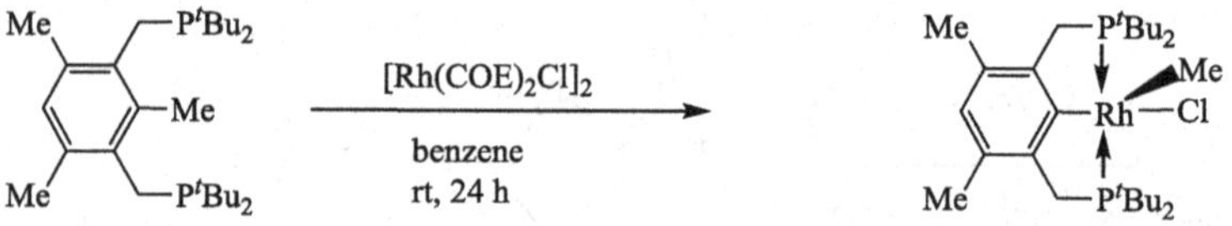
[Rh(COE)2Cl]2
benzene
rt, 24 h

98%

(5.43)
[39,40]

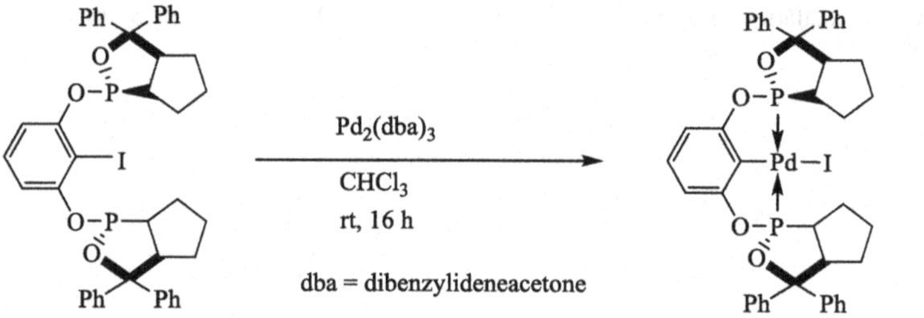
Pd2(dba)3
CHCl3
rt, 16 h
dba = dibenzylideneacetone

86% (dr = 98 : 2)

55% (dr > 99 : 1 after recry)

(5.44)
[41]

Fig 5.7 (continued)

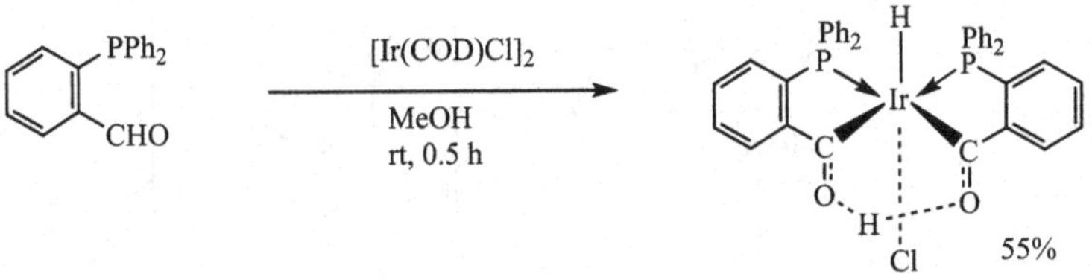

(5.45)

[41]

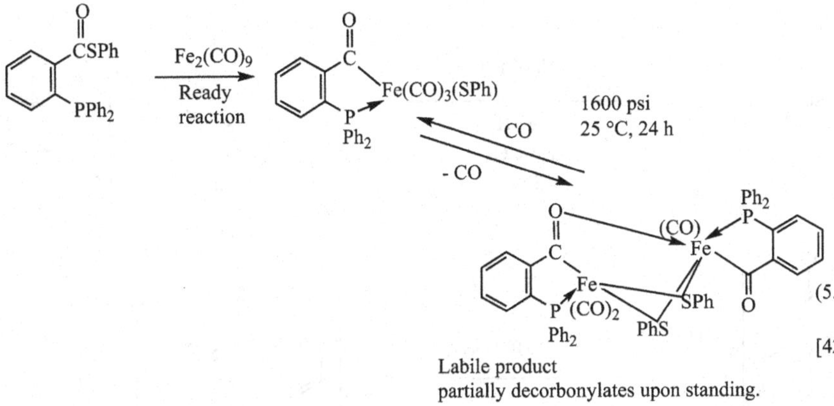

55%

Labile product
partially decorbonylates upon standing.

(5.46)

[42]

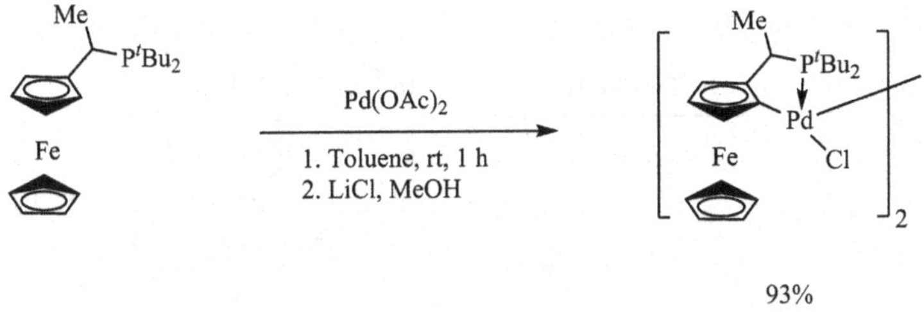

93%

(5.47)

[43]

Fig 5.7 (continued)

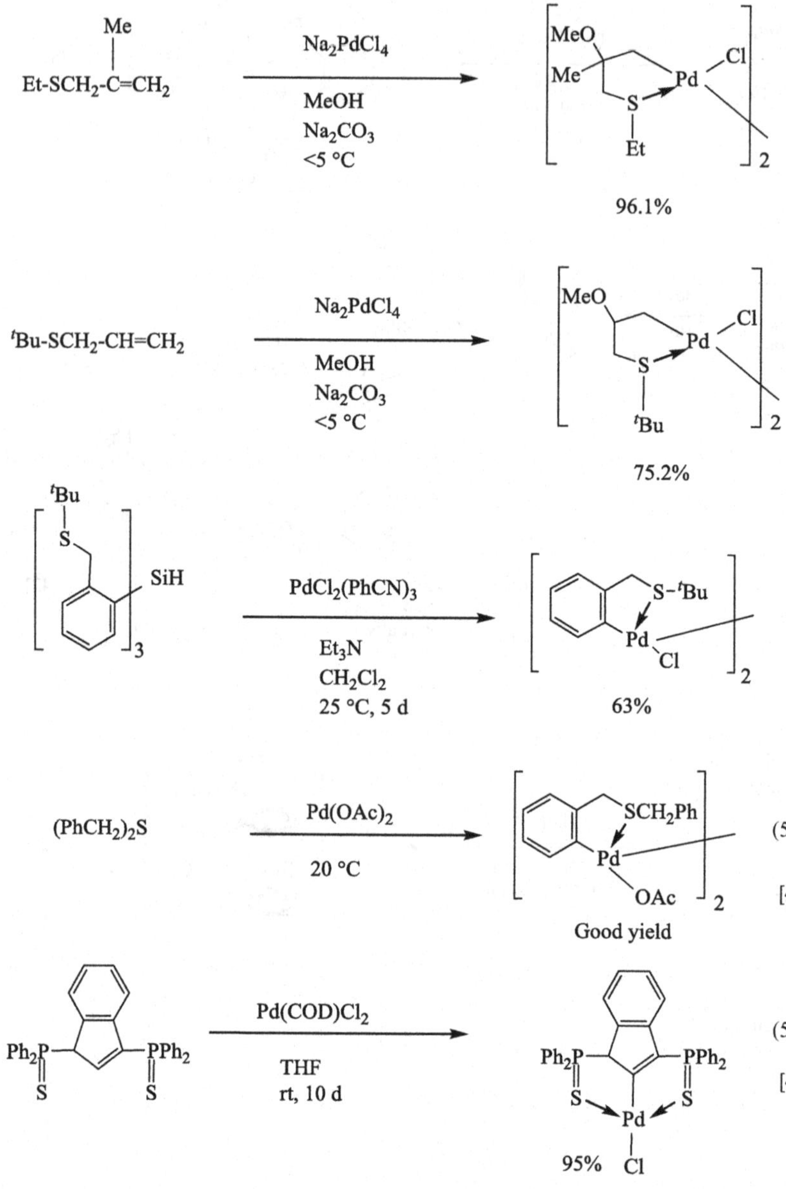

(5.48) [46]

(5.49) [46]

(5.50) [47]

(5.51) [48]

(5.52) [49]

Fig. 5.8 Cyclometalation reactions with sulfur compounds at room temperature

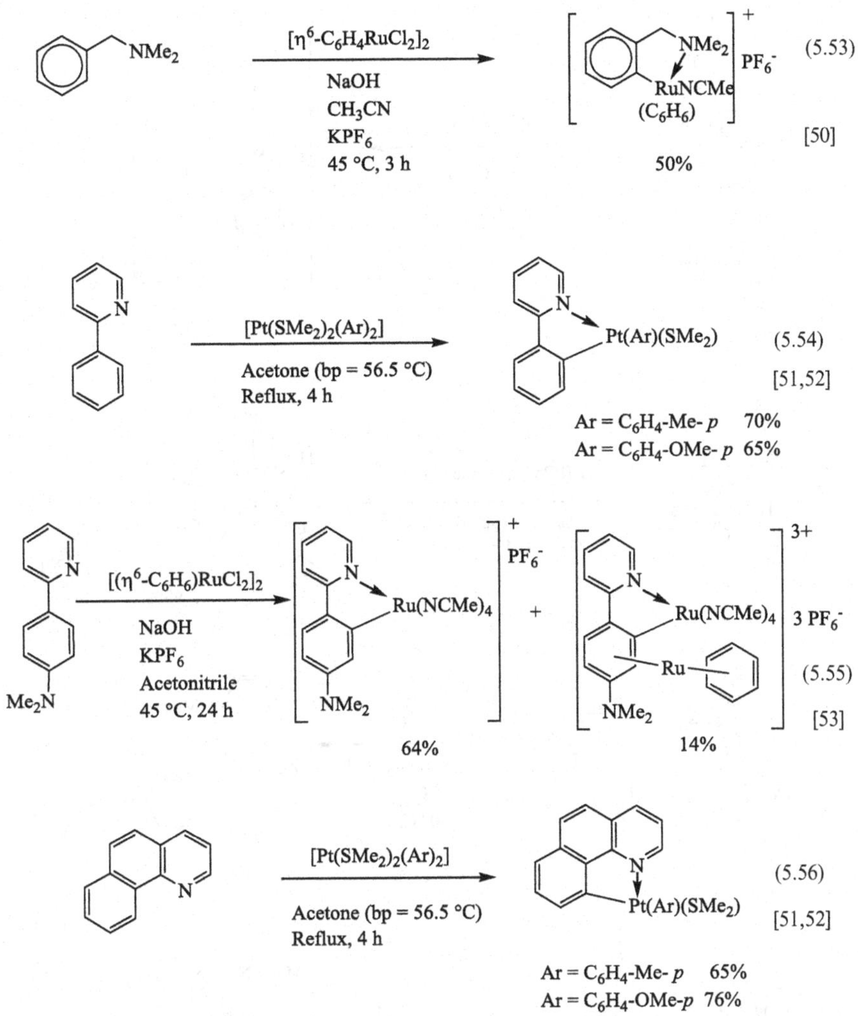

Fig. 5.9 Cyclometalation reactions with other compounds under mild reaction conditions (60° and below)

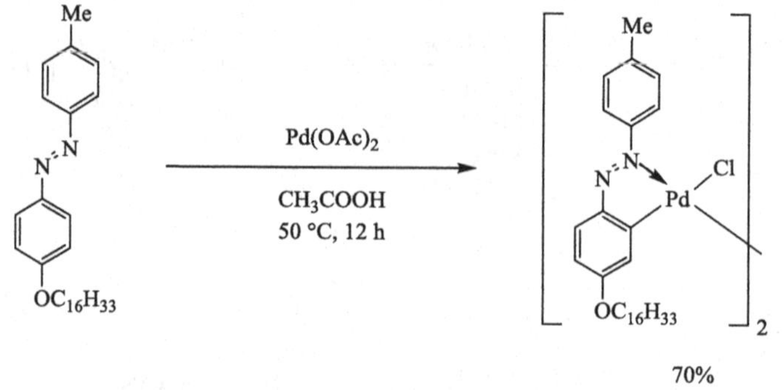
Me
N
N
OC16H33
Pd(OAc)2
CH3COOH
50 °C, 12 h
Cl
Pd
2
70%

(5.57)

[54]

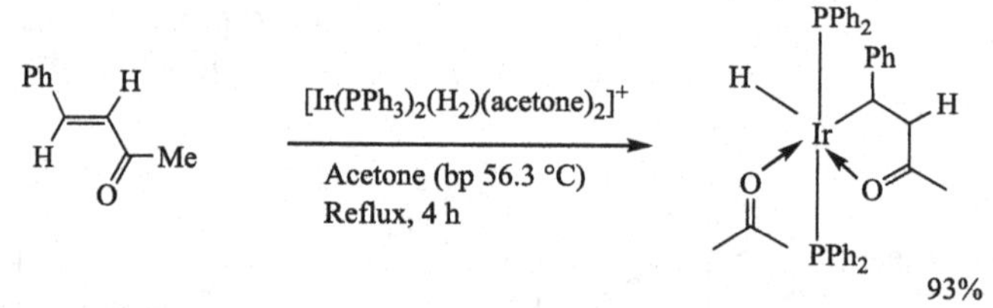
Ph
H
Me
O
[Ir(PPh3)2(H2)(acetone)2]+
Acetone (bp 56.3 °C)
Reflux, 4 h
PPh2
Ir
93%

(5.58)

[55]

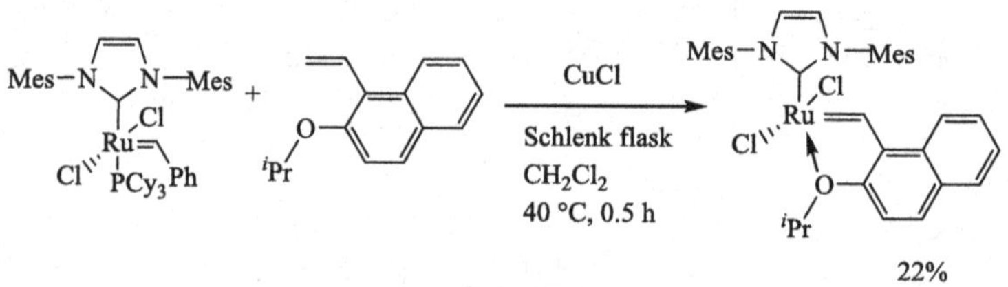
Mes−N
N−Mes
Cl
Ru
PCy3
Ph
iPr
CuCl
Schlenk flask
CH2Cl2
40 °C, 0.5 h
22%

(5.59)

[56]

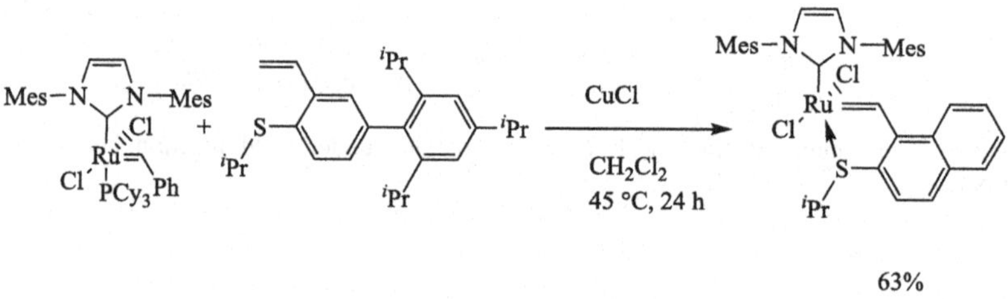
Mes−N
N−Mes
Cl
Ru
PCy3
Ph
S
iPr
CuCl
CH2Cl2
45 °C, 24 h
63%

(5.60)

[57]

Fig 5.9 (continued)

5.3 Various Kinds of Metal Atoms

Since cyclometalation reactions proceed extremely easily, many such reactions have been tried with many metal elements, and the number of metal elements tested for the production of the organometallic intramolecular-coordination five-membered

Table 5.4 Numbers of metal elements employed to produce organometallic intramolecular-coordination compounds [1]

Publication year	Number of metal elements	References
1986	31	[47]
1997	48	[40]
2004	57	[41]
2007	68	[43, 44]
2010	69	[45]

The metal elements employed in cyclometalation reactions to produce organometallic intramolecular-coordination five-membered ring compounds are shown in Fig. 5.10 [1]

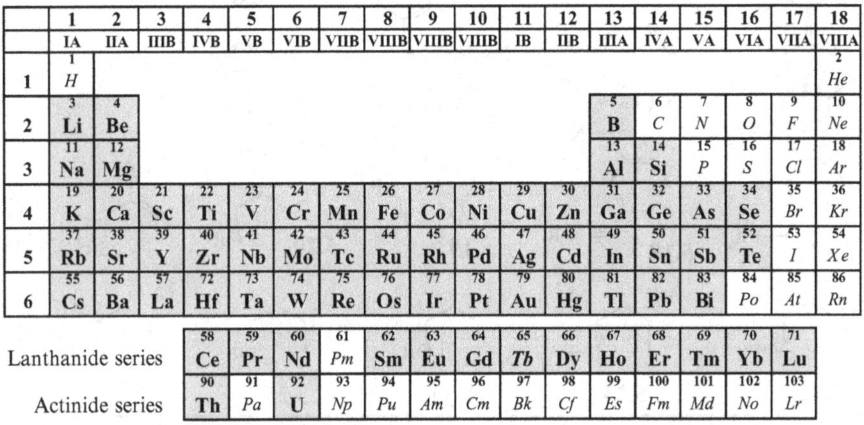

Fig. 5.10 Metal elements employed to produce organometallic intramolecular-coordination five-membered ring compounds [1]

ring compounds has increased rapidly, as shown in Table 5.4 [1]. Almost all of the transition metal elements and main group metal elements, a total of 69 metal compounds, are now used in forming organometallic intramolecular-coordination five-membered ring compounds.

5.4 Various Kinds of Substrates

Since cyclometalation reactions proceed extremely easily, many reactions have been tried with many substrates, and the number of reaction substrates employed has increased rapidly as a result. Many representative substrates are shown in Fig. 5.11 [58]. These substrates are derived from bidentate compounds or tridentate compounds. The tridentate substrates are usually called pincer compounds. Tridentate pincer compounds are used as catalysts, with an emphasis on their stability as catalysts.

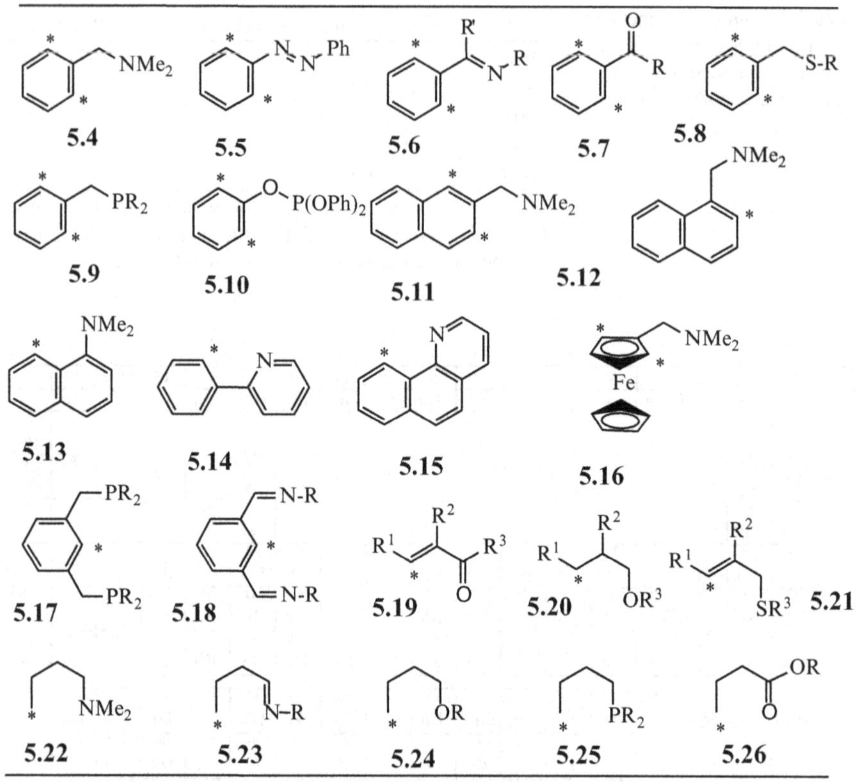

Asterisks (*) indicate the reaction sites of the metal compounds.
These are generally at the γ-position to acoordinating atom such as N, P, O or S.

Fig. 5.11 Representative substrates for synthesis of intramolecular five-membered ring compounds [58]

In the former review [2], the number of the articles reported in the Cambridge Structural Database regarding the major aryl and other substrates employed in intramolecular five-membered ring compounds is shown in Table 5.5 [2] and Table 5.6 [2], respectively.

Of particular note, *N,N*-dimethylbenzylamines **5.4** are the most popular with 350 articles reported, followed by such other popular compounds as benzyl methyl derivatives **5.30** with 168, 2-phenhylpyridines **5.14** with 135, dimethylaminopropyl (or propenyl) compounds **5.22** with 159, alkyl (or aryl) ethyl (or ethenyl) ketones **5.19** with 108, diallyl (or diaryl or propenyl) phosphines **5.25** with 206, and alkyl (or aryl) propionic acid esters **5.26** with 164.

The reported metal elements applied with *N,N*-dimethylbenzylamines **5.4** were Li (G1: G1 shows the first group's elements in the periodic table of elements), Mg (G2), Sc, Y, La, Nd, Er, Lu (G3), Ti, Zr (G4), V, Ta (G5), Cr, Mo, W (G6), Mn (G7), Co, Rh (G9), Ni, Pd, Pt (G10), Cu, Au, Zn (G11), Cd, Hg, B (G12), B, Al, Ga, In,

Table 5.5 The major aryl types of intramolecular five-membered ring compounds reported in the Cambridge Structural Database, together with the number of articles reported therein for each type [2]

Number of articles reported in the Cambridge Structural Database

Structure	No.	Articles	Structure	No.	Articles
NMe$_2$ / M	**5.4**	350	R^1 R^2 / NMe$_2$ / M	**5.30**	168
N=N–Ph / M	**5.5**	89	(pyridine ring) N / M	**5.31**	67
NMe$_2$ / M / NMe$_2$	**5.27**	100	PR$_2$ / M	**5.9**	73
R / O / M	**5.7**	32	Me$_2$N→M	**5.13**	26
OR / M	**5.28**	45	NMe$_2$ / M	**5.12**	64
R^1 / N–R^2 / M	**5.6**	41	R^{1}R^{2}C–M / N	**5.32**	51
S–R / M	**5.8**	52	M / N	**5.15**	13
NMe$_2$ / M	**5.29**	28	N / M	**5.14**	135

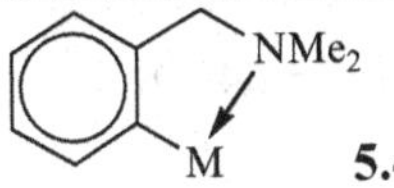

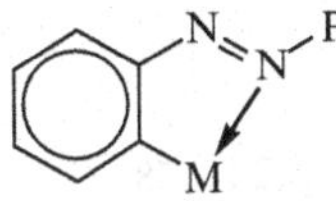

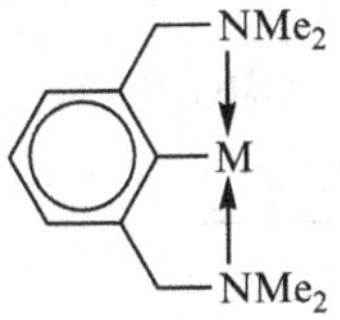

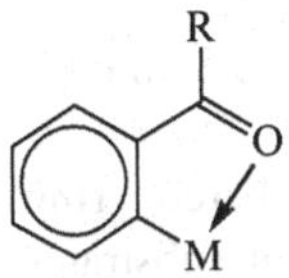

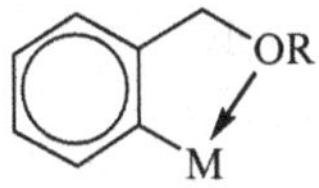

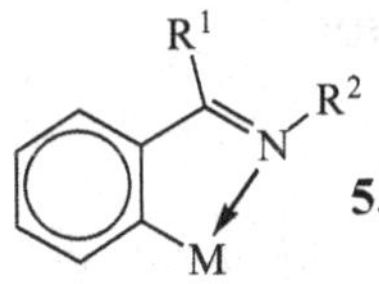

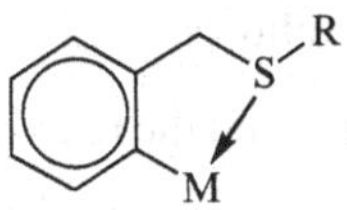

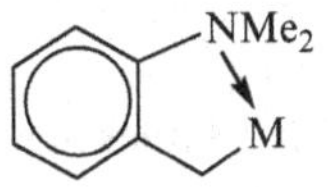

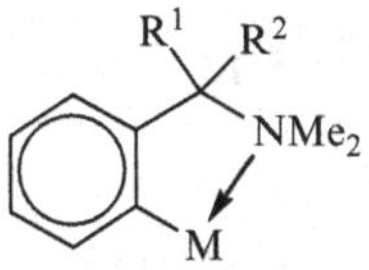

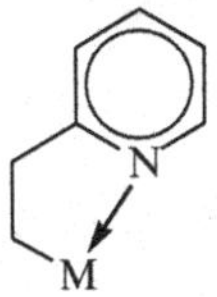

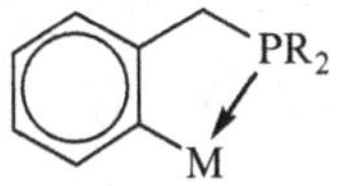

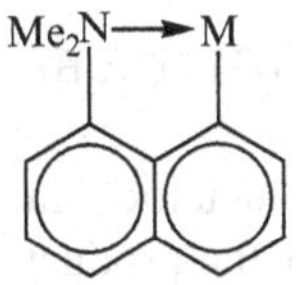

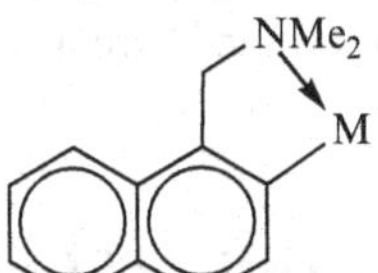

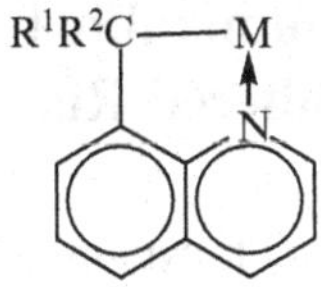

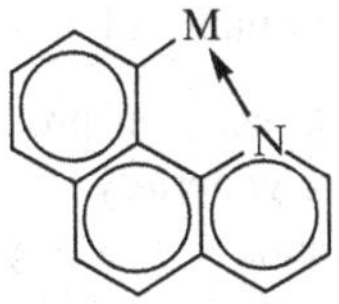

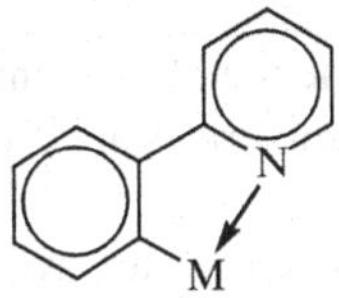

Table 5.6 Other types of intramolecular five-membered ring compounds reported in the Cambridge Structural Database, together with the number of article reported therein for each type [2]

Number of articles reported in the Cambridge Structural Database

(Fe)⟋⟍NMe$_2$ M	**5.16** 42		M⟵OR	**5.24** 99
M⟵NMe$_2$	**5.22** 159		M⟵PR^{1}R^2	**5.25** 206
M⟵N∼R	**5.23** 82		M⟵SR	**5.21** 58
M⟵O (R)	**5.19** 108		M⟵O (O-R)	**5.26** 164

Tl (G13), Si, Sn, Pb (G14), P, Bi (G15), and Te (G16), among the total of 39 metals reported in 2003 [59].

As for organometallic intramolecular-coordination five-membered ring compounds used for the production of organic electronic devices, their substrates and their ancillary ligands are also shown in Figs. 9.2 and 9.3, respectively.

5.5   Characteristics of Cyclometalation Reactions for Organometallic Intramolecular-Coordination Five-Membered Ring Compounds

The characteristics of cyclometalation reactions for producing organometallic intramolecular-coordination five-membered ring compounds are as follows:

1. Cyclometalation reactions proceed very easily and regioselectively, and the reaction products are generally stable compounds.
2. Chiral compounds are very easily synthesized using chiral substrates (see Sect. 8.2)
3. Almost all of the metal elements are used (see Fig. 5.10).
4. Many kinds of substrates are used for these reactions (see Fig. 5.11 and Tables 5.5 and 5.6).
5. Many kinds of catalysts are available, because many kinds of metal elements and many kinds of substrates can be used in the reactions.
6. Since some cyclometalation reactions proceed too rapidly, the reactions are also used for synthesis of the derivatives as cyclometalation reaction intermediates (see Sect. 7.4).

References

1. Omae I (2010) Appl Organomet Chem 24:347
2. Omae I (2004) Coord Chem Rev 248:995
3. Cope AC, Friedrich EC (1968) J Am Chem Soc 90:909
4. Chooi SYM, Ranford JD, Leung P-H, Mok KF (1994) Tetrahedron Asymmetr 5:805
5. Allen DG, McLaughlin GM, Robertson GB, Steffen WL, Salem G, Wild SB (1982) Inorg Chem 21:1007
6. Davies DL, Al-Duaij O, Farwcett J, Giardiello M, Hilton ST, Russel DR (2003) Dalton Trans 4132
7. Albert J, Crespo M, Granell J, Rodríguez J, Zafrilla J, Calvet T, Font-Bardia M, Solans X (2010) Organometallics 29:214
8. Larocque TG, Lavoie GG (2012) J Organomet Chem 715:26
9. Mariño M, Martínez J, Caamaño M, Pereira MT, Ortigueira JM, Gayoso E, Fernández A, Vila JM (2012) Organometallics 31:890
10. Volpe EC, Wolczanski PT, Lobkovsky EB (2010) Organometallics 29:364
11. Onoue H, Minami K, Nakagawa K (1970) Bull Chem Soc Jpn 43:3480
12. Anderson CM, Crespo M, Jennings MC, Lough AJ, Ferguson G, Puddephatt RJ (1991) Organometallics 10:2672
13. Ceder RM, Granell J, Muller G, Font-Bardía M, Solans X (1995) Organometallics 14:5544
14. Martín R, Crespo M, Font-Bardia M, Calvet T (2009) Organometallics 28:587
15. Chen S, Li Y, Zhao J, Li X (2009) Inorg Chem 48:1198
16. Guari Y, Sabo-Etienne S, Chaudret B (1998) J Am Chem Soc 120:4228
17. Toner AJ, Gründemann S, Clot E, Limbach H-H, Donnadieu B, Sabo-Etienne S, Chaudret B (2000) J Am Chem Soc 122:6777
18. Rahaman SMW, Dinda S, Ghatak T, Bera JK (2012) Organometallics 31:5533
19. Scheeren C, Maasarani F, Hijazi A, Djukic J-P, Pfeffer M, Zarić SD, Goff X-FL, Ricard L (2007) Organometallics 26:3336
20. Mdleleni MM, Bridgewater JS, Watts RJ, Ford PC (1995) Inorg Chem 34:2334
21. Lavin M, Holt EM, Crabtree RH (1989) Organometallics 8:99
22. Clot E, Eisenstein O, Dubé T, Faller JW, Crabtree RH (2002) Organometallics 21:575
23. Canty AJ, Hoare JL, Patel J, Pfeffer M, Skelton BW, White AH (1999) Organometallics 18:2660
24. Albéniz AC, Schulte G, Crabtree RH (1992) Organometallics 11:242
25. Patra SK, Bera JK (2006) Organometallics 25:6054
26. Blackburn OA, Coe BJ (2011) Organometallics 30:2212
27. Liniger M, Gschwend B, Neuburger M, Schaffner S, Pfaltz A (2010) Organometallics 29:5953
28. Yoshida Y, Miyamoto R, Nakato A, Santo R, Kuwamura N, Gobo K, Nishioka T, Hirotsu M, Ichimura A, Hashimoto H, Kinoshita I (2011) Bull Chem Soc Jpn 84:600
29. Esterulas MA, Lledós A, Oliván M, Oñate E, Tajada MA, Ujaque G (2003) Organometallics 22:3753
30. Eguillor B, Esteruelas MA, Oliván M, Oñate E (2004) Organometallics 23:6015
31. Liu X-C, Hubbard JL, Scouten WH (1995) J Organomet Chem 493:91
32. Li X, Voge T, Incarvito CD, Crabtree RH (2005) Organometallics 24:62
33. Cheney AJ, Shaw BL (1972) J Chem Soc Dalton Trans 754
34. Burford RJ, Piers WE, Parvez M (2012) Organometallics 31:2949
35. Benhamou L, César V, Lugan N, Lavigne G (2007) Organometallics 26:4673
36. Maimone TJ, Milner PJ, Kinzel T, Zhang Y, Takase MK, Buchwald SL (2011) J Am Chem Soc 133:18106
37. Stallinger S, Reitsamer C, Schuh W, Kopacka H, Wurst K, Peringer P (2007) Chem Commun 510
38. Churruca F, SanMartin R, Tellitu I, Domínguez E (2006) Tetrahedron Lett 47:3233
39. Rybtchinski B, Vigalok A, Ben-David Y, Milstein D (1996) J Am Chem Soc 118:12406
40. Vigalok A, Rybtchinski B, Shimon LJW, Ben-David Y, Milstein D (1999) Organometallics 18:895

41. Li J, Lutz M, Spek AL, van Klink GPM, van Koten G, Gebbink RJMK (2010) Organometallics 29:1379
42. Garralda MA, Hernández R, Ibarlucea L, Pinilla E, Torres MR (2003) Organometallics 22:3600
43. Royer AM, Rauchfuss TB, Gray DL (2009) Organometallics 28:3618
44. Royer AM, Salomone-Stagni M, Rauchfuss TB, Meyer-Klaucke W (2010) J Am Chem Soc 132:16997
45. Dunina VV, Gorunova ON, Linantsov MV, Grishin YK, Kuz'mina LG, Kataeva NA, Churakov AV (2000) Inorg Chem Commun 3:354
46. Takahashi Y, Tokuda A, Sakai S, Ishii Y (1972) J Organomet Chem 35:415
47. Takeda N, Watanabe D, Nakamura T, Unno M (2010) Organometallics 29:2839
48. Rüger R, Rittner W, Jones PG, Isenberger W, Sheldrick GM (1981) Angew Chem Int Ed 20:382
49. Oulié P, Nebra N, Saffon N, Maron L, Martin-Vaca B, Bourissou D (2009) J Am Chem Soc 131:3493
50. Fernandez S, Pfeffer M, Ritleng V, Sirlin C (1999) Organometallics 18:2390
51. Nabavizadeh SM, Haghighi MG, Esmaeilbeig AR, Raoof F, Mandegani Z, Jamali S, Rashidi M, Puddephatt RJ (2010) Organometallics 29:4893
52. Nabavizadeh SM, Amini H, Shahsavari HR, Namdar M, Rashidi M, Kia R, Hemmateenejad B, Nekoeinia M, Ariafard A, Hosseini FN, Gharavi A, Khalafi-Nezhad A, Sharbati MT, Panahi F (2011) Organometallics 30:1466
53. Djukic J-P, Fetzer L, Czysz A, Iali S, Sirlin C, Pfeffer M (2010) Organometallics 29:1675
54. Choudhury TD, Shen Y, Rao NVS, Clark NA (2012) J Organomet Chem 712:20
55. Li X, Chen P, Faller JW, Crabtree RH (2005) Organometallics 24:4810
56. Barbasiewicz M, Szadkowska A, Makal A, Jarzembska K, Woźniak K, Grela K (2008) Chem Eur J 14:9330
57. Aharoni A, Vidavsky Y, Diesendruck CE, Ben-Asuly A, Goldberg I, Lemcoff NG (2011) Organometallics 30:1607
58. Omae I (2007) J Organomet Chem 692:2608
59. Omae I (1998) Kagaku Kogyo 40:303

Chapter 6
Reasons Why Organometallic Intramolecular-Coordination Five-Membered Ring Compounds Are Extremely Easily Synthesized Through Cyclometalation Reactions

Abstract There are three reasons why organometallic intramolecular-coordination five-membered ring compounds are extremely easily synthesized through cyclometalation reactions: The first is metal activation by the coordination of lone electron pair of a hetero atom to a metal atom. The second is the chelate effect caused by the formation of a five-membered ring. The third is also metal activation by ligands, such as hetero atom groups (e.g., bipyridines, phosphines, and carboxylates), unsaturated groups (e.g., aryl, allyl, Cp, and Cp*), carbonyl groups, and halogen atoms (F, Cl, Br, Cl), which are bonded to a central metal atom. Many articles including title words, such as C–H activation, C–X activation, C–H functionalization, and chelation-assisted reactions, related to the abovementioned metal activation and chelate effect in cyclometalation reactions have been published recently.

Keywords C–H activation • C–H functionalization • Chelate effect • Chelation-assisted • C–X activation • Cyclometalation • Five-membered ring • Intramolecular coordination • Metal activation

6.1 High Stability of Five-Membered Ring Compounds in Organometallic Intramolecular-Coordination Compounds

As described in Chap. 2, we found that direct reactions of halo-β-carbonyl compounds with tinfoil proceed surprisingly easily and promptly and afford mostly simple five-membered ring products as compared with the reactions of alkyl halides

I. Omae, *Cyclometalation Reactions: Five-Membered Ring Products as Universal Reagents*, DOI 10.1007/978-4-431-54604-7_6, © Springer Japan 2014

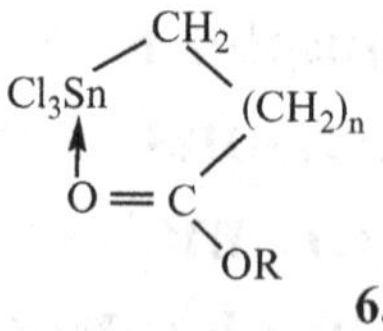

Fig. 6.1 Trichlorocarboxylic
acid ester tin compounds **6.1**
[1–3]

$$Cl_3Sn \overset{CH_2}{\underset{O=C}{\overset{}{\big\backslash}}} (CH_2)_n \\ OR$$

6.1

R = Me, ^{i}Pr; n = 1
R = Et, n = 2

in the presence of the same catalysts. This selectivity is recognizable in Table 5.3 in the previous chapter, where the ratio of five-membered rings is about 95 %.

In 1994, about 30 years after direct reactions of halo-β-carbonyl compounds with tinfoil were first conducted, the author sought to show the reasons why cyclometalation reactions for the production of organometallic intramolecular-coordination five-membered ring compounds proceed extremely quickly. To verify the above, the five-membered ring structure was demonstrated to be the most stable as compared with four- and six-membered ring structures.

Based on three X-ray diffraction data for trichlorocarboxylic acid ester tin compounds **6.1** as shown in Fig. 6.1 [1–3], the strain energies for the cyclic structures of alkoxycarbonylalkyltin trichlorides were calculated utilizing the MM2 force field and the Dreiding force field among molecular force field methods, the PM3 semiempirical molecular orbital method and the ab initio molecular orbital method making use of the Gaussian 94 program.

The results showed that the orders of stability in these ring systems are five-membered rings > six-membered rings by the MM2 method, five- > six- > seven- > four-membered rings by the Dreiding method, and five- > six- > four-membered rings by the PM3 semiempirical molecular orbital method and ab initio molecular orbital method. Hence, all the results of calculations distinctly demonstrated the five-membered ring to be the most stable as compared with rings with four or six members.

For the ab initio molecular orbital method employing Gaussian 94, basis set: 3–21+g**, the ring formation energies of five-, six-, and four-membered rings are 10.83, 3.88, and −16.23 kcal/mol, respectively, and the stability of the five-membered ring is thus some three times that of the six-membered ring, while the four-membered ring shows a rather unstable value as compared with a linear structure.

These results were published in *The Proceedings of the 69th Annual Meeting of the Chemical Society of Japan, 1995* [4], and in the journal *Kagaku Kogyo* in 1998 [5].

In 1997, Gómez, Granell, and Martinez [6] also reported on similarly easy five-membered ring formation. The cyclometalation reactions of imines with palladium acetate have been studied in toluene solutions, as shown in Fig. 6.2.

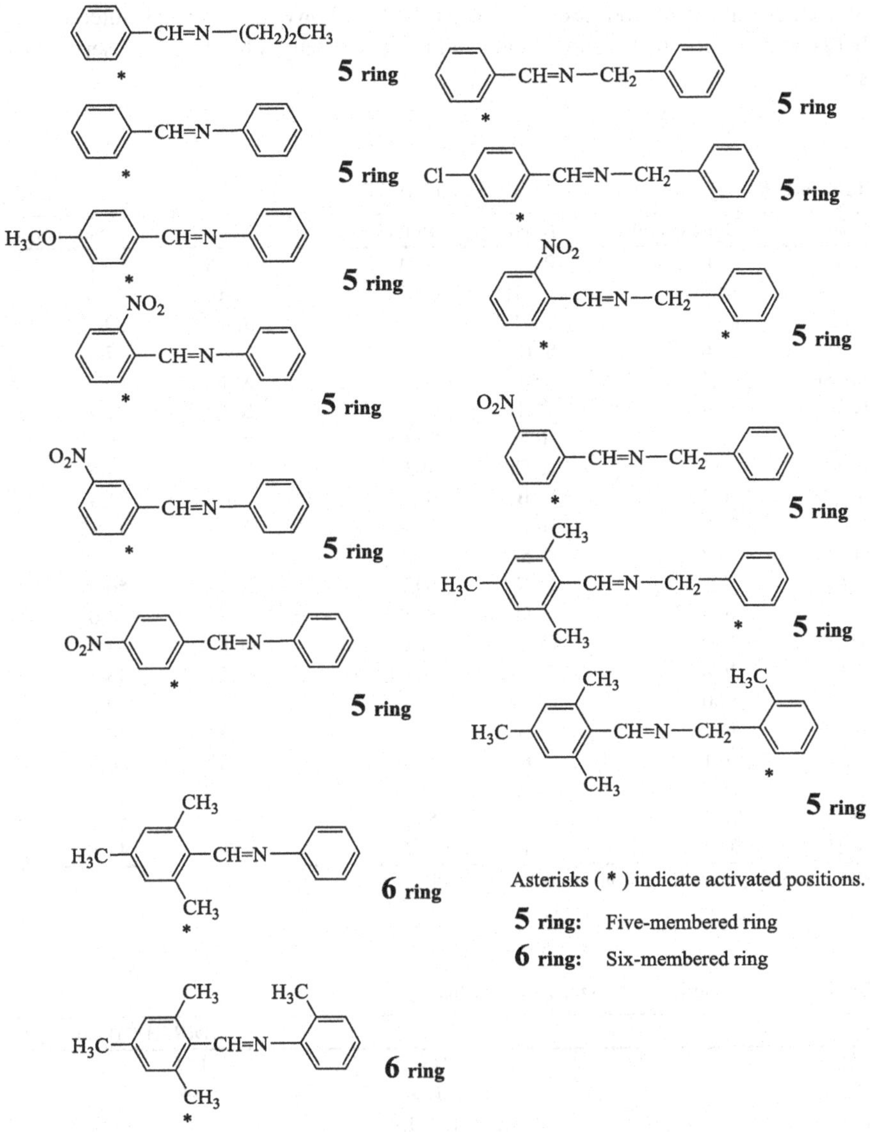

Fig. 6.2 Formation of ring compounds in cyclometalation reactions of imines with palladium acetate [6]

6.2 Metal Activation, Agostic Interactions, C–H Activation, and the Chelate Effect in Cyclometalation Reactions

The energy values for single bonds are shown in Tables 6.1 [7] and 6.2 [8]. Alkanes, aromatic compounds, and hydrogen have high bond energies, and these compounds are generally considered to be inert. In such inert compounds as hydrogen molecules, alkanes, and aromatic C–H bonds, however, M–C–H interactions between transition metal compounds and these substances have been observed for some time.

Table 6.1 Energy values for single bonds (kcal/mol) [7][a]

Bond	Bond energy	Bond	Bond energy	Bond	Bond energy
H–H	104.2	P–H	76.4	Si–Cl	85.7
C–C	83.1	As–H	58.6	Si–Br	69.1
Si–Si	42.2	O–H	110.6	Si–I	50.9
Ge–Ge	37.6	S–H	81.1	Ge–Cl	97.5
Sn–Sn	34.2	Se–H	66.1	N–F	64.5
N–N	38.4	Te–H	57.5	N–Cl	47.7
P–P	51.3	H–F	134.6	P–Cl	79.1
As–As	32.1	H–Cl	103.2	P–Br	65.4
Sb–Sb	30.2	H–Br	87.5	P–I	51.4
Bi–Bi	25.0	H–I	71.4	As–F	111.3
O–O	33.2	C–Si	69.3	As–Cl	69.9
S–S	50.9	C–N	69.7	As–Br	56.5
Se–Se	44.0	C–O	84.0	As–I	41.6
Te–Te	33.0	C–S	62.0	O–F	44.2
F–F	36.6	C–F	105.4	O–Cl	48.5
Cl–Cl	58.0	C–Cl	78.5	S–Cl	59.7
Br–Br	46.1	C–Br	65.9	S–Br	50.7
I–I	36.1	C–I	57.4	Cl–F	60.6
C–H	98.8	Si–O	88.2	Br–Cl	52.3
Si–H	70.4	Si–S	54.2	I–Cl	50.3
N–H	93.4	Si–F	129.3	I–Br	42.5

[a]Bond energy values for diatomic molecules of alkali metals

Table 6.2 Characteristics of some hydrocarbons [8]

RH	$\rightarrow$	R•	D(R–H)* (kcal mol^{-1})
CH_4		CH_3•	104
C_2H_6		C_2H_5•	98
C_3H_8		n-C_3H_7•	97
		iso-C_3H_7•	94
$cyclo$-C_6H_{12}		$cyclo$-C_6H_{11}•	94
C_6C_6		C_6H_5•	109
$CH_2{=}CH_2$		$CH_2{=}CH$•	106
$CH{\equiv}CH$		$CH{\equiv}C$•	120
$C_6H_5CH_3$		$C_6H_5CH_2$•	85
CH_3CN		$NCCH_2$•	79
H_2		H•	104
H_2O		HO•	118

Energy of C–H bonds

As shown in Eq. (6.1), for example, Vaska and DiLuzio reported in 1962 that hydrogen molecules react with iridium phosphine compounds at room temperature to produce a hydrogen complex [9, 10].

1962

$$Ir^I Cl(CO)(PPh_3)_2 \quad + \quad H_2 \xrightarrow[\substack{Benzene \\ 25\ ^\circ C, 1d}]{H_2\,(1\ atom)} \quad Ir^{III}H_2Cl(CO)(PPh_3)_2 \qquad (6.1)$$

As shown in Eq. (6.2), it was indeed found in 1969 that Pt^{II} could react with C–H bonds in methane molecules; H/D exchange between methane and D_2O was observed in the presence of Pt^{II}. This reaction was suggested to proceed via the mechanism of an electrophilic substitution [10, 11].

1969

$$CH_4 \quad + \quad D_2O \xrightarrow[D_2O\ /\ CH_3COOD]{K_2PtCl_4} \quad CH_3D \quad + \quad HDO$$

$$(6.2)$$

Electrophilic substitution

$$(\ Pt^{II} \quad + \quad R\text{-}H \longrightarrow Pt^{II}\text{-}R \quad + \quad H^+)$$

In 1970, Green and Knowles reported that benzene or toluene reacts with dicyclopentadienyltungsten to give oxidative addition products **6.2**, as shown in Eq. (6.3) [12].

1970

$$Cp_2W \quad + \quad HC_6H_4\text{-}R \xrightarrow{120\ ^\circ C,\ 3\ d} \quad Cp_2W \begin{smallmatrix} H \\ \diagdown \end{smallmatrix} \langle\!\langle \text{aryl} \rangle\!\rangle R \qquad (6.3)$$

6.2

These M–C–H interactions were named "agostic interactions" by Brookhart and Green [13] in 1983. They reported in detail on agostic interactions in their 1988 review entitled "Carbon–Hydrogen–Transition Metal Bonds [14]."

Carbon–hydrogen bonds, especially those with saturated (sp^3) carbon centers, are normally considered to be chemically inert. There is a rapidly increasing body of evidence, however, showing that C–H bonds can act as ligands to transition metal centers by the formation of three-centered, two-electron bonds (3c–2e) and that the interactions occur to such an extent that they show marked effects on the molecular and electronic structures and hence on the reactivity of the molecules.

Table 6.3 Ratios of the ligand groups in these 157 types of agostic interactions [14]

Ligand group	Ratio (%)
Phosphines	About 40
CO	About 30
Cp + Cp*	About 30
π-ally	About 20
Alkenyl, Cl, N	About 10
C:, O, diene, I, Ph, etc.	1–6

Brookhart and Green reported about 200 agostic interactions in their review. The ratios of the ligand groups in these 157 types of agostic interactions are shown in Table 6.3:

Examples for simple alkyl compounds with an agostic hydrogen atom in various kinds of agostic interactions are shown in Fig. 6.3 [14].

Agostic interactions are caused by the connection of these ligands, such as phosphines, carbonyl, cyclopentadienyl (Cp), pentamethylcyclopentadienyl (Cp*), π-allyl, η^4-dienyl, nitrogen, oxygen, chlorine, bromine, and iodine, to metals.

Based on the data for about 200 agostic interactions in their review and the various types of agostic interactions shown in Fig. 6.3 [14], it was determined that these ligands initiate agostic interactions in metal atoms. In other words, it is considered that interactions of ligands with metal atoms activate the inert C–H bonds.

Most of the ligands in these agostic interactions are phosphines. The lone electron pair in phosphorus atom coordinate to the metal atom. The number of electrons transferred from the phosphines to the metal varies with the phosphine moiety, e.g., an alkylphosphine or phenylphosphine. Metal atoms usually receive electrons from phosphine ligands. Under certain circumstances, however, the metal atoms pass electrons to the phosphines. Phosphines are especially well matched to low-valence transition metal compounds and form stable complexes as a result. During catalytic reactions, on the other hand, the ligand adjusts the oxidation state of the central metal atom or provides structural support, and the reaction is promoted as a result [15].

It is considered that these ligands should cause metal activation, agostic interaction, and C–H activation.

These inert compounds are able to react with transition metal compounds containing some ligands under mild reaction conditions, that is, without being exposed to high temperatures or high pressure. Most importantly the reactions proceed with transition metal compounds activated by these ligands, such as phosphines, cyclopentadienyl, carbonyl, π-allyl, and halogen atoms, bonding with the transition metal atom.

In 2003, van der Boom and Milstein [16] reported that the following three mechanistic pathways account for most metal-based aromatic C–H activation processes in a review entitled "Cyclometalated Phosphine-Based Pincer Complexes: Mechanistic Insight in Catalysis, Coordination, and Bond Activation."

1. Oxidative addition
2. Electrophilic metalation
3. Agostic intermediate

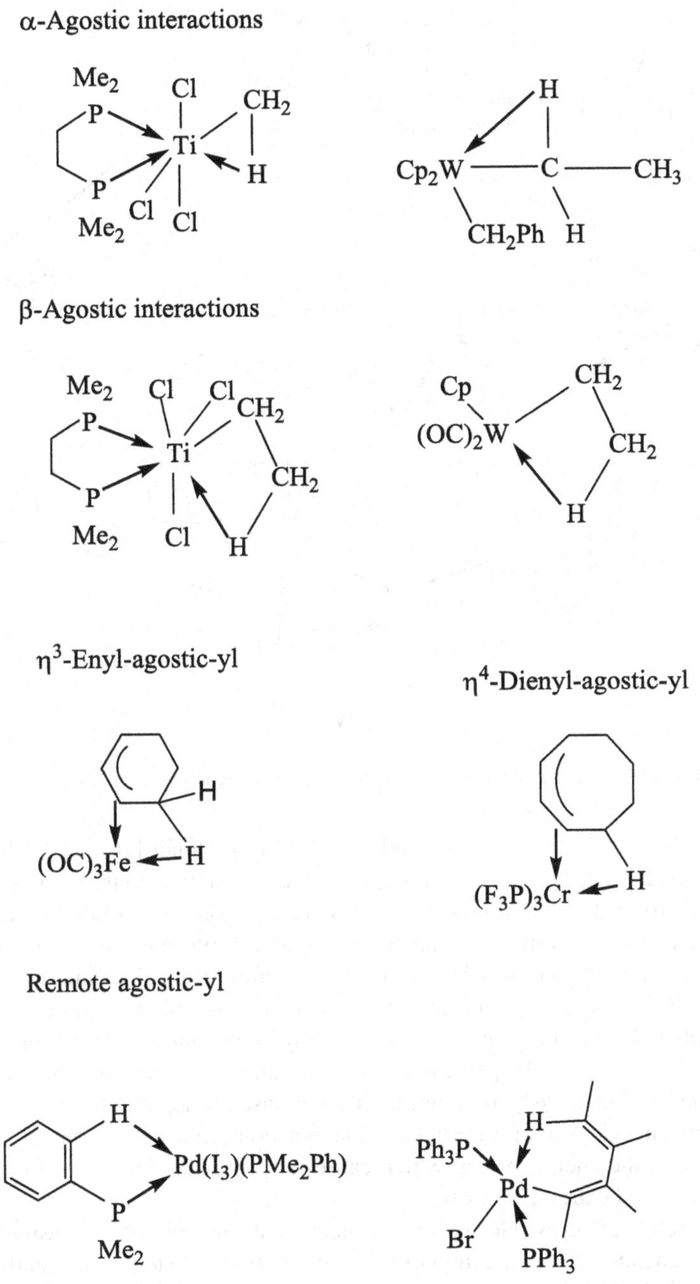

α-Agostic interactions

β-Agostic interactions

η^3-Enyl-agostic-yl

η^4-Dienyl-agostic-yl

Remote agostic-yl

Fig. 6.3 Various types of agostic interactions [14]

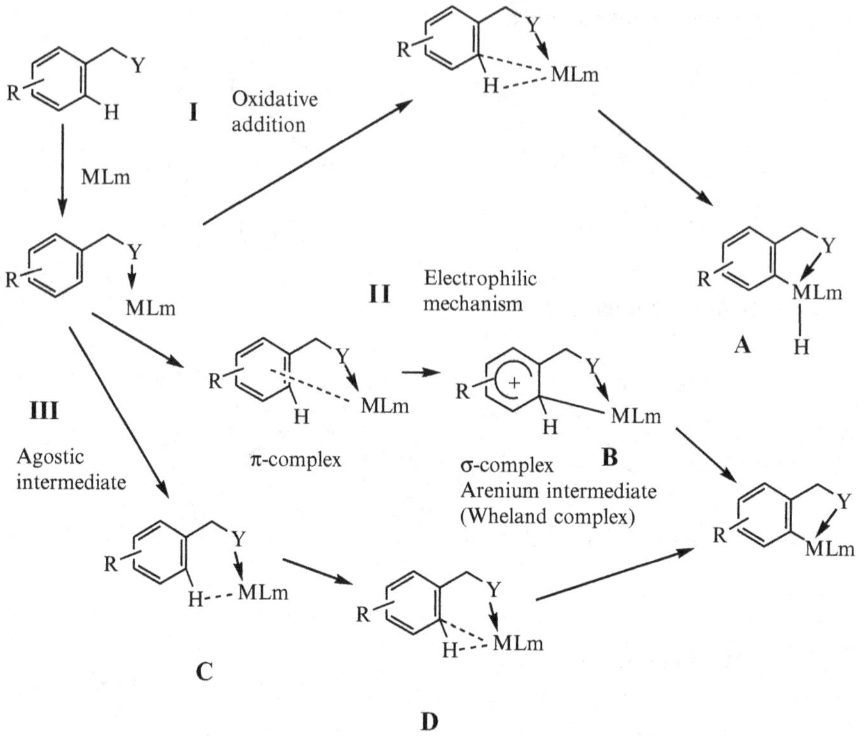

Scheme 6.1 Three mechanistic pathways of cyclometalation reactions [16, 17]

Based on a review entitled "Activation of Aryl C–H Bonds [16, 17]," Scheme 6.1 is proposed for these three pathways of cyclometalation reactions, as follows:

Route I (oxidative addition) involves a concerted oxidative addition process with the formation of metal-hydride species **A**. Alternatively, an electrophilic attack by the metal center on the aryl *ipso*-carbon may afford a metal arenium (Wheland) complex **B** followed by proton loss. In the agostic C–H bond activation route, the six-membered transition state **C** including a hydrogen–metal interaction has been found to initiate the C–H activation process, leading to an agostic intermediate **D** and acting simultaneously as an intramolecular base for deprotonation.

In this mechanism of cyclometalation reactions, the author places a heavy emphasis on the chelate effect, which causes strong activation enabling the reactions to proceed extremely easily.

The chelate effect is relevant to five-membered rings because it easily forms a bidentate chelate ring with extremely low strain energy, enabling the central metal atom to bond with both the coordinating atom and a γ-carbon atom.

Basola and Johnson [18] reported in a monograph entitled *Coordination Chemistry* that chelating ligands, in general, form more stable complexes than those of monodentate analogs. This is known as the *chelate effect*, and it is explained in terms of favorable entropy for the chelation process.

Shilow and Shul'pin [8] also pointed out for the oxidative addition process shown in Scheme 6.1 that, due to the chelate effect, the intramolecular cleavage of the C–H bond occurs much more readily than intermolecular activation and gives rise to a more stable σ-organyl hydride complex **A**.

Most reactions among cyclometalation reactions are orthometalation reactions. These reactions are used for the synthesis of orthometalation products or derivatives of orthometalation products. Pfeffer et al. [19] named these synthetic reactions "chelation-assisted reactions" in 2002.

Chen et al. [20], for example, reported on chelation-assisted reactions in an article entitled "Chelation-Assisted Carbon–Halogen Bond Activation by a Rhodium(I) Complex" in 2009. These reactions proceed by C–Br bond activation via an oxidative addition mechanism. They take place in reactions of $[Rh(PPh_3)_2(acetone)_2]$ $^+PF_6^-$ with 2-(2-bromophenyl)pyridine at room temperature to give the cyclometalated rhodium bromide shown in Eq. (6.4).

$$ (6.4) $$

Iwasawa et al. [21] also reported chelation-assisted reactions in an article entitled "Rhodium(I)-Catalyzed Direct Carboxylation of Arenes with CO_2 via Chelation-Assisted C–H Bond Activation," in which the cyclometalation reactions proceed easily and form cyclometalation intermediates. The metal atoms are active centers in their intermediates. Hence, the active metal atom reacts easily with inert carbon dioxide to give carboxylic acid derivatives. Examples include the cyclometalation of 2-phenylpyridine as a substrate in the presence of a rhodium intermediate. Carbon dioxide can be inserted into the rhodium–phenyl carbon bond, and a methyl ester is formed with $TMSCH_2N_2$ from a rhodium carboxylate, as shown in Eq. (6.5). The reaction mechanism is proposed as shown in Scheme 6.2 [21].

$$ (6.5) $$

COE = Cyclooctene

Chatani et al. [22] reported on another chelation-assisted reaction, moreover, in an article entitled "Nickel-Catalyzed Chelation-Assisted Transformations Involving

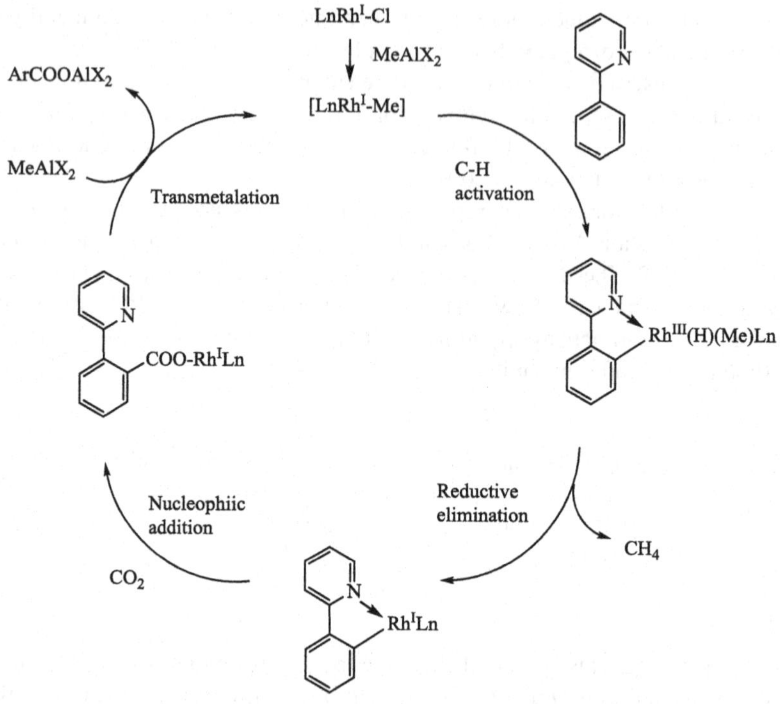

Scheme 6.2 Chelation-assisted reaction mechanism [21]

Ortho C–H Bond Activation: Regioselective Oxidative Cycloaddition of Aromatic Amides to Alkynes." Cyclometalation reactions with nickel phosphine COD complexes used an amido nitrogen atom as the coordinating atom. The insertion and cyclization with alkynes is then proposed to proceed via the cyclometalation nickel intermediate as an active center to give the six-membered isoquinolone derivatives shown in Eq. (6.6). In 2013, Chatani et al. [22] also reported on these chelation-assisted transformations in details as the review articles.

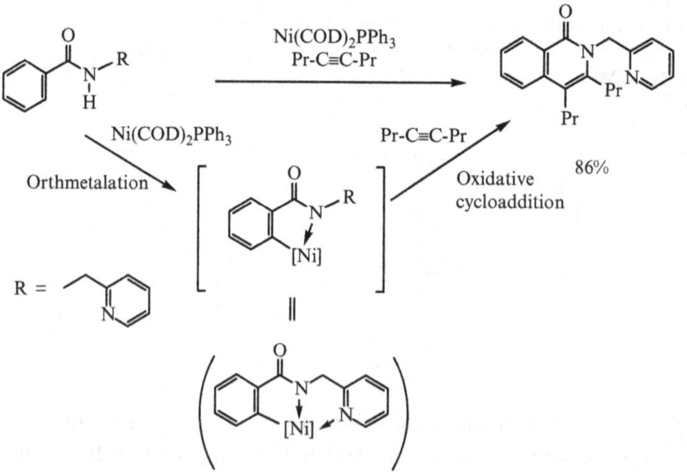

(6.6)

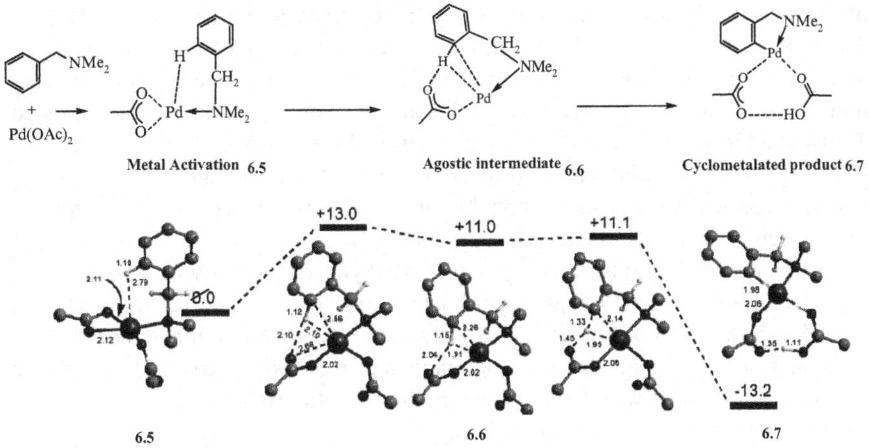

Scheme 6.3 Computed reaction profile (kcal/mol) and key distances (Å) for the cyclometalation of Pd(OAc)$_2$ with *N,N*-dimethylbenzylamine via an agostic interaction [24]

In cyclometalation reactions, as shown in Eq. (6.7), a central atom is activated by the coordination of lone electron pair of the coordinating atom, such as nitrogen, phosphorus, oxygen, or sulfur. Metal activation **6.3** and agostic interactions **6.4** consequently occur, and bond formation between the γ-carbon atom and the metal atom follows. Final cyclization proceeds via agostic interactions and CH activation through the chelate effect to cause bidentate coordination of the central metal atom with the coordinating atom and the γ-carbon atom [23].

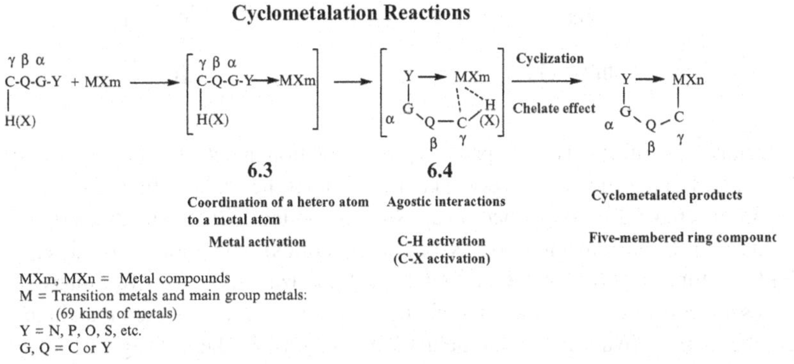

Cyclometalation Reactions

(6.7)

MXm, MXn = Metal compounds
M = Transition metals and main group metals:
    (69 kinds of metals)
Y = N, P, O, S, etc.
G, Q = C or Y

As for the computer chemistry of cyclometalation reactions, the reaction with the most representative substrate, *N,N*-dimethylbenzylamine in a palladium compound, was studied. This reaction proceeds very easily, and its intermediate state, or agostic interaction, is therefore not actually isolated. As shown in Scheme 6.3, however, the activation energy for the agostic interaction is only 13 kcal/mol. It may be pointed out that the acyl group assists the formation of the agostic interaction in the reaction, as exhibited by the agostic intermediate **6.6** shown in Scheme 6.3 [24].

In the cyclometalation reaction of *N,N*-dimethylbenzylamine with palladium acetate, palladium acetate is thought to play the dual roles of electrophilic activation of the arene (Wheland complex **B** in Scheme 6.1) and an intramolecular base for deprotonation (base-assisted metalation [25]), as shown in agostic intermediate **6.6** in Scheme 6.3 [24]. As the acetate, carboxylate acts as the key to activating the reaction, as Ackermann [25] reported in his review entitled "Carboxylate-Assisted Transition-Metal-Catalyzed C–H Bond Functionalizations: Mechanism and Scope" in 2011. This carboxylate-assisted reaction is essentially the same as a chelation-assisted reaction, because the carboxylate strongly assists the cyclometalation reaction, which is to say, the metal cyclization reaction.

In the case of reactions of halo-β-monocarboxylic acid esters with tinfoil, on the other hand, the carbonyl oxygen coordinates to the tin atom as the first step. It is considered that the coordination to the metal causes metal activation in the tin atom. C–Br activation and a cyclization reaction then proceed with the oxidative addition reactions of C and Br with the tin atom, as shown in Eq. (6.8) [26].

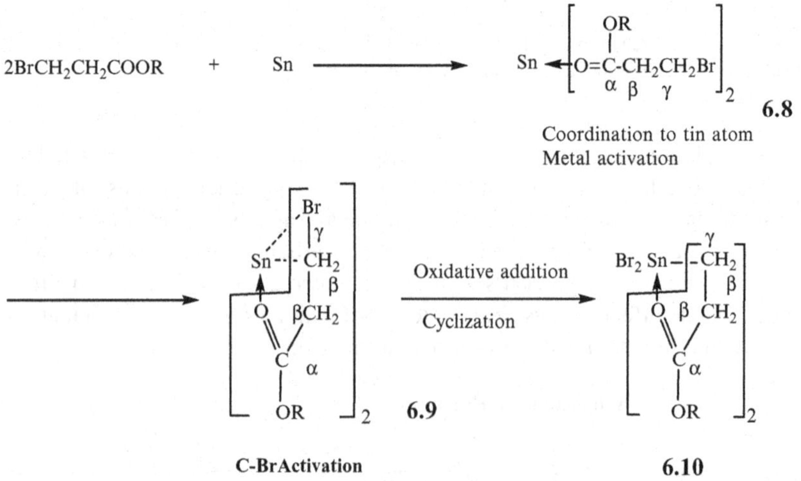

$$2BrCH_2CH_2COOR \quad + \quad Sn \quad \longrightarrow \quad Sn \longleftarrow \left[\begin{array}{c} OR \\ | \\ O=C\text{-}CH_2CH_2Br \\ \alpha \ \ \beta \ \ \gamma \end{array} \right]_2$$

6.8

Coordination to tin atom
Metal activation

$$(6.8)$$

Oxidative addition

Cyclization

6.9   C-BrActivation

6.10

The mechanisms of the two steps in cyclometalation reactions are shown in Eq. (6.7). The first step is metal activation, and the second is the chelate effect. Recently, numerous articles have been published on agostic interactions, C–H activations (C–H bond activations), C–X activations (C–X bond activations), chelation-assisted reactions, and C–H functionalizations (C–H bond functionalizations). However, many of these articles are concerned with cyclometalation reactions. It is considered that, in the first stage, the metal activation in cyclometalation reactions is related to agostic interactions, C–H activations, and C–X activations and that, in the second stage, the chelate effect is related to chelation-assisted reactions and C–H functionalizations.

In Scheme 6.3, for example, the titles of Eqs. (6.4), (6.5), and (6.6) presented above are as follows:

Scheme 6.3: Easy cyclometalation reactions with a benzylamine proceed via *agostic interaction* as shown in agostic intermediate **6.6**

Equation (6.4): "Chelation-Assisted Carbon–Halogen Bond Activation by a Rhodium(I) Complex"

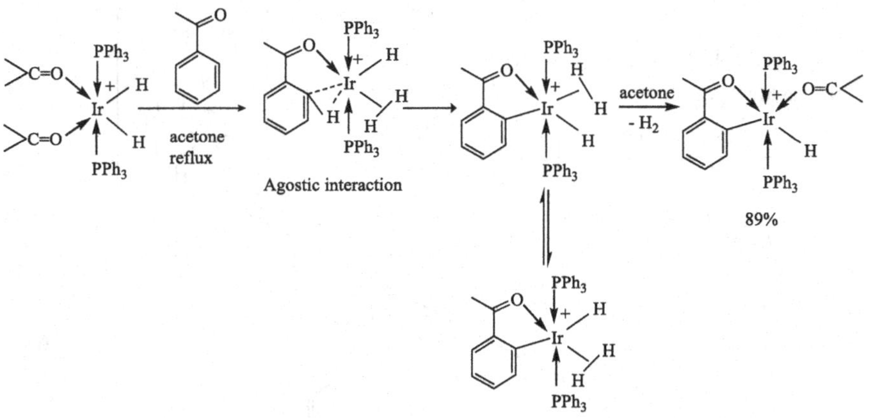

Scheme 6.4 Proposed mechanism for the formation of hydride [27]

Equation (6.5): "Rhodium(I) Catalyzed Direct Carboxylation of Arenes with CO_2 via Chelation-Assisted C–H Bond Activation"

Equation (6.6): "Nickel-Catalyzed Chelation-Assisted Transformations Involving Ortho C–H Bond/Activation: Regioselective Oxidation Cycloaddition of Aromatic Amides to Alkynes"

In an earlier review the author published on the agostic interaction entitled "Agostic Bonds in Cyclometalation" in 2011 [23], cyclometalation reactions proceed extremely easily with a one-step reaction between metal compounds and substrates containing a heteroatom such as O, S, N, P, or As. Under mild reaction conditions, however, many agostic compounds, which are intermediates in these cyclometalation reactions with both transition metal and main group metal compounds, can be isolated.

Crabtree et al. reported on the cyclometalation reactions of acetophenones, for which a plausible mechanism is shown in Scheme 6.4 [27]. The acetophenone coordinates with a carbonyl group when an acetone ligand is substituted. The carbonyl coordination guides the *ortho* C–H bonds to interact with iridium, probably via a C–H agostic intermediate. C–H activation is then proposed to generate an iridium aryl hydride dihydrogen species; and the loss of dihydrogen and substitution by acetone give a final orthometalation product. The agostic products are not isolated because the cyclometalation of acetophenones proceeds very easily, since the activation of the *ortho* C–H bonds in acetophenones easily gives orthometalated products [27, 28].

In the cyclometalation reaction of 2-(dimethylamino)pyridine with an iridium complex ([$H_2Ir(OCMe_2)_2L_2$]BF_4 (L = PPh_3) as shown in Eq. (6.9), the agostic intermediate **6.11** observed by NMR is predicted (DFT(B3PWW91) computation) to give C–H oxidative addition to form an alkyl intermediate **6.12**. Loss of H_2 leads to the fully characterized alkyl product **6.13**, which loses acetone to give the alkylidene product **6.14** by rapid reversible α-elimination [29].

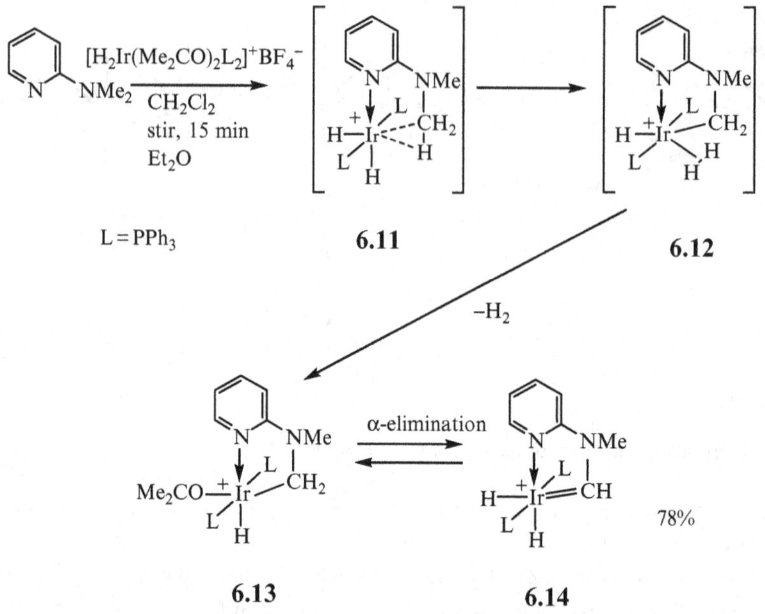

$$(6.9)$$

Other examples of agostic interactions in cyclometalation reactions are shown in Eqs. (6.10)–(6.17), as follows [30–41]:

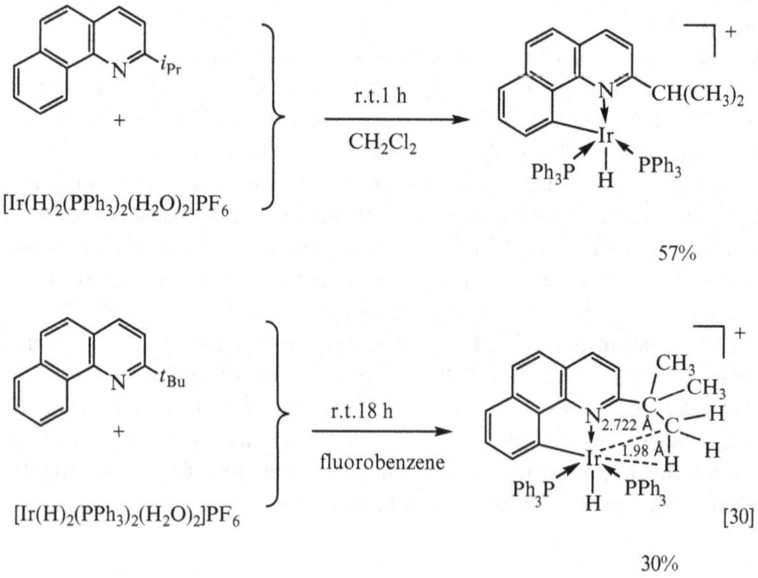

$$(6.10)$$

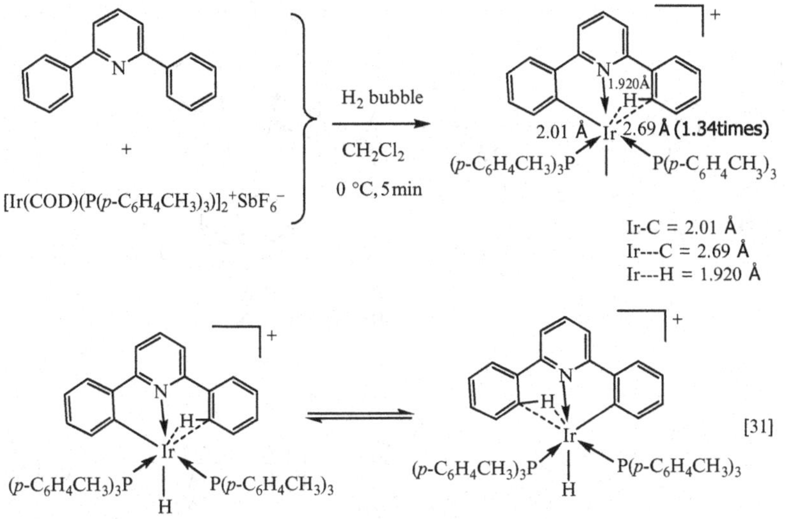
+
[Ir(COD)(P(p-C6H4CH3)3)]2+SbF6−
H2 bubble
CH2Cl2
0 °C, 5 min
1.920Å
2.01 Å
2.69 Å (1.34times)
(p-C6H4CH3)3P
P(p-C6H4CH3)3
Ir-C = 2.01 Å
Ir---C = 2.69 Å
Ir---H = 1.920 Å
[31]

(6.11)

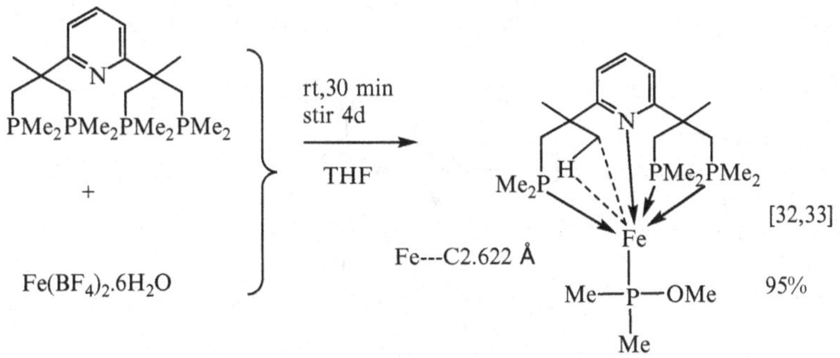
PMe2PMe2PMe2PMe2
+
Fe(BF4)2.6H2O
rt,30 min
stir 4d
THF
Me2P
PMe2PMe2
Fe---C2.622 Å
Me—P—OMe
Me
[32,33]
95%

(6.12)

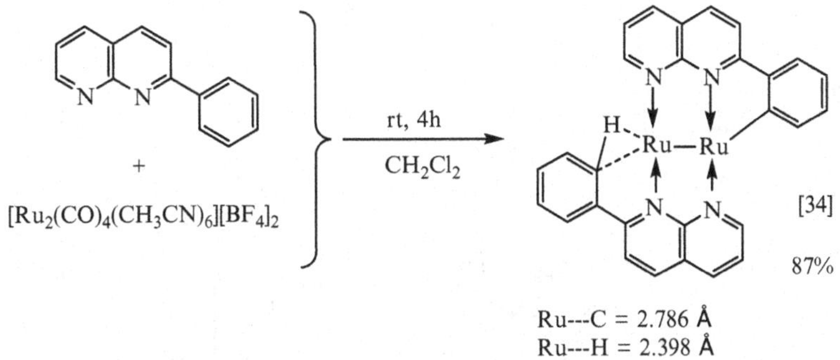
+
[Ru2(CO)4(CH3CN)6][BF4]2
rt, 4h
CH2Cl2
[34]
87%
Ru---C = 2.786 Å
Ru---H = 2.398 Å

(6.13)

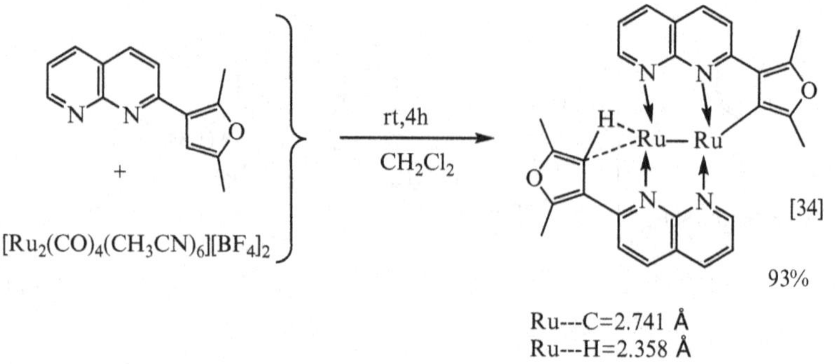

$$[Ru_2(CO)_4(CH_3CN)_6][BF_4]_2$$

rt, 4h
CH_2Cl_2

[34]

(6.14)

93%

Ru---C=2.741 Å
Ru---H=2.358 Å

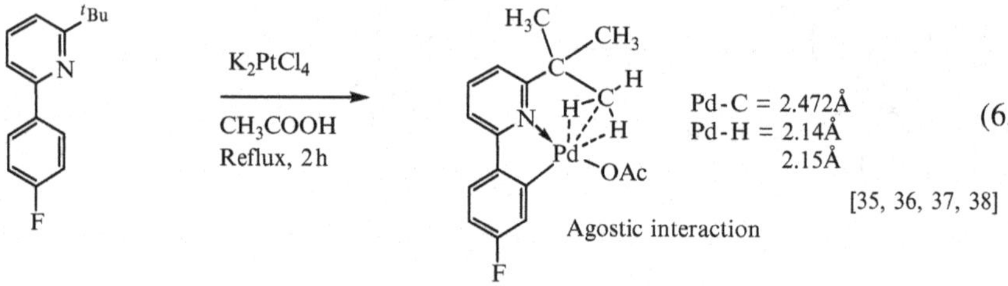

K_2PtCl_4

CH_3COOH
Reflux, 2 h

Pd-C = 2.472Å
Pd-H = 2.14Å
       2.15Å

(6.15)

[35, 36, 37, 38]

Agostic interaction

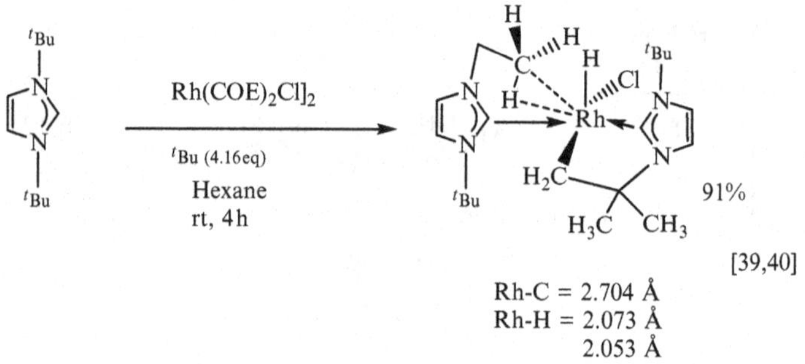

$Rh(COE)_2Cl]_2$

^tBu (4.16eq)
Hexane
rt, 4h

91%

(6.16)

[39,40]

Rh-C = 2.704 Å
Rh-H = 2.073 Å
       2.053 Å

MeO   PPh'$_2$

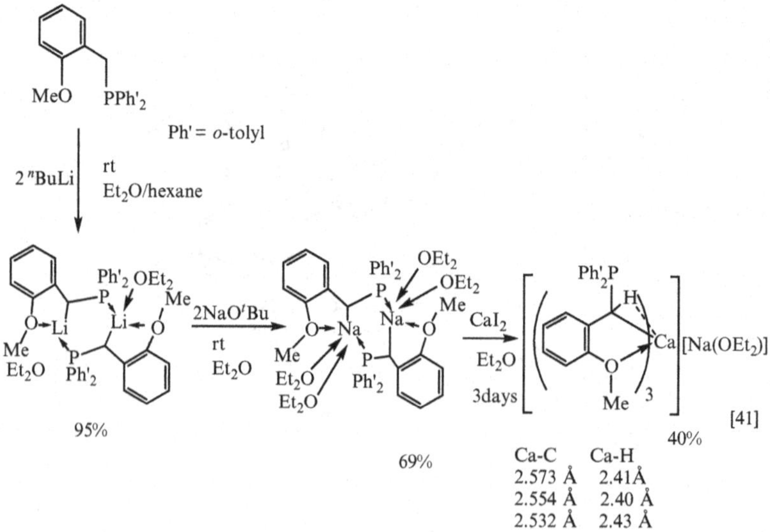

Ph' = o-tolyl

2 ^{n}BuLi   rt
Et$_2$O/hexane

2NaO^{t}Bu
rt
Et$_2$O

CaI$_2$
Et$_2$O
3days

[Na(OEt$_2$)]

(6.17)

[41]

95%

69%

40%

Ca-C   Ca-H
2.573 Å   2.41Å
2.554 Å   2.40 Å
2.532 Å   2.43 Å

Other terms related to the cyclometalation reactions described above, that is, C–H activations (C–H bond activations), C–X activations (C–X bond activations), chelation-assisted reactions, and C–H functionalizations (C–H bond functionalizations), are employed as title words in article titles, as shown in the following sequence.

First, articles employing C–H activations as title words are shown in Eqs. (6.18)–(6.29).

Intermolecular Amidation of Unactivated sp^2 and sp^3 C–H Bonds via Palladium-Catalyzed Cascade *C–H Activation*/Nitrene Insertion [42]

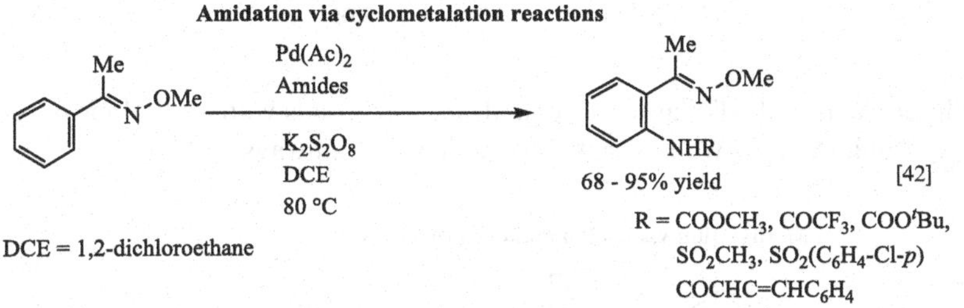

Amidation via cyclometalation reactions

$$(6.18)$$

Double *C–H Activation* in Osmium and Ruthenium Centers: Carbene versus Olefin Products

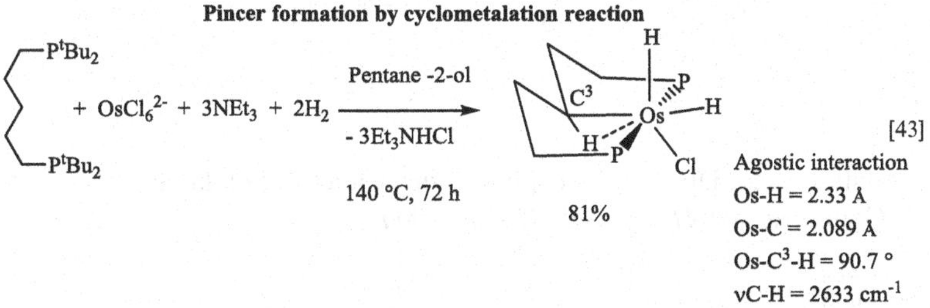

Pincer formation by cyclometalation reaction

$$(6.19)$$

C–H Activation Induced by Water. Monocyclometalated to Dicyclometalated: CNC Tridentate Platinum Complexes

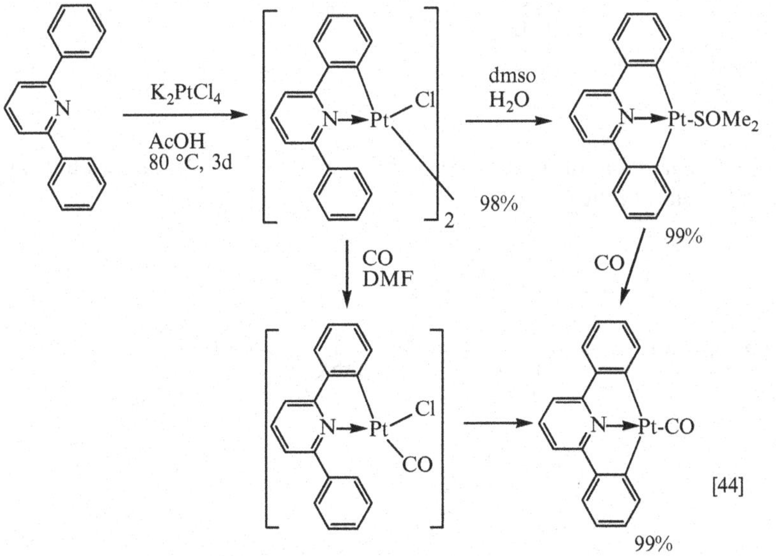

$$(6.20)$$

Transcyclometalation Processes with Late Transition Metals: C_{aryl}–H Bond *Activation* via Noncovalent C–H……Interaction

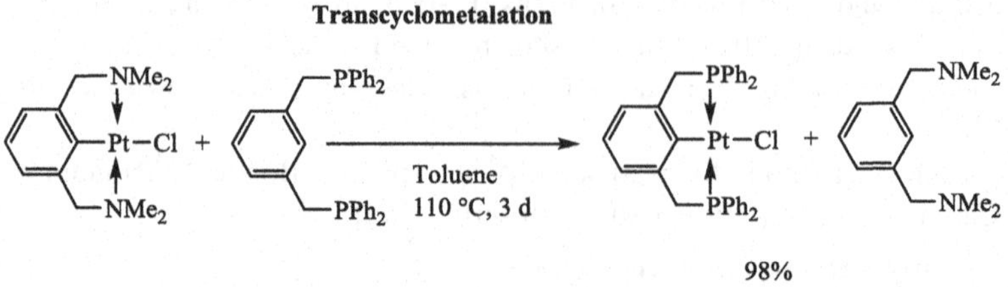

Transcyclometalation

$$(6.21)$$

Toluene
110 °C, 3 d

[45]

98%

Palladium-Catalyzed Alkylation of sp^2 and sp^3 C–H Bonds with Methylboroxine and Alkylboronic Acids: Two Distinct *C–H Activation* Pathways

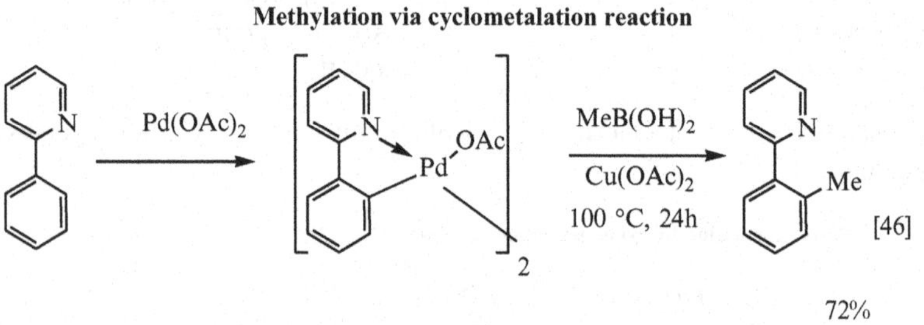

Methylation via cyclometalation reaction

$Pd(OAc)_2$

$MeB(OH)_2$

$Cu(OAc)_2$

100 °C, 24h

$$(6.22)$$

[46]

72%

Cyclometalation of Dimesitylphosphine in Cationic Palladium(II) and Platinum(II) Complexes: *P–H* versus *C–H Activation*

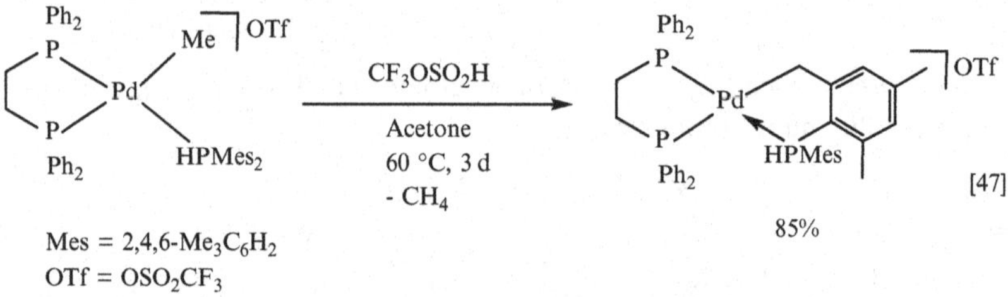

CF_3OSO_2H

Acetone
60 °C, 3 d
- CH_4

$$(6.23)$$

[47]

85%

Mes = $2,4,6\text{-Me}_3C_6H_2$
OTf = OSO_2CF_3

C–H Bond Activation of Heteroarenes Mediated by a Half-Sandwich Iron Complex of *N*-Heterocyclic Carbene

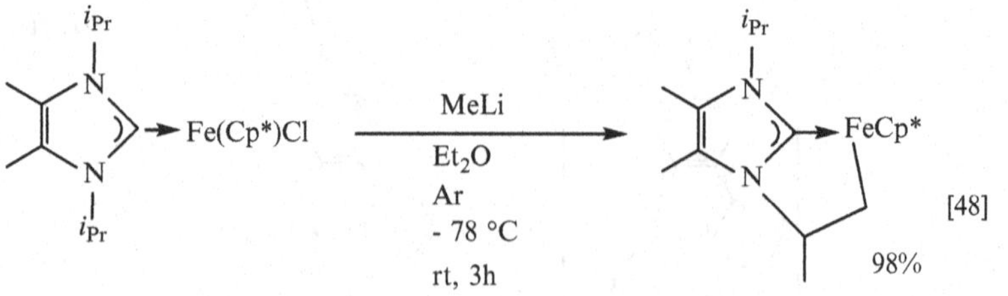

$Fe(Cp^*)Cl$

MeLi

Et_2O
Ar
- 78 °C
rt, 3h

$$(6.24)$$

[48]

98%

Cyclopalladation of (S)-4-*tert*-Butyl-2-methyl-2-oxazoline: An Unprecedented Case of (sp³)C–H Bond Activation Resulting in *Exo*-Palladacycle Formation

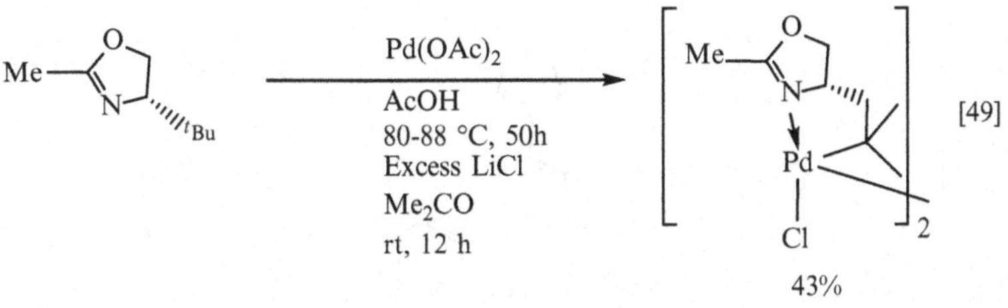

$$(6.25)$$

Unexpected Formation of Chiral Pincer CNN Nickel Complexes with β-Diketiminato-Type Ligands via C–H Activation: Synthesis, Properties, Structures, and Computational Studies

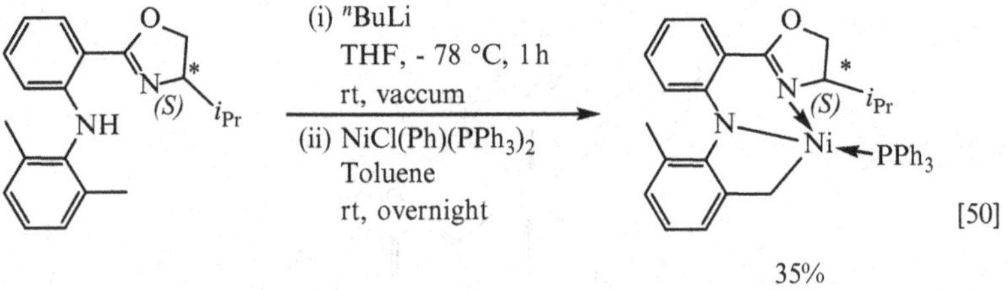

$$(6.26)$$

Reaction of 16-Electron Ruthenium and Iridium Amide Complexes with Acidic Alcohols: Intramolecular C–H Bond Activation and the Isolation of Cyclometalated Complexes

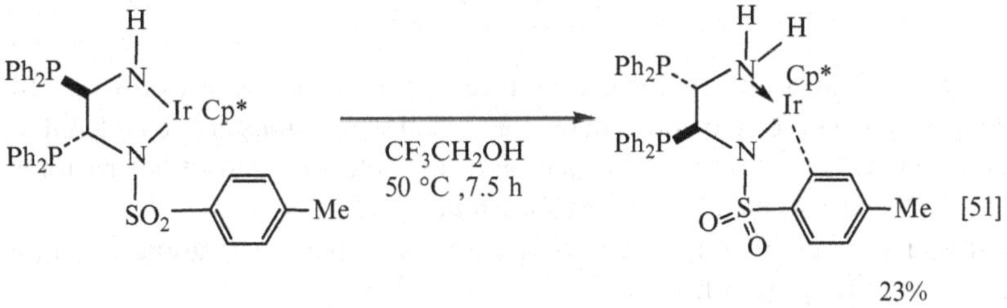

$$(6.27)$$

Understanding C–H Bond Activation on a Diruthenium(I) Platform

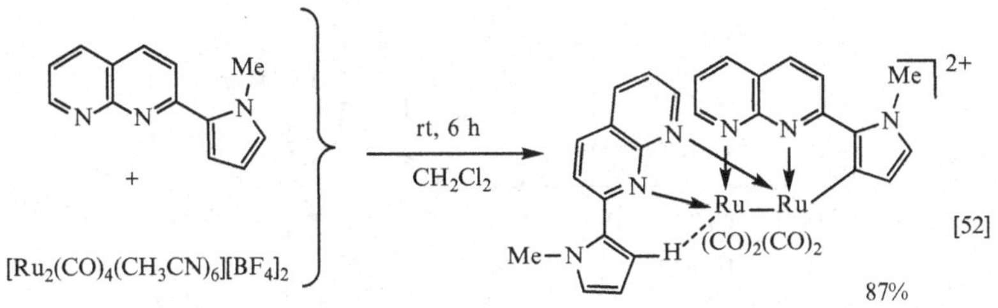

$$(6.28)$$

Regioselective *C–H Activation* of Cyclometalated Bis-Tridentate Ruthenium Complexes

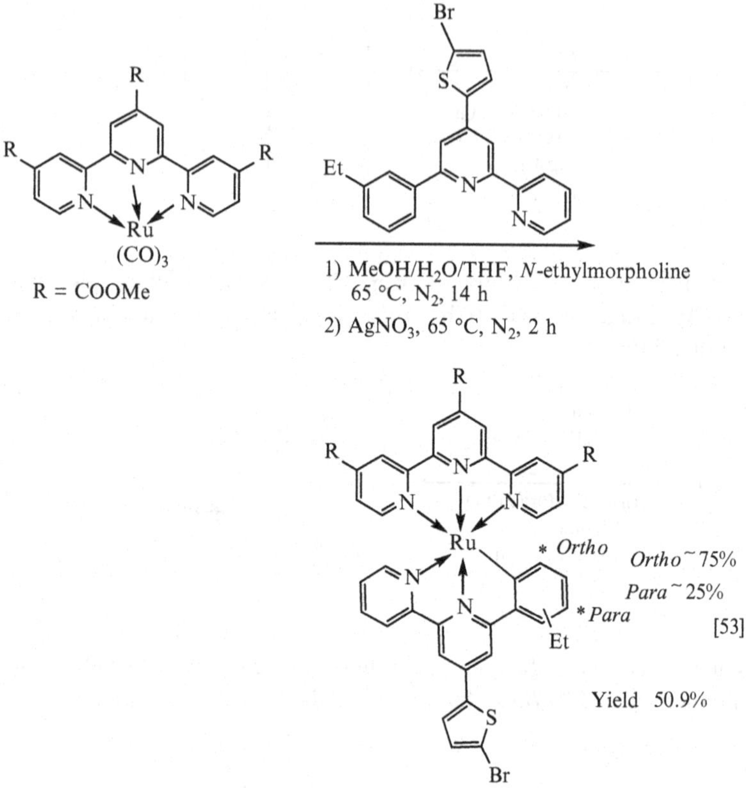

(6.29)

These articles concerning cyclometalation reactions employ title words related to C–H activations. These activations in cyclometalation reactions are considered to be caused by metal activation by coordination of lone electron pair of heteroatoms such as N, P, O, and S to metal atoms, as shown in Eq. (6.7).

Secondly articles employing C–X activations (C–X bond activations) as title words are shown in Eqs. (6.30)–(6.41).

Activation of C–F and CH– Bonds by Platinum in Trifluorinated [C,N,N′] Ligands. Crystal Structures of [PtFMe$_2$\{Me$_2$NCH$_2$CH$_2$NHCH(CH$_2$COMe)(2,4-C$_6$H$_2$F$_2$)\}] and [PtMe\{Me$_2$NCH$_2$CH$_2$N=CH(2,3,4-C$_6$HF$_3$)\}]

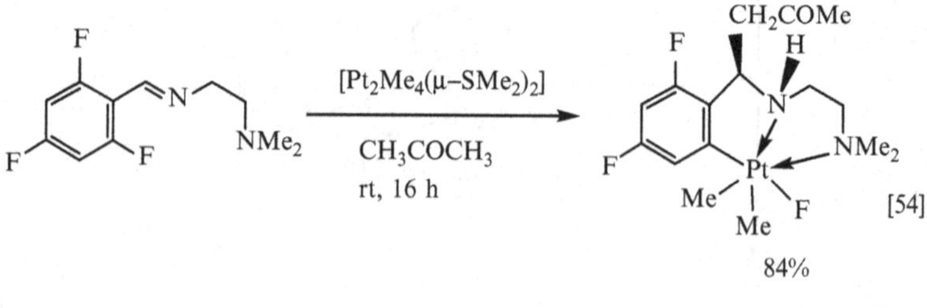

(6.30)

Kinetico-Mechanistic Studies on Intramolecular *C–X Bond Activation* (*X=Br*, *Cl*) of Amino–Imino Ligands on Pt(II) Compounds. Prevalence of a Concerted Mechanism in Nonpolar, Polar, and Ionic Liquid Media

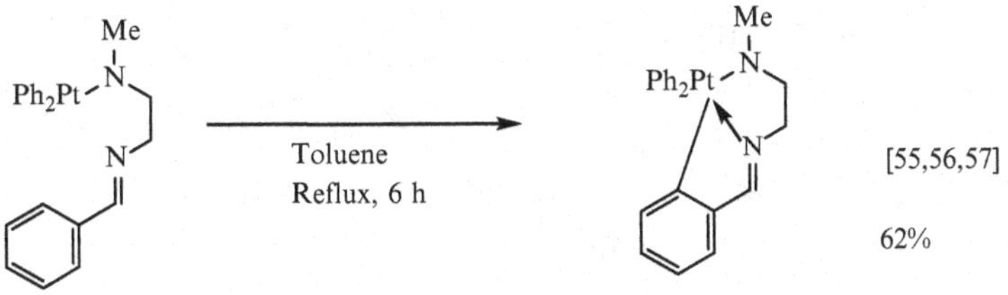

Me
Ph_2Pt
N
N
Toluene
Reflux, 6 h
Me
Ph_2Pt
N
N
[55,56,57]
62%

(6.31)

Cyclometalation Reactions Involving *C–Cl Bond Activation* of *Ortho*-Chlorinated Substrated with Imines as Anchoring Groups by Cobalt Complexes

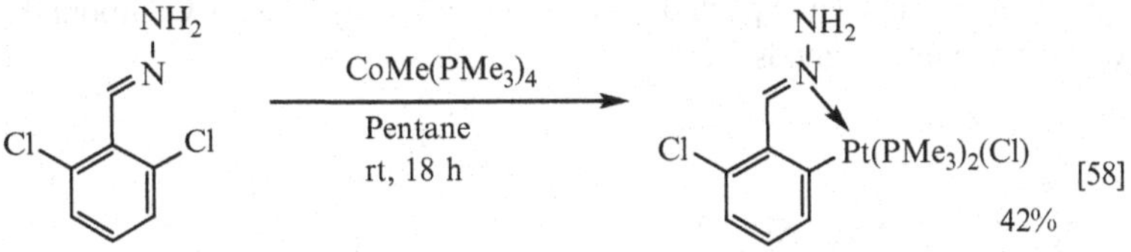

NH_2
N
Cl
Cl
$CoMe(PMe_3)_4$
Pentane
rt, 18 h
NH_2
N
Cl
$Pt(PMe_3)_2(Cl)$
[58]
42%

(6.32)

C–Cl Bond Activation of *Ortho*-Chlorinated Imines with Iron Complexes in Low Oxidation States

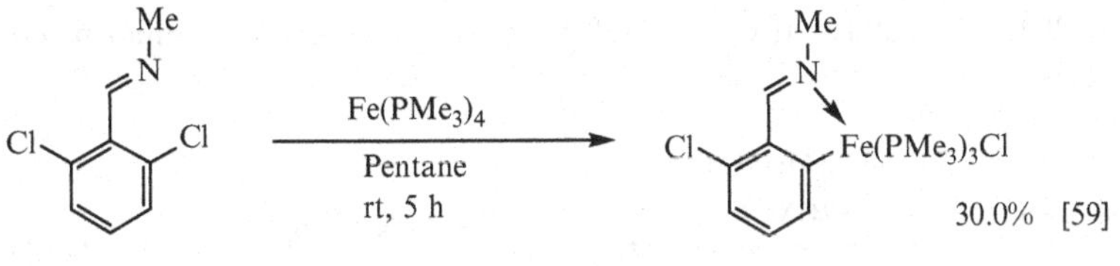

Me
N
Cl
Cl
$Fe(PMe_3)_4$
Pentane
rt, 5 h
Me
N
Cl
$Fe(PMe_3)_3Cl$
30.0% [59]

(6.33)

A PCN Ligand System. Exclusive *C–C Activation* with Rhodium(I) and C–H Activation with Platinum(II)

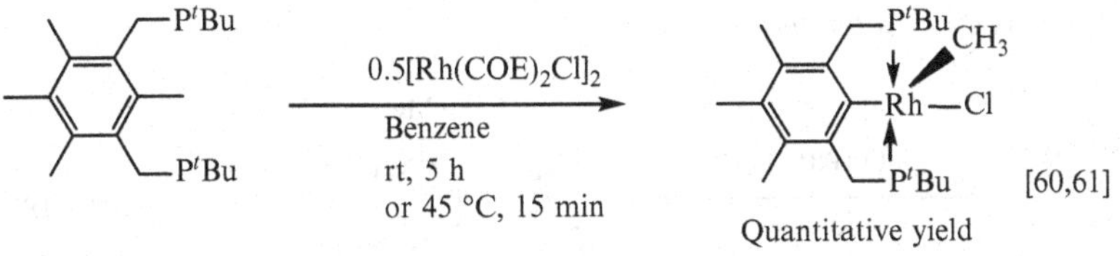

P^tBu
P^tBu
$0.5[Rh(COE)_2Cl]_2$
Benzene
rt, 5 h
or 45 °C, 15 min
P^tBu
CH_3
Rh—Cl
P^tBu
[60,61]
Quantitative yield

(6.34)

Exploration of the Mechanism of Platinum(II)-Catalyzed *C–F Activation*: Characterization and Reactivity of Platinum(IV) Fluoroaryl Complexes Relevant to Catalysis

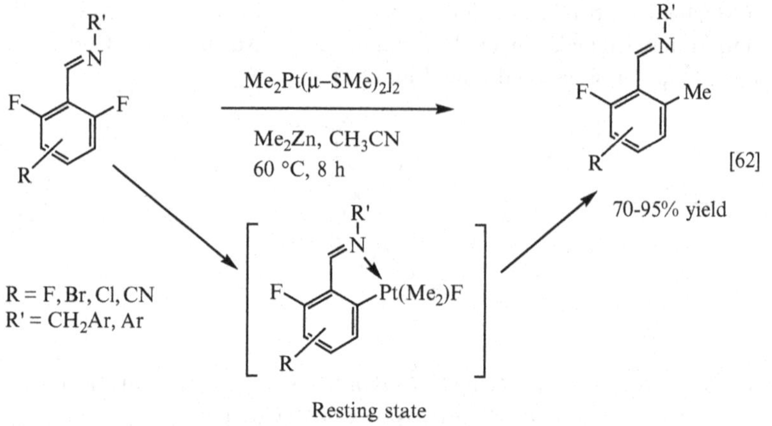

$$[62] \quad (6.35)$$

70-95% yield

R = F, Br, Cl, CN
R' = CH$_2$Ar, Ar

Resting state

C–F Bond Cleavage and Unexpected *C–CN Activation* by Cobalt Compounds Supported with Phosphine Ligands

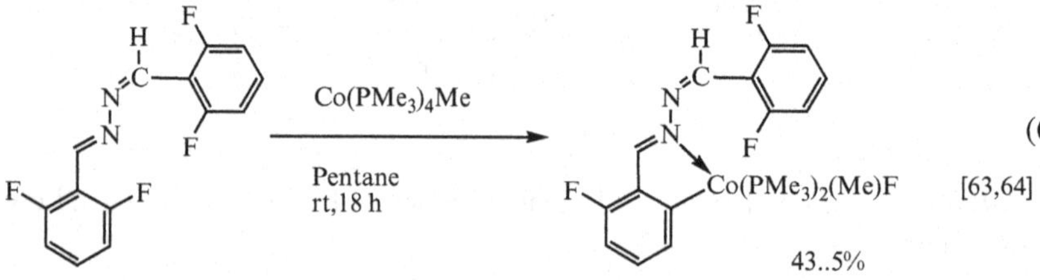

$$(6.36)$$

[63,64]

43..5%

Selective Platinum-Catalyzed *C–F Bond Activation* as a Route to Fluorinated Aryl Methyl Ethers

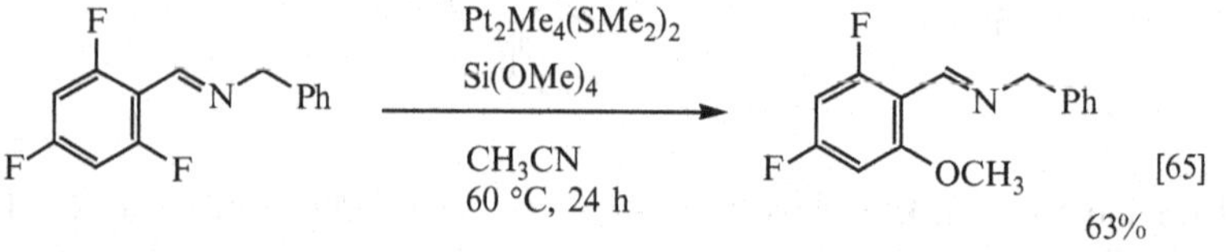

$$(6.37)$$

[65]

63%

Reactions of a Hexahydride–Osmium Complex with Aromatic Ketones: C–H Activation versus *C–F Activation*

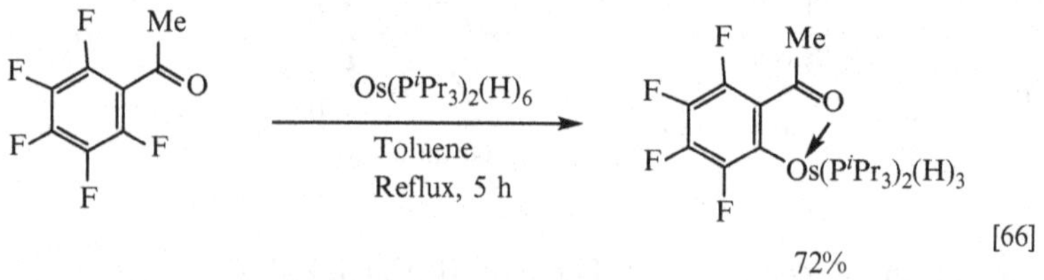

$$(6.38)$$

[66]

72%

Formation of Difluoromethylene–Arenium Complexes by Consecutive Aryl–CF$_3$ *C–C Bond Activation* and C–F Bond Cleavage

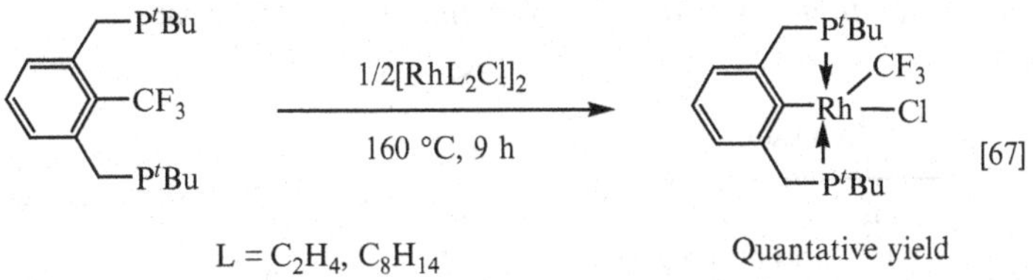

$$\text{[structure]} \xrightarrow[160\ °C,\ 9\ h]{1/2[RhL_2Cl]_2} \text{[structure]} \quad [67] \tag{6.39}$$

L = C_2H_4, C_8H_{14} Quantative yield

Preparation of Five-Membered Nickelacycles of N-Donor Ligands by *Activation of C–X Bonds* (*X = F, Cl, or Br*). X-ray Crystal Structure of [NiBr{2-(CH=NCH_2Ph) C_6H_4}(2,4,6-Me_3C_5H_2N)]

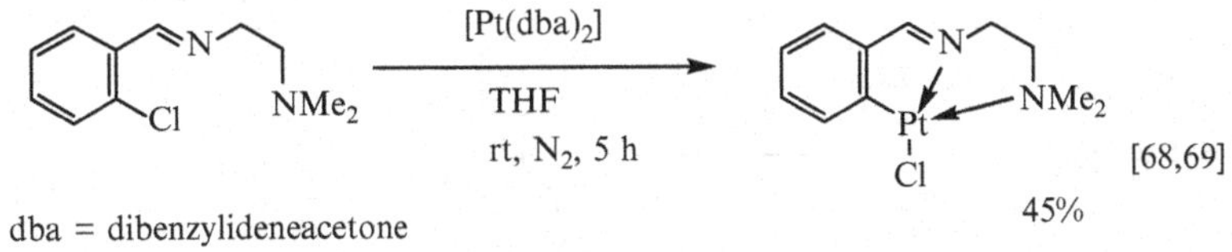

$$\text{[structure]} \xrightarrow[\substack{\text{THF}\\ \text{rt, } N_2,\ 5\ h}]{[Pt(dba)_2]} \text{[structure]} \quad [68,69] \tag{6.40}$$

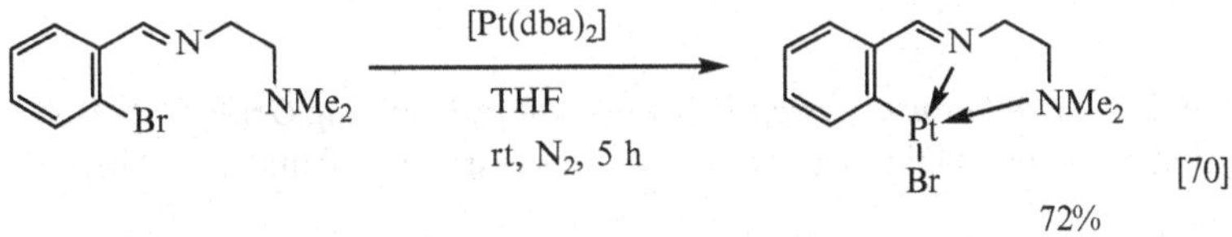

dba = dibenzylideneacetone   45%

$$\text{[structure]} \xrightarrow[\substack{\text{THF}\\ \text{rt, } N_2,\ 5\ h}]{[Pt(dba)_2]} \text{[structure]} \quad [70] \tag{6.41}$$

72%

Third, articles employing chelation-assisted reactions as title words are shown in Eqs. (6.42)–(6.51).

Chelate-Assisted Oxidative Coupling Reaction of Arylamides and Unactivated Alkenes: Mechanistic Evidence for Vinyl C–H Bond Activation Promoted by an Electrophilic Ruthenium Hydride Catalyst

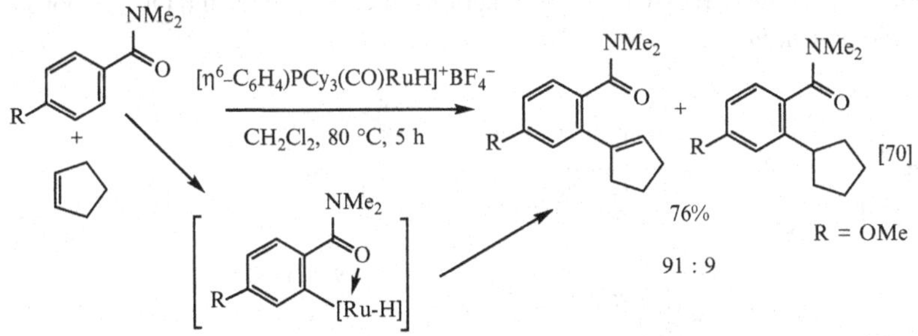

$$\text{[structure]} \xrightarrow[CH_2Cl_2,\ 80\ °C,\ 5\ h]{[\eta^6\text{-}C_6H_4)PCy_3(CO)RuH]^+BF_4^-} \text{[structures]} \quad [70] \tag{6.42}$$

76%

91 : 9   R = OMe

Cobalt-Catalyzed Hydroarylation of Alkynes Through *Chelation-Assisted C–H Bond Activation*

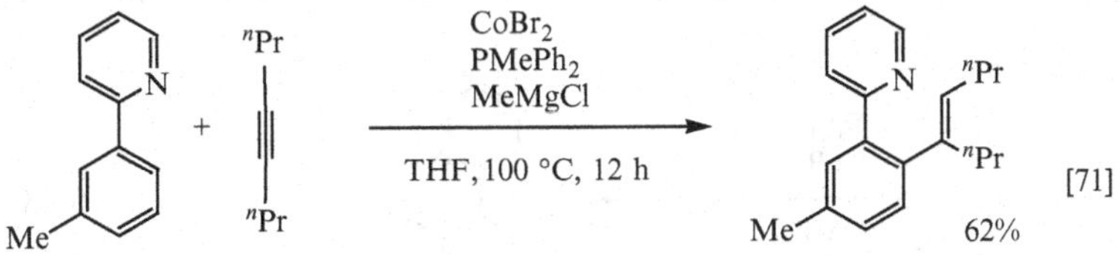

$$\text{[structure]} \xrightarrow[THF,\ 100\ °C,\ 12\ h]{\substack{CoBr_2\\ PMePh_2\\ MeMgCl}} \text{[structure]} \quad [71] \tag{6.43}$$

62%

Copper-Mediated *Chelation-Assisted Ortho Nitration* of (Hetero)arenes

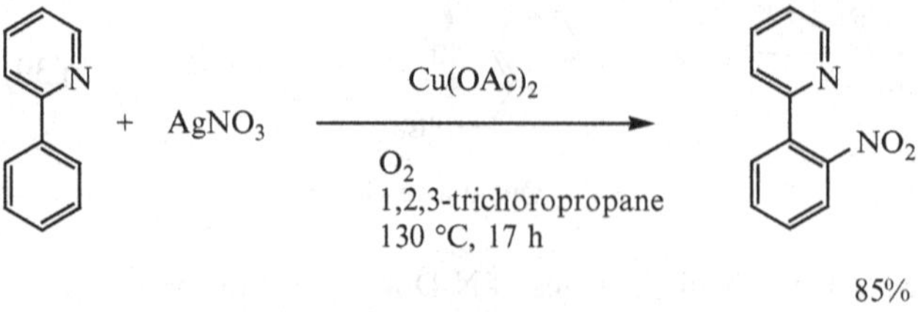

N
+ AgNO3
Cu(OAc)2
O2
1,2,3-trichoropropane
130 °C, 17 h
N
NO2
85%

(6.44) [72]

Chelation-Assisted Palladium-Catalyzed Cascade Bromination/Cyanation Reaction of 2-Arylpyridine and 1-Arylpyrazole C–H Bonds

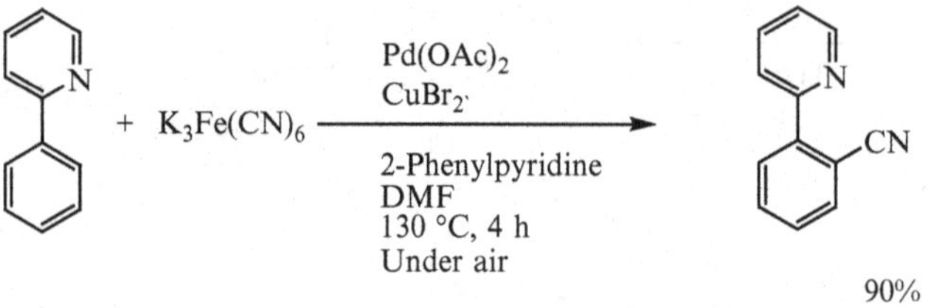

N
+ K3Fe(CN)6
Pd(OAc)2
CuBr2·
2-Phenylpyridine
DMF
130 °C, 4 h
Under air
N
CN
90%

(6.45) [73]

Nickel-Catalyzed *Chelation-Assisted Transformations* Involving *Ortho* C–H Bond Activation: Regioselective Oxidative Cycloaddition of Aromatic Amides to Alkynes

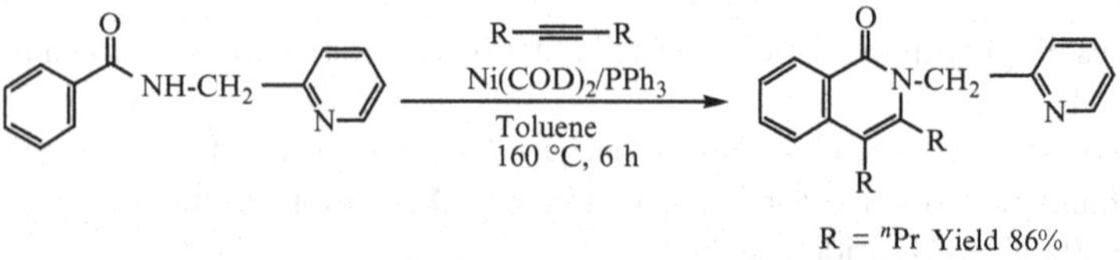

O
NH-CH2
N
R—≡—R
Ni(COD)2/PPh3
Toluene
160 °C, 6 h
O
N-CH2
N
R
R
R = nPr Yield 86%

(6.46) [22]

Rhodium-Catalyzed Cross-Coupling Reactions of Carboxylate and Organoboron Compounds via *Chelation-Assisted C–C Bond Activation*

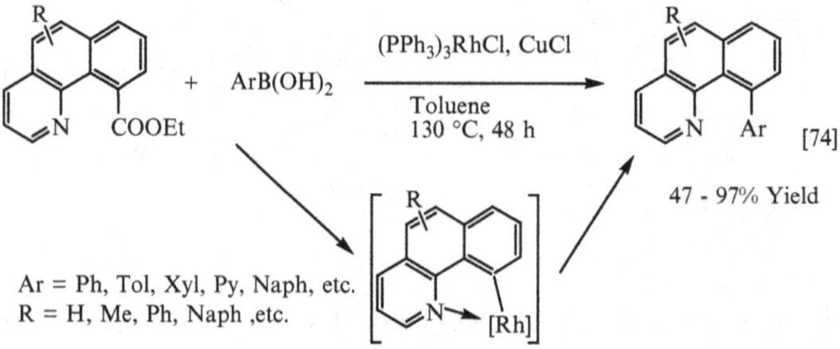

R
N
COOEt
+ ArB(OH)2
(PPh3)3RhCl, CuCl
Toluene
130 °C, 48 h
R
N
Ar
47 - 97% Yield
R
N
[Rh]
Ar = Ph, Tol, Xyl, Py, Naph, etc.
R = H, Me, Ph, Naph ,etc.

(6.47) [74]

Chelation-Assisted Palladium-Catalyzed Direct Cyanation of 2-Arylpyridine C–H Bonds

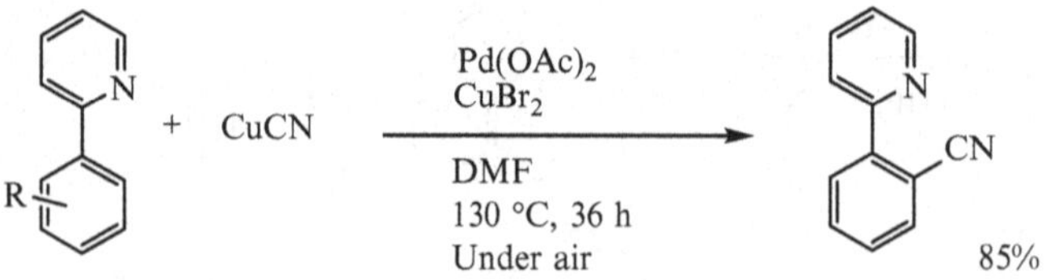

N
R
+ CuCN
Pd(OAc)2
CuBr2
DMF
130 °C, 36 h
Under air
N
CN
85%

(6.48) [75]

Rhodium-Catalyzed Intermolecular Amidation of Arenes with Sulfonyl Azides via *Chelation-Assisted C–H Bond Activation*

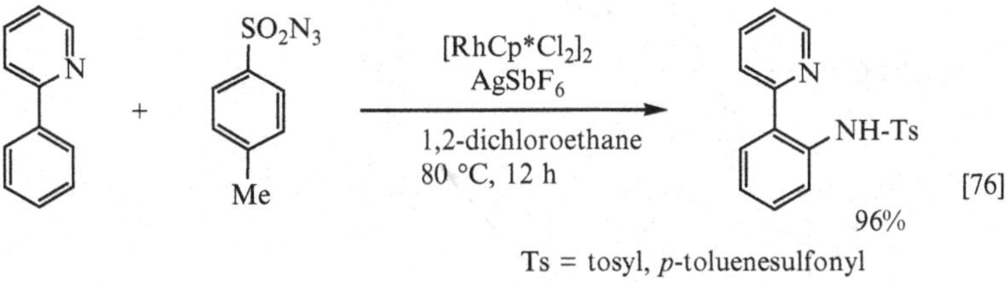

(6.49)

Ts = tosyl, *p*-toluenesulfonyl

Snapshot of a *Chelation-Assisted C–H/Alkyne Coupling*: A Ruthenium Complex Caught in the Act of C–C Bond Formation

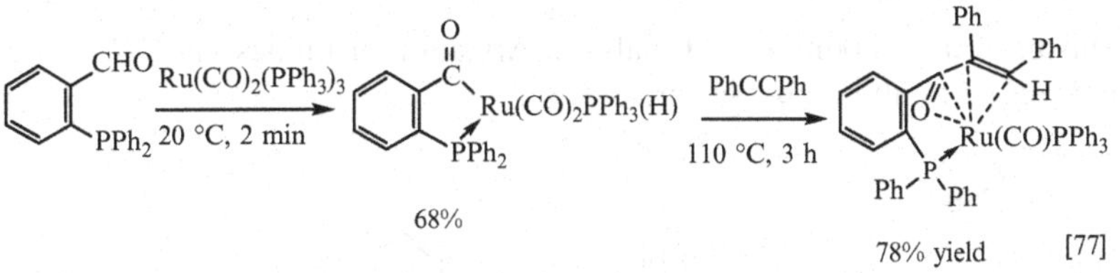

(6.50)

Double-Chelation-Assisted Rh-Catalyzed Intermolecular Hydroacylation Between Salicylaldehydes and 1,4-Peneta- or 1,5-Hexadienes

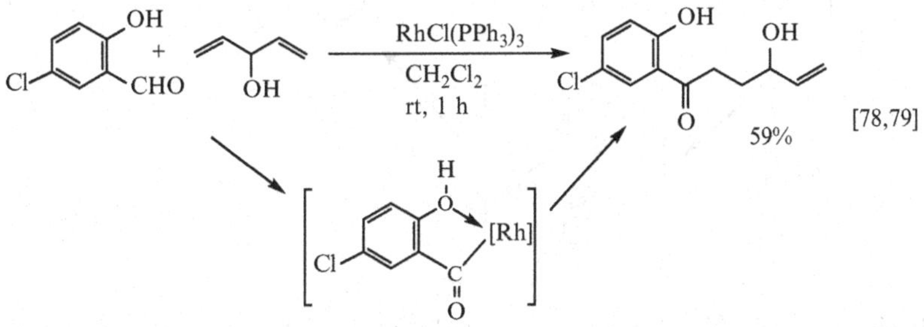

(6.51)

Fourth, articles employing C–H functionalizations (C–H bond functionalizations) as title words are shown in Eqs. (6.52)–(6.64).

Rhodium-Catalyzed Regioselective *C–H Functionalization* via Decarbonylation of Acid Chlorides and C–H Bond Activation Under Phosphine-Free Conditions

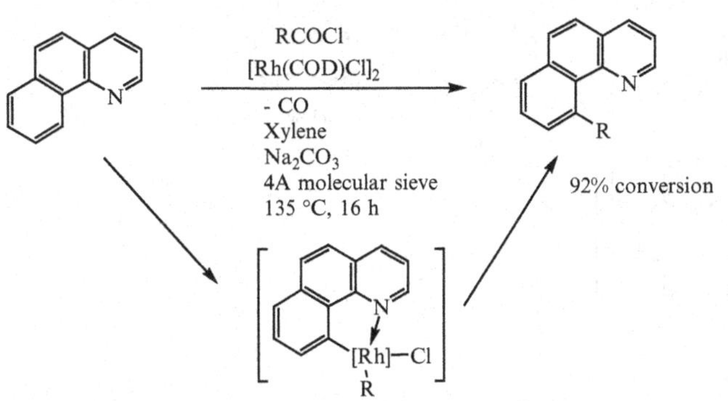

(6.52)

Palladium-Catalyzed *C–H Bond Functionalization* with Arylsulfonyl Chlorides

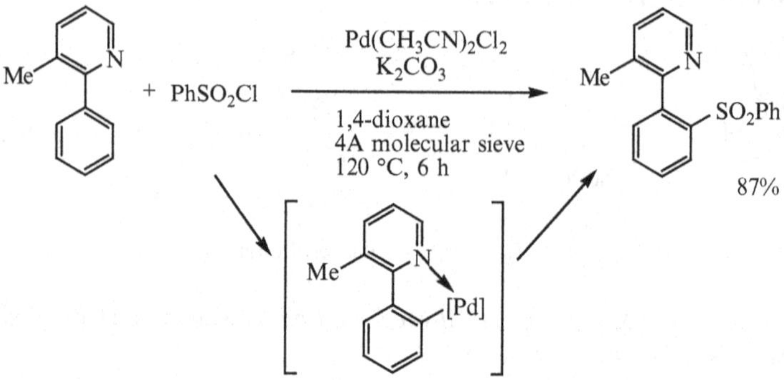

$$(6.53)$$

Mechanism of the Rhodium(III)-Catalyzed Arylation of Imines via *C–H Bond Functionalization*: Inhibition by Substrate

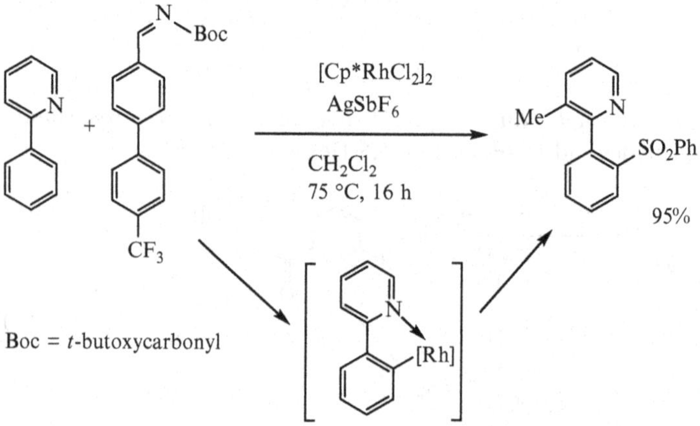

Boc = *t*-butoxycarbonyl

$$(6.54)$$

Synthesis of 7,7′-Dihydroxy-8,8′-biquinolyl(azaBINOL) via Pd-Catalyzed Directed Double *C–H Functionalization* of 8,8′-Biquinolyl: Emergence of an *Atropos* from a *Tropos* State

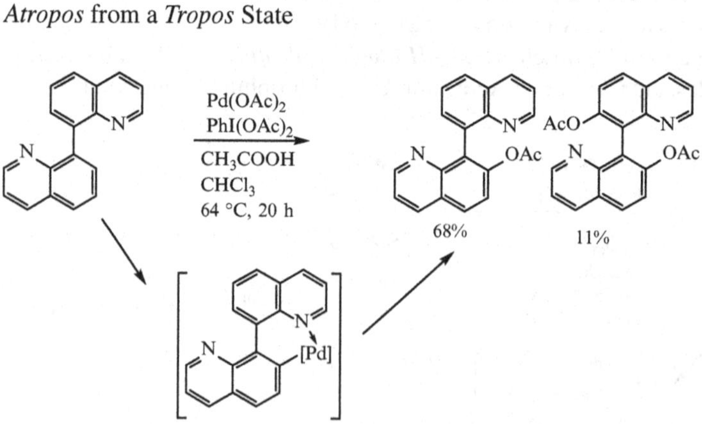

$$(6.55)$$

Diversity Synthesis via *C–H Bond Functionalization*: Concept-Guided Development of New C-Arylation Methods for Imidazoles

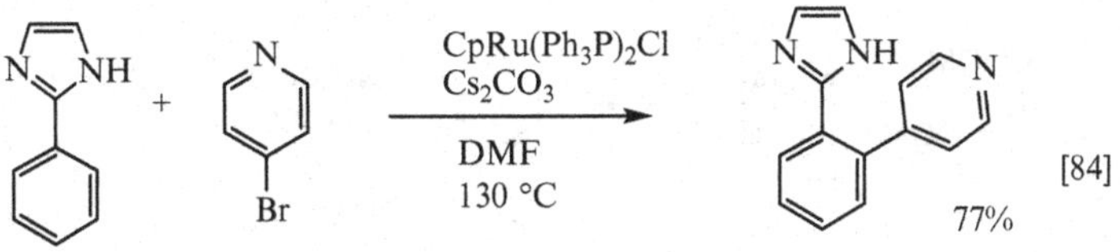

Ruthenium(II)-Catalyzed Regio- and Stereoselective Hydroarylation of Alkynes via Directed *C–H Functionalization*

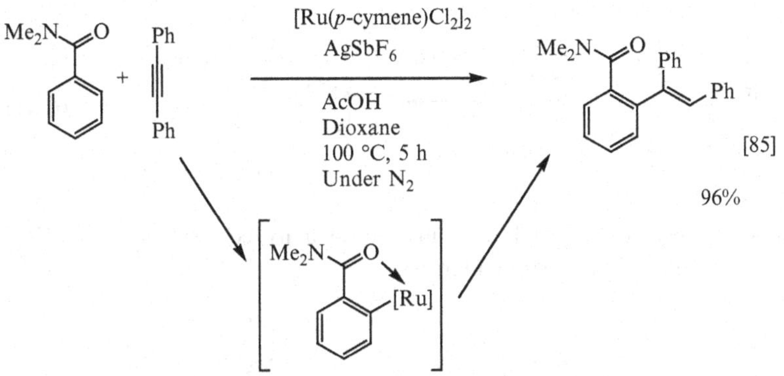

Rollover Cyclometalation Pathway in Rhodium Catalysis: Dramatic NHC Effects in the *C–H Bond Functionalization*

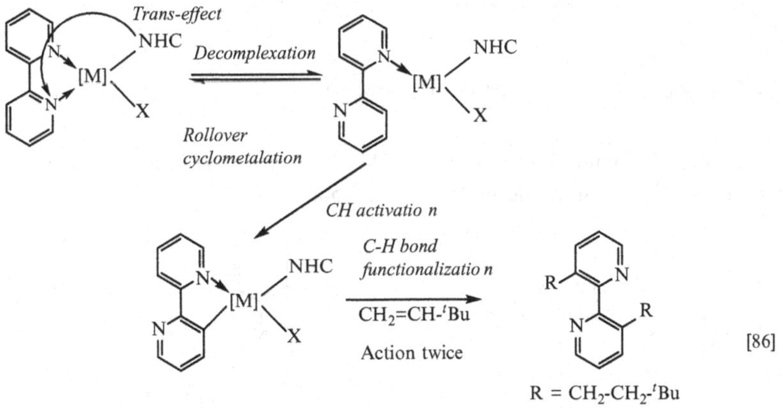

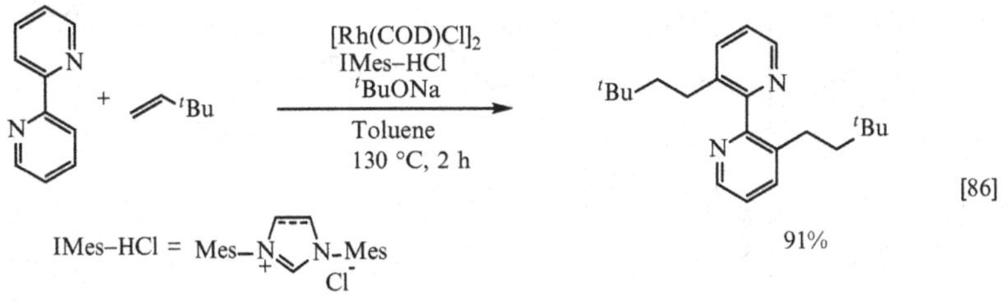

Rh-Catalyzed Intermolecular Carbenoid *Functionalization of Aromatic C–H Bonds* by α-Diazomalonates

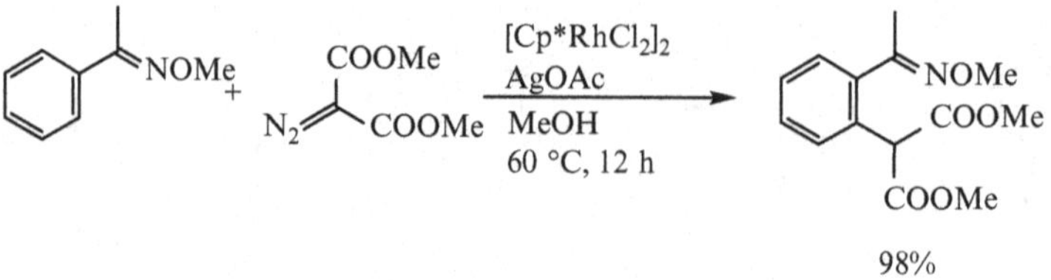

(6.60)

[87]

98%

Ruthenium-Catalyzed *C–H/N–O Bond Functionalization*: Green Isoquinolone Synthesis in Water

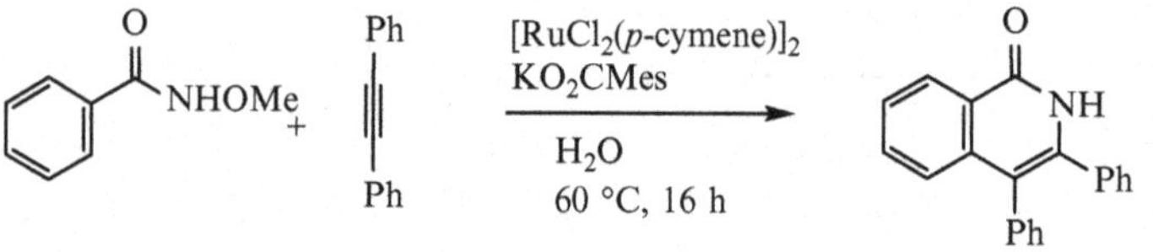

(6.61)

[88]

Phosphine Oxides as Preligands in Ruthenium-Catalyzed *Arylations* via *C–H Bond Functionalization* Using Aryl Chlorides

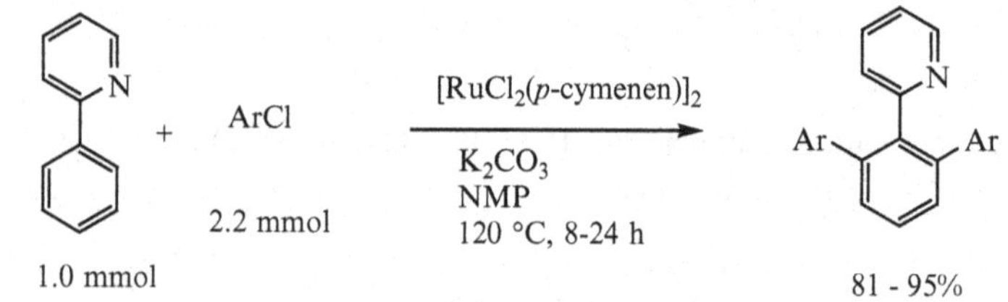

(6.62)

[89]

81 - 95%

Ar = Ph, 4-OOCC$_6$H$_4$, 3-NCC$_6$H$_4$, 4-MeCOC$_6$H$_4$, 4-MeOC$_6$H$_4$

Direct Access to Acylated Azobenzenes via Pd-Catalyzed *C–H Functionalization* and Further Transformation into an Indazole Backbone

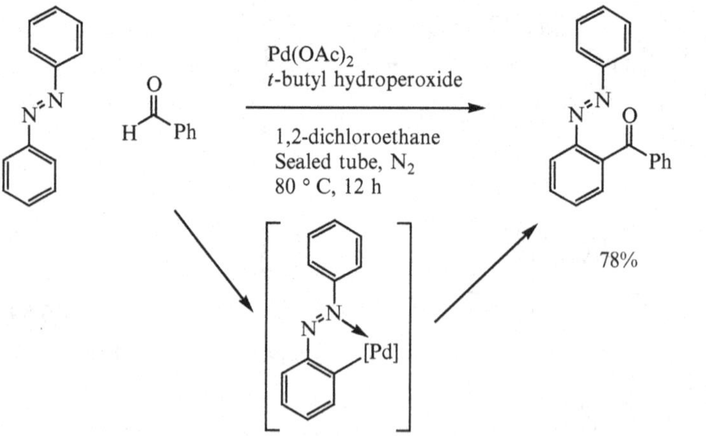

[90]

(6.63)

78%

Shot Synthesis of Alkyl-Substituted Acenes Using Carbonyl-Directed *C–H and C–O Functionalization*

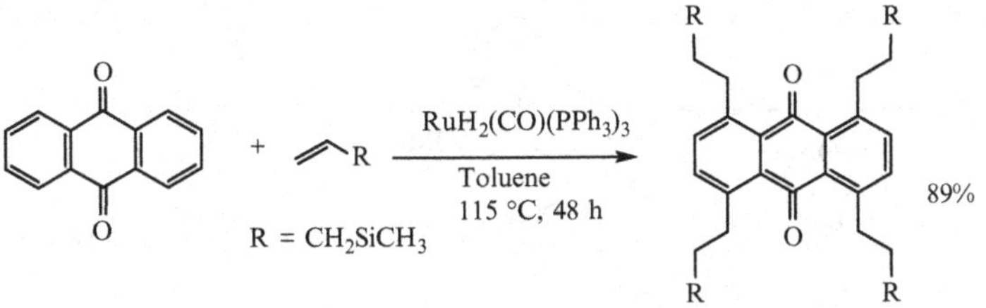

$$\text{(6.64)}$$

89%   [91]

R = CH$_2$SiCH$_3$

Rhodium-Catalyzed Oxidation *Ortho*-Acylation of Benzamides with Aldehydes: *Direct Functionalization of the sp^2 C–H Bond*

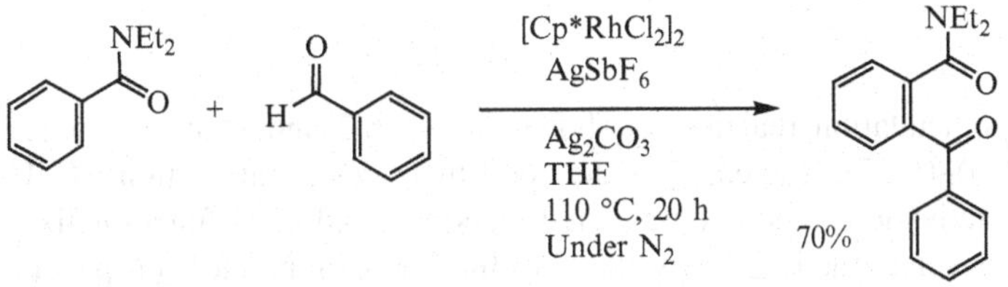

$$\text{(6.65)}$$

70%   [92]

Rhodium(III)-Catalyzed Arylation of Boc-Imines via *C–H Bond Functionalization*

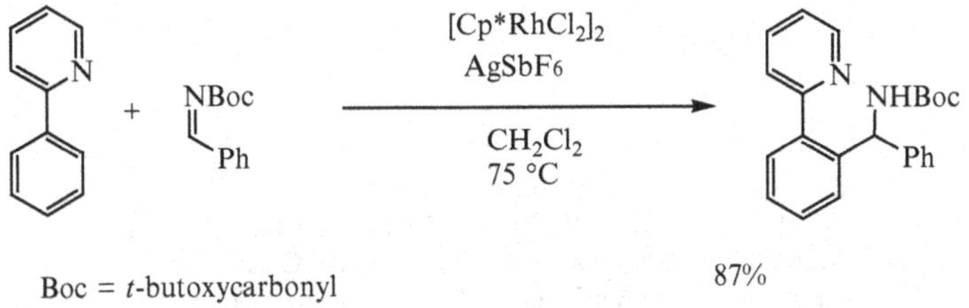

$$\text{(6.66)}$$

Boc = *t*-butoxycarbonyl   87%   [93]

Hydroxyl-Directed   Ruthenium-Catalyzed   *C–H   Bond   Functionalization*: Versatile Access to Fluorescent Pyrans

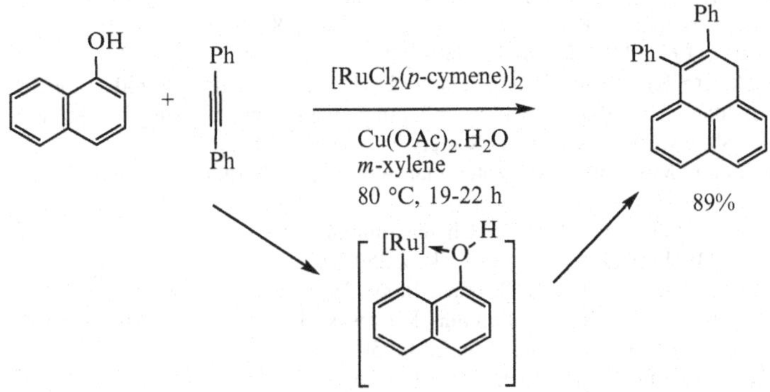

89%   [94]   (6.67)

A Highly Selective Catalytic Method for the *Oxidative Functionalization of C–H Bonds*

$$(6.68)$$

From the cyclometalation reaction as shown in Eq. 6.7 and many examples shown in Eqs. (6.9)–(6.67), it is easily understood that agostic interactions, C–H activations, C–X activations, chelation-assisted reactions, and C–H functionalizations are caused by metal activation by the coordination of lone electron pair of hetero atoms such as N, P, O, and S to the central metal atom and the chelate effect, in cyclometalation reactions.

References

1. Harrison PG, King TJ, Healy MA (1979) J Organomet Chem 182:17
2. Howie RA, Paterson ES, Wardell JL, Burley JW (1986) J Organomet Chem 304:301
3. Howie RA, Paterson ES, Wardell JL, Burley JW (1983) J Organomet Chem 259:71
4. Aoki A, Horiuchi K, Omae I (1995) The 69th Annual Meeting of the Chemical Meeting of Japan, vol 1, Kyoto, p 480
5. Omae I, Aoki A, Horiuchi K (1998) Kagaku Kogyo 49:469
6. Gómez M, Granell J, Martinez M (1997) Organometallics 16:2539
7. Pauling L (1959) The nature of the chemical bond, 3rd edn. Cornell University Press, Ithaca
8. Shilov A, Shul'pin AE (2000) Activation and catallytic reactions of saturated hydrocarbons in the presence of metal complexes, Kluwer. Academic, London pp 8–20, pp 129–199
9. Vaska L, DiLuzio JW (1962) J Am Chem Soc 84:679
10. Goldman AS, Goldberg KI (2004) Organometallic C–H bond activation: introduction. In: Goldberg KI, Goldman AS (eds) Activation and Functionalization of C–H Bonds, ACS Symposium Series 885. American Chemical Society, Washington, DC, p 2
11. Gol'dshleger NF, Tyabin MB, Shilov AE, Shteinman AA (1969) Zhurnal Fizicheskoi Khimii 43:2174
12. Green MLH, Knowles PJ (1970) J Chem Soc Chem Commun 24:1677
13. Brookhart M, Green MLH (1983) J Organomet Chem 250:395
14. Brookhart M, Green MLH, Wong L-L (1988) Prog Inorg Chem 36:1
15. Ohshima T, Kawabata T, Takeuchi Y, Kakinuma T, Iwasaki T, Yonezawa T, Murakami H, Nishiyama H, Mashima K (2011) Angew Chem Int Ed 50:6296
16. van der Boom ME, Milstein D (2003) Chem Rev 103:1759

17. Albrecht M (2008) C–H bond activation. In: Dupont J, Pfeffer M (eds) Palladacycles. Synthesis, characterization and applications. Wiley-VCH, Weinheim, p 13
18. Basolo F, Johnson R (1964) Coordination chemistry. The Benjamin/Cummings Publishing Company, London
19. Ritleng V, Sirlin C, Pfeffer M (2002) Chem Rev 102:1731
20. Chen S, Li Y, Zhao J, Li X (2009) Inorg Chem 48:1198
21. Mizuno H, Takaya J, Iwasawa N (2011) J Am Chem Soc 133:1251
22. Shiota H, Ano Y, Aihara Y, Fukumoto Y, Chatani N (2011) J Am Chem Soc 133:14952 Chatani N (2013) Yuki Gosei Kagaku Kyokaishi 71:406
23. Omae I (2011) J Organomet Chem 696:1128
24. Davies DL, Donald SMA, Macgregor SA (2005) J Am Chem Soc 127:13754
25. Ackermann L (2011) Chem Rev 111:1315
26. Matsuda S, Kikkawa S, Nomura M (1966) Kogyo Kagaku Zasshi 69:649
27. Li X, Chen P, Faller JW, Crabtree RH (2005) Organometallics 24:4810
28. Li X, Vogel T, Incarbito CD, Crabtree RH (2005) Organometallics 24:62
29. Clot E, Chen J, Lee D-H, Sung SY, Appelhans LN, Faller JW, Crabtree RH, Eisenstein O (2004) J Am Chem Soc 126:8795
30. Clot E, Eisenstein O, Dubé T, Faller JW, Crabtree RH (2002) Organometallics 21:575
31. Albéniz AC, Schulte G, Crabtree RH (1992) Organometallics 11:242
32. Kohl SW, Heinemann FW, Hummert M, Bauser W, Grohmann A (2006) Chem Eur J 12:4313
33. Kohl SW, Heinemann FW, Hummert M, Bauer W, Grohmann A (2006) Dalton Trans 5583
34. Patra SK, Bera JK (2006) Organometallics 25:6054
35. Crosby SH, Clarkson GJ, Rourke JP (2009) J Am Chem Soc 131:14142
36. Crosby SH, Clarkson GJ, Rourke JP (2010) Organometallics 29:1966
37. Crosby SH, Clarkson GJ, Rourke JP (2011) Organometallics 30:3603
38. Thomas HR, Deeth RJ, Clarkson GJ, Rourke JP (2011) Organometallics 30:5641
39. Dorta R, Stevens ED, Nolan SP (2004) J Am Chem Soc 126:5054
40. Scott NM, Dorta R, Stevens ED, Correa A, Cavallo L, Nolan SP (2005) J Am Chem Soc 127:3516
41. Knapp V, Müller G (2001) Angew Chem Int Ed 40:183
42. Thu H-Y, Yu W-Y, Che C-M (2006) J Am Chem Soc 128:9048
43. Gusev DG, Lough AJ (2002) Organometallics 21:2601
44. Cave GWV, Fanizzi FP, Deeth RJ, Errington W, Rourke JP (2000) Organometallics 19:1355
45. Arbrecht M, Dani P, Lutz M, Spek AL, van Koten G (2000) J Am Chem Soc 122:11822
46. Chen X, Goodhue CE, Yu J-Q (2006) J Am Chem Soc 128:12634
47. Zhuravel MA, Grewal NS, Glueck DS, Lam K-C, Rheingold AL (2000) Organometallics 19:2882
48. Ohki Y, Hatanaka T, Tatsumi K (2008) J Am Chem Soc 130:17174
49. Keuseman KJ, Smoliakova IP, Dunina VV (2005) Organometallics 24:4159
50. Lu Z, Abbina S, Sabin JR, Nemykin VN, Du G (2013) Inorg Chem 52:1454
51. Koike T, Ikariya T (2005) Organometallics 24:724
52. Sinha A, Majumdar M, Sarkar M, Ghatak T, Bera JK (2013) Organometallics 32:340
53. Muise SSR, Severin HA, Koivisto BD, Robson KCD, Schott E, Berlinguette CP (2011) Organometallics 30:6628
54. López O, Crespo M, Font-Bardía M, Solan X (1997) Organometallics 16:1233
55. Calvet T, Crespo M, Font-Bardía M, Jansat S, Martínez M (2012) Organometallics 31:4367
56. Crespo M, Font-Bardia M, Solans X (2004) Organometallics 23:1708
57. Martín R, Crespo M, Font-Bardia M, Calvet T (2009) Organometallics 28:587
58. Chen Y, Sun H, Flörke U, Li X (2008) Organometallics 27:270
59. Shi Y, Li M, Hu Q, Li X, Sun H (2009) Organometallics 28:2206
60. Gandelman M, Vigalok A, Shimon LJW, Milstein D (1997) Organometallics 16:3981
61. Rybtchinski B, Vigalok A, Ben-David Y, Milstein D (1996) J Am Chem Soc 118:12406
62. Wang T, Keyes L, Patrick BO, Love JA (2012) Organometallics 31:1397

63. Li X, Sun H, Yu F, Flörke U, Klein J-F (2006) Organometallics 25:4695
64. Zheng T, Sun H, Chen Y, Li X, Dürr S, Radius U, Harms K (2009) Organometallics 28:5771
65. Buckley HL, Wang T, Tran O, Love JA (2009) Organometallics 28:2356
66. Barrio P, Castarlenas R, Esteruelas MA, Lledós A, Maseras F, Oñate E, Tomàs J (2001) Organometallics 20:442
67. van der Boom ME, Ben-David Y, Milstein D (1999) J Am Chem Soc 121:6652
68. Ceder RM, Granell J, Muller G, Font-Bardía M, Solans X (1995) Organometallics 14:5544
69. Crespo M, Granell J, Solans X, Font-Bardia M (2002) Organometallics 21:5140
70. Kwon K-H, Lee DW, Yi CS (2010) Organometallics 29:5748
71. Gao K, Lee P-S, Fujita T, Yoshikai N (2010) J Am Chem Soc 132:12249
72. Zhang L, Liu Z, Li H, Fang G, Barry B-D, Belay TA, Bi X, Liu Q (2011) Org Lett 13:6536
73. Jia X, Yang D, Wang W, Luo F, Cheng J (2009) J Org Chem 74:9470
74. Wang J, Liu B, Zhao H, Wang J (2012) Organometallics 31:8598
75. Jia X, Yang D, Zhang S, Cheng J (2009) Org Lett 11:4716
76. Kim JY, Park SH, Ryu J, Cho SH, Kim SH, Chang S (2012) J Am Chem Soc 134:9110
77. Benhamou L, César V, Lugan N, Lavigne G (2007) Organometallics 26:4673
78. Imai M, Tanaka M, Tanaka K, Yamamoto Y, Imai-Ogata N, Shimowatari M, Nagumo S, Kawahara N, Suemune H (2004) J Org Chem 69:1144
79. Tanaka M, Imai M, Yamamoto Y, Tanaka K, Shimowatari M, Nagumo S, Kawahara N, Suemune H (2003) Org Lett 5:1365
80. Zhao X, Yu Z (2008) J Am Chem Soc 130:8136
81. Zhao X, Dimitrijević E, Dong VM (2009) J Am Chem Soc 131:3466
82. Tauchert ME, Incarvito CD, Rheingold AL, Bergman RG, Ellman JA (2012) J Am Chem Soc 134:1482
83. Wang C, Flanigan DM, Zakharov LN, Blakemore PR (2011) Org Lett 13:4024
84. Sezen B, Sames D (2003) J Am Chem Soc 125:10580
85. Hashimoto Y, Hirano K, Satoh T, Kakiuchi F, Miura M (2012) Org Lett 14:2058
86. Kwak J, Ohk Y, Jung Y, Chang S (2012) J Am Chem Soc 134:17778
87. Chan W-W, Lo S-F, Zhou Z, Yu W-Y (2012) J Am Chem Soc 134:13565
88. Ackermann L, Fenner S (2011) Org Lett 13:6548
89. Ackermann L (2005) Org Lett 7:3123
90. Li H, Li P, Wang L (2013) Org Lett 15:620
91. Matsumura D, Kitazawa K, Terai S, Kochi T, Ie Y, Nitani M, Aso Y, Kakiuchi F (2012) Org Lett 14:3882
92. Park J, Park E, Kim A, Lee Y, Chi K-W, Kwak JH, Kwak YH, Jung YH, Kim IS (2011) Org Lett 13:4390
93. Tsai AS, Tauchert ME, Bergman RG, Ellman JA (2011) J Am Chem Soc 133:1248
94. Trirunavukhasasu VS, Donati M, Ackermann L (2012) Org Lett 14:3416
95. Dick AR, Hull KL, Sanford MS (2004) J Am Chem Soc 126:2300

Chapter 7
Applications of Cyclometalation Reactions and Five-Membered Ring Products for Synthetic Purposes

Abstract There are two applications for cyclometalation reactions and five-membered ring products for synthetic purposes. The first is synthesis of five-membered ring products by cyclometalation reactions. The second is synthesis of the derivatives of five-membered ring products or their intermediates during cyclometalation reactions. Pincer products are also used for synthesis of their derivatives.

Keywords Alkenylation • Alkynylation • Arylation • Carbonylation • Cross-coupling • Cyclometalation • Five-membered ring • Rollover

7.1 Introduction

As applications of cyclometalation reactions for producing organometallic intramolecular-coordination five-membered ring compounds, two categories of uses are considered. These are applications of cyclometalation reactions and syntheses of the derivatives of their five-membered ring products.

These two categories have the following characteristics:

1. Applications of cyclometalation reactions for producing organometallic intramolecular-coordination five-membered ring products
   (a) Very easy regioselective reactions (Tables 5.1, 5.2, and 5.3; Figs. 5.1, 5.2, 5.3, 5.4, 5.5, 5.6, 5.7, 5.8, and 5.9)
   (b) Use of almost all kinds of metal elements (Fig. 5.10)
   (c) Use of many substrates (Fig. 5.11, Tables 5.5 and 5.6; Fig 9.2)

I. Omae, *Cyclometalation Reactions: Five-Membered Ring Products as Universal Reagents*, DOI 10.1007/978-4-431-54604-7_7, © Springer Japan 2014

2. Applications of five-membered ring products for synthetic uses
   (a) Very easy synthesis of the derivatives by substituting active metal atoms
   (b) Many kinds of reactions available in terms of both reaction products and reaction intermediates
   (c) Also, use of pincer products, which are tridentate compounds, as more stable cyclometalation products

As concerns the first category of applications, the author has published reports on these applications in many reviews [1–20] and three monographs [21–23], mainly regarding their coordinating atoms such as nitrogen, phosphorus, oxygen, sulfur, and arsine, as described in Chap. 1.

As the second category of applications, or applications of the reaction products, on the other hand, the author has published only two reviews, in 2004 and in 2007. The title of the former is "Intramolecular Five-Membered Ring Compounds and Their Applications [16]" and that of the latter is "Three Types of Reactions with Intramolecular Five-Membered Ring Compounds in Organic Synthesis [18]."

This monograph presents recent types of new cyclometalation reactions in the first category. Reactions with new types of substrates, cyclometalation reactions via mild oxidation, and rollover cyclometalation reactions are presented, for example.

As concerns the second category, this monograph presents applications for both five-membered ring products and their intermediates produced by cyclometalation reactions from the author's two reviews as well as from recent articles.

7.2   Applications for Cyclometalation Reactions

7.2.1   Introduction

The author has already reported in detail on cyclometalation reactions in many reviews and three monographs, as described in Chap. 1 and in the previous section. Applications for cyclometalation reactions in this section, therefore, describe only the following three new cyclometalation reactions:

First, cyclometalation reactions with new types of substrates that are formed during the reactions
Second, cyclometalation reactions with substrates that are formed by mild oxidation
Third, rollover cyclometalation reactions

7.2.2   Cyclometalation Reactions with Reaction Products of Amines and Aldehydes or Alcohols as Substrates

Jun et al. [24–28] have published reports recently on hydroacylations via new types of cyclometalation reactions. These cyclometalation reactions proceed with substrates formed by reactions of amines such as 2-aminopyridine with aldehydes or

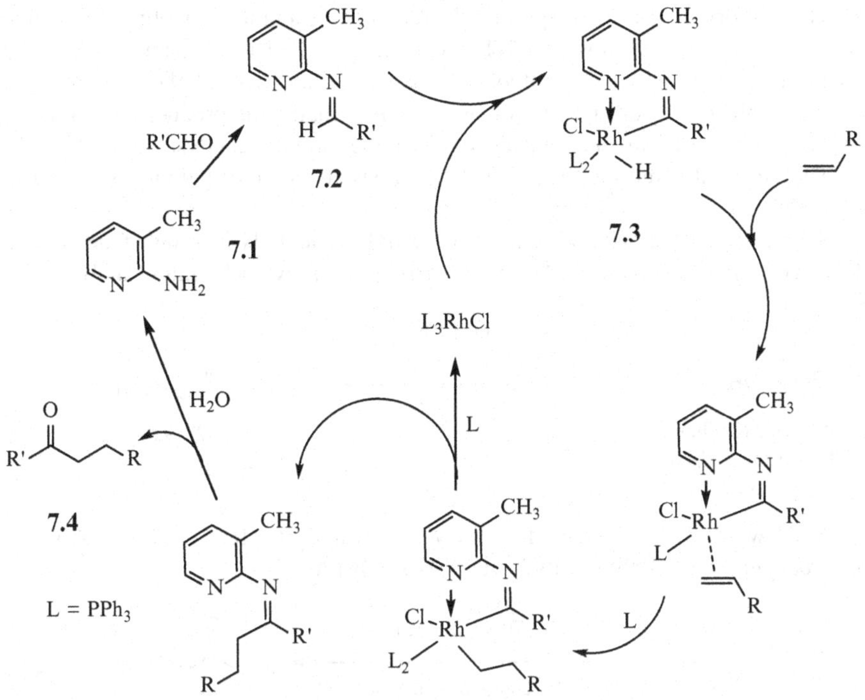

Scheme 7.1 Proposed mechanism for the formation of hydride [24]

alcohols. These amines have no γ-carbon atom for the formation of five-membered rings, as in the conventional substrates shown in Fig. 5.11, Tables 5.5 and 5.6. The amines form conventional substrates, "imines," during reactions with aldehydes or alcohols (see Scheme 7.1). In these reactions, the catalysts are rhodium compounds, the substrates are imines such as 2-amino-3-picoline, and ketones are synthesized by the hydroacylation of olefins with aldehydes or alcohols. For example, chelation-assisted hydroacylation is shown in Eq. (7.1) [24].

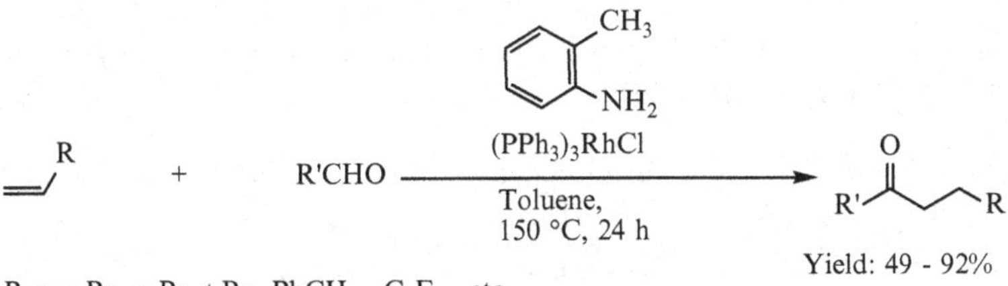

$$R = n\text{-}Bu, n\text{-}Pr, t\text{-}Bu, PhCH_2\text{-}, C_6F_5\text{-}, \text{etc.}$$
$$R' = Ph, C_6H_5\text{-}OMe\text{-}p; C_6H_5\text{-}CH_2CH_2\text{-}, \text{Cyclohexyl}$$

Jun et al. [24] reported the above reaction in an article entitled "Chelation-Assisted Intermolecular Hydroacylation: Direct Synthesis of Ketone from Aldehyde and 1-Alkene" in 1997 [24]. They also reported on the reaction mechanism, as shown in Scheme 7.1 [24].

The authors used a new type of substrate, e.g., 2-amino-3-picoline **7.1**, which produces conventional substrates **7.2**, which, in turn, form a five-membered ring **7.3** during a reaction with one component, such as an aldehyde. Other components, such as 1-alkenes, coordinate to the metal atom. Finally, the product **7.4** is formed by the coupling of 1-alkenes and aldehydes at the rhodium catalyst.

They have also reported these chelation-assisted hydroacylations, as shown in Eqs. (7.2)–(7.5).

Synthesis of Aliphatic Ketones from Allylic Alcohols Through Consecutive Isomerization and *Chelation-Assisted Hydroacylation* by a Rhodium Catalyst [25]

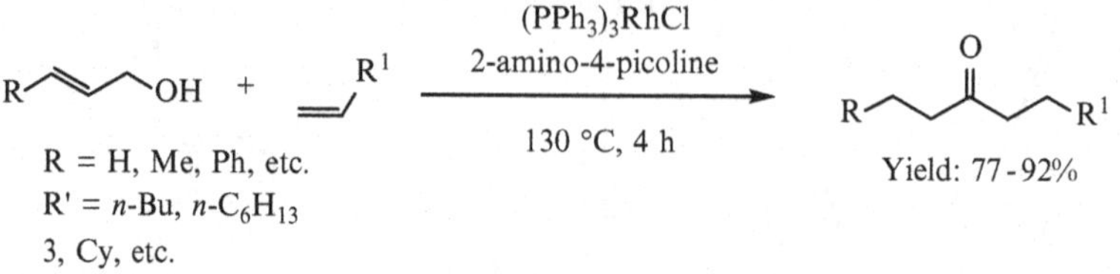

$$R = H, Me, Ph, etc.$$
$$R' = n\text{-Bu}, n\text{-}C_6H_{13}$$
3, Cy, etc.

$$(7.2)$$

Yield: 77-92%

A New Solvent System for Recycling Catalysts for *Chelation-Assisted Hydroacylation* of Olefins with Primary Alcohols [26]

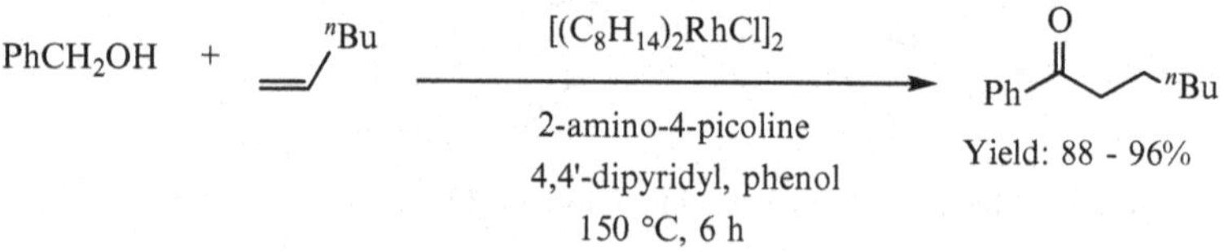

Yield: 88 - 96%

$$(7.3)$$

Recyclable Self-Assembly-Supported Catalyst for *Chelation-Assisted Hydroacylation* of an Olefin with a Primary Alcohol [27]

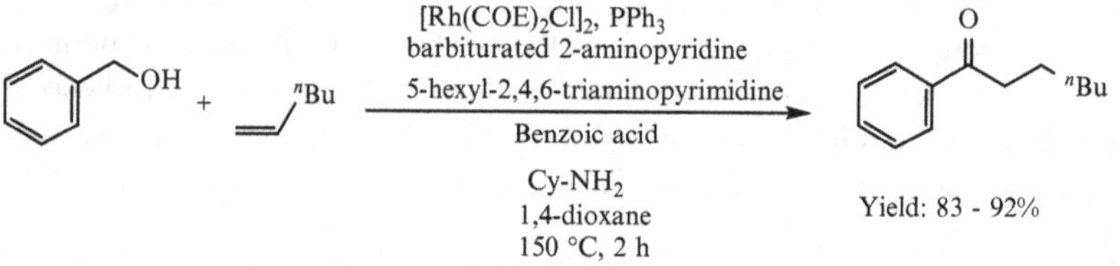

Yield: 83 - 92%

$$(7.4)$$

Dual Functionalities of Hydrogen-Bonding Self-Assembled Catalysts in *Chelation-Assisted Hydroacylation* [28]

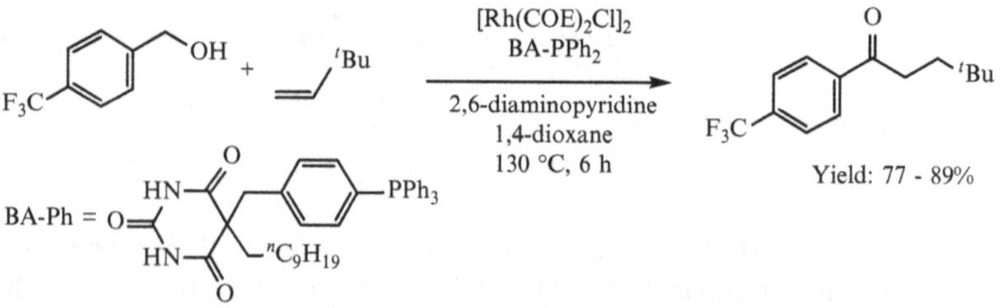

Yield: 77 - 89%

$$(7.5)$$

The hydroacylation examples shown in Eqs. (7.2)–(7.5) are presented as new types of cyclometalation reactions using new types of substrates.

7.2.3   Cyclometalation Reactions with Substrates Formed by Mild Oxidation

Padilla et al. [29] reported that benzoylcarboxylic acid ester is prepared in high yield by the reaction of acetylenic diester with diphenyliridium containing hydrotris(3,5-dimethylpyrazolyl)borate and dinitrogen under oxygen pressure. The phenylbenzoylcarboxylic acid ester is prepared via the intermediate **7.5**, when a phenyl group bonds with acetylenic diester and decomposes to form benzoyl derivatives under mild oxidation reaction conditions. The final product is a conventional cyclometalation reaction compound with a benzoyl group as the conventional substrate as shown in Eq. (7.6).

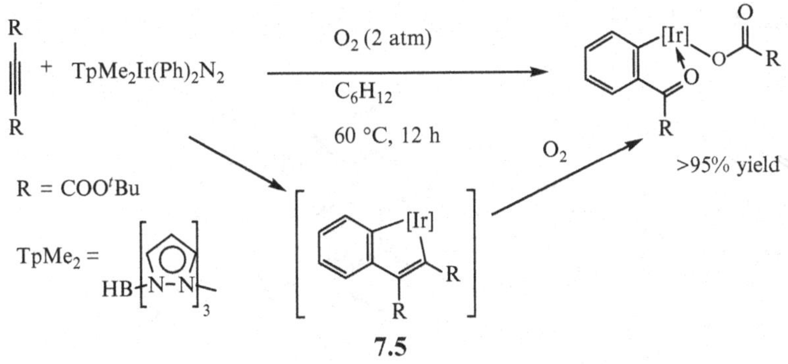

$$\text{(7.6)}$$

7.2.4   Rollover Cyclometalation Reactions

Generally, 2,2′-bipyridines are representative N,N'-chelating ligands that are similar to ethylenediamine and 1,10-phenanthroline as inorganic chelate ligands [30]. Only recently has a new coordinating behavior called a "rollover cyclometalation reaction" appeared in the literature. A rollover reaction involves γ-C–H bond activation and formation of a γ-C–M bond, as shown in Eq. (7.7) [31].

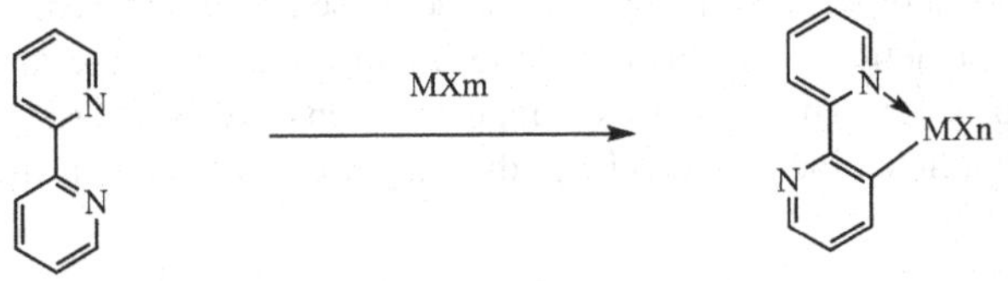

$$\text{(7.7)}$$

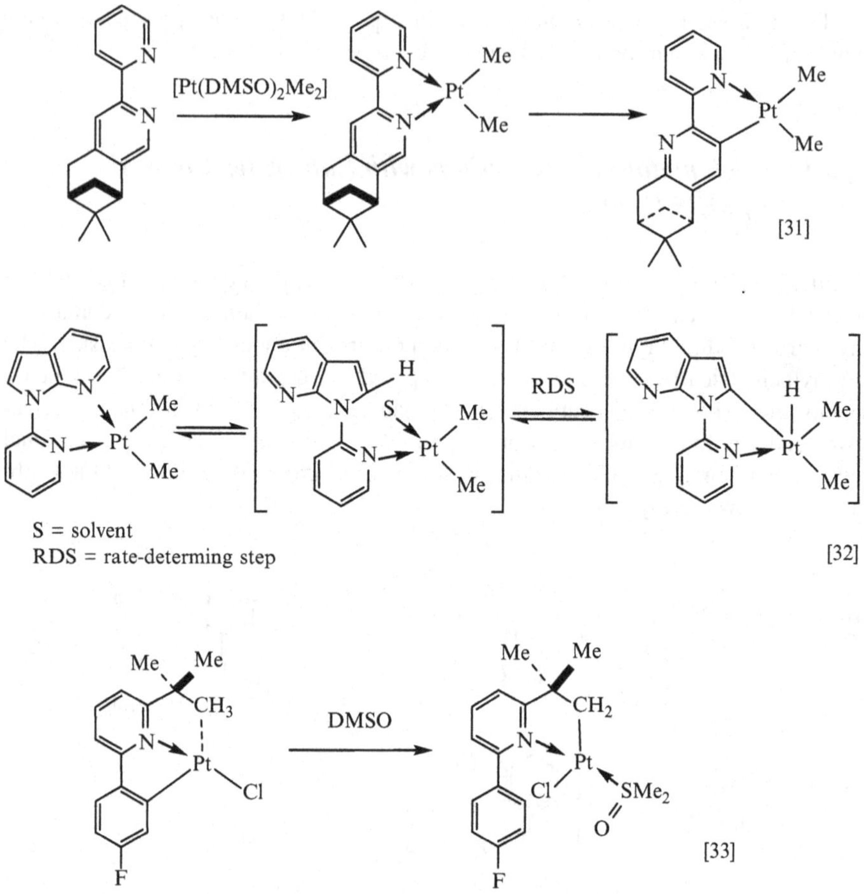

Fig. 7.1 Some substrates in rollover cyclometalation reactions

The reaction is not restricted to 2,2'-bipyridines but is also applicable to (5S,7S)-5,7-methane-6,6-dimethyl-2-(pyridin-2-yl)-5,6,7,8-tetrahydroquinoline [31], N-(2'-pyridyl)-7-azaindole [32] and 2-phenylpyrines [33–36], as shown in Fig. 7.1.

Metal atoms such as Pt [31–49], Pd [45, 50–52], Ir [53, 54], Ni [45], Au [55, 56], and Cu [57] have been reported for rollover cyclometalation reactions.

A platinum complex with a bipyridine as the most widely used ligand is easily prepared in high yields by reaction with Pt(Me)$_2$(DMSO)$_2$ in toluene to give an air- and moisture-stable product. The ligand exchange reactions proceed easily with other ligands such as phosphine, CO, and pyridine derivative, as shown in Eq. (7.8) [48].

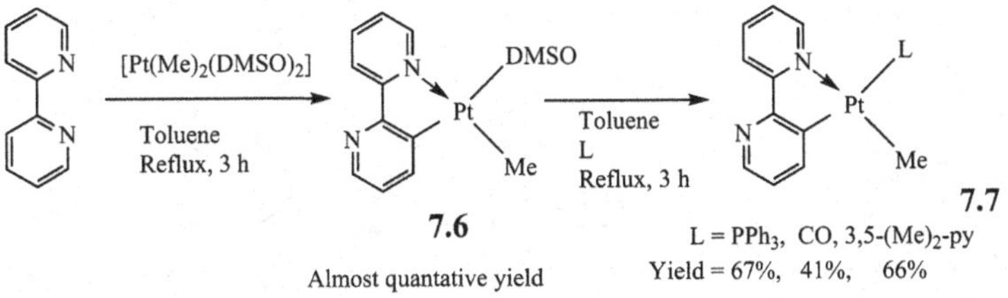

$$(7.8)$$

L = PPh₃, CO, 3,5-(Me)₂-py
Yield = 67%, 41%, 66%

Treatment of the platinum species [Pt(bipy-H)(Me)(DMSO)] **7.6** with acids containing poorly coordinating anions, such as [18-crown-6-H₃O][BF₄], forms a *mesoionic* complex [Pt(bipy*)(Me)(DMSO)]⁺ **7.8** derived from a rollover cyclometalation reaction as shown in Eq. (7.9) [38].

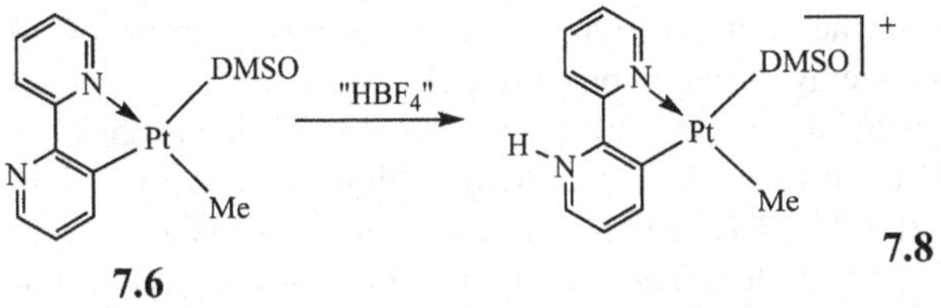

$$(7.9)$$

The protonation of the triphenylphosphine derivative **7.7** (L=PPh₃) shown in Eq. (7.8) with [18-crown-6-H₃O][BF₄] also proceeds smoothly to the corresponding cationic complex [Pt(bipy*)(Me)(PPh₃)][BF₄] **7.9**, as shown in Eq. (7.10) [38].

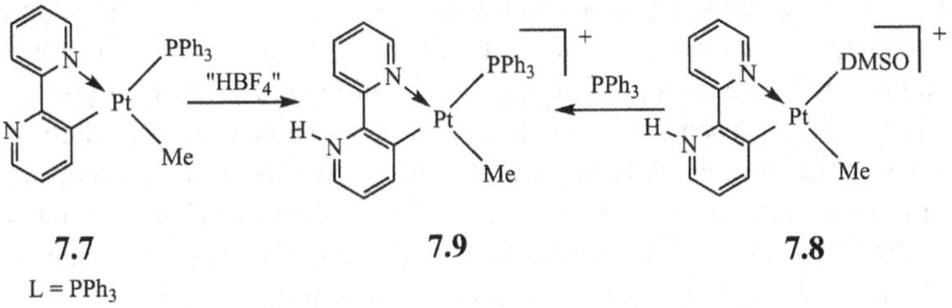

$$(7.10)$$

When the cationic complex [Pt(bipy*)(Me)(PPh₃)][BF₄] **7.9** is left in solution for several days, it is slowly converted into a new species **7.10**, which is an isomerized product, with a completion of ca. 30 days. The reaction rate may be influenced by addition of DMSO to the solution as shown in Eq. (7.11) [38].

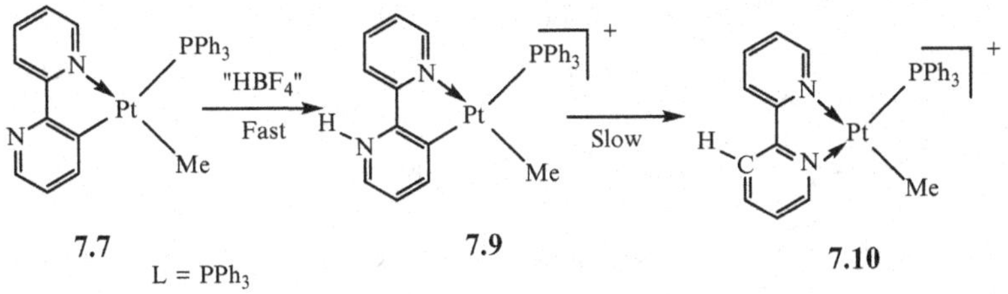

$$(7.11)$$

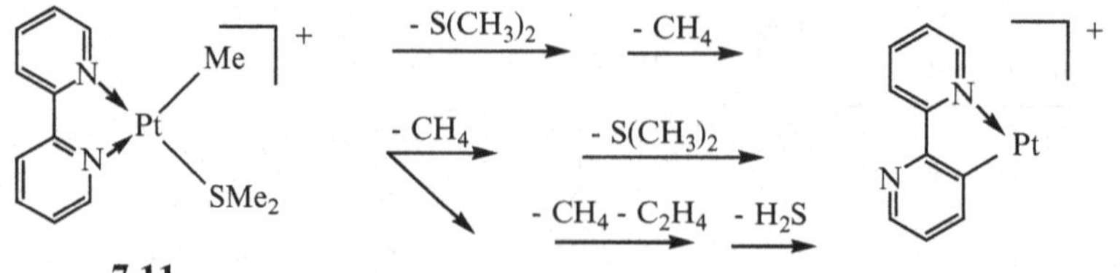

Scheme 7.2 Gas-phase generation of cyclometalated 2,2'-bipyridine [Pt(bipy-H)]$^+$ [46]

Butschke et al. [46] reported in a combined experimental/computational investigation report on the gas phase of cationic [Pt(bipy)-(Me)(SMe$_2$)]$^+$ **7.11** as shown in Scheme 7.2. Losses of CH$_4$ and SMe$_2$ from **7.11** result in the formation of a cyclometalated 2,2'-bipyrid-3-yl species [Pt(bipy-H)]$^+$ **7.12**. The process reveals a rollover cyclometalation path in the course of which a hydrogen atom from γ-C is combined with the Pt-bound methyl group to produce CH$_4$. The activation of a C–H bond of the SMe$_2$ ligand occurs as well, but this is less favored (35 versus 65 %) as compared with the γ-C–H bond activation of bipy. In addition, the thermal ion/molecule reactions of [Pt-(bipy-H)] **7.12** with SMe$_2$ have been examined along with the major pathway, that is, the dehydrogenative coupling of the two methyl groups to form C$_2$H$_4$, for which a mechanism compatible with the experimental and computation findings is suggested.

Recently, Kwak et al. [52] reported on hydroallylation with these rollover cyclometalated complexes as the catalysts. The bipyridine rhodium complex containing an NHC ligand **7.13** is labile because of the strong *trans*-effect of *N*-heterocyclic carbene (NHC), thus weakening the rhodium–pyridyl bond that is *trans* to the bound NHC ligand. Subsequent rollover cyclometalation leads to C–H activation to form the isomerized complex **7.15**, as shown in Eq. (7.12). In the reaction of bipyridine with an ally *t*-butyl in the presence of a rhodium complex (Rh(acac)$_3$) as a catalyst, the monohydroallylation product **7.16** is formed as the derivative from the intermediate **7.15** in the first cyclometalation reaction, and the dihydroallylation product **7.17** is also formed in the second cyclometalation reaction. The second product is prepared in a high yield under the reaction conditions shown in Eq. (7.13) [52].

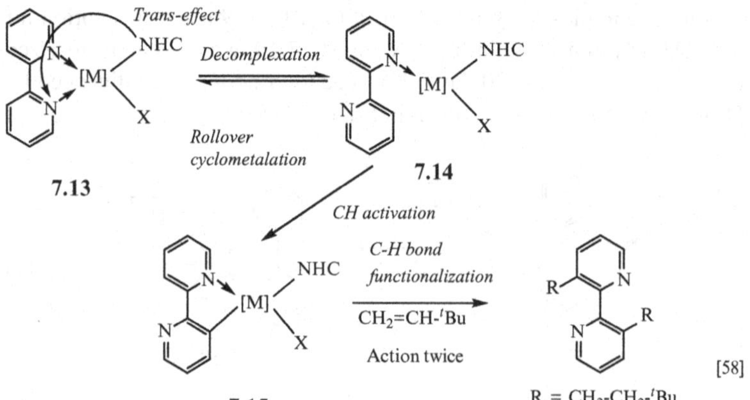

(7.12)

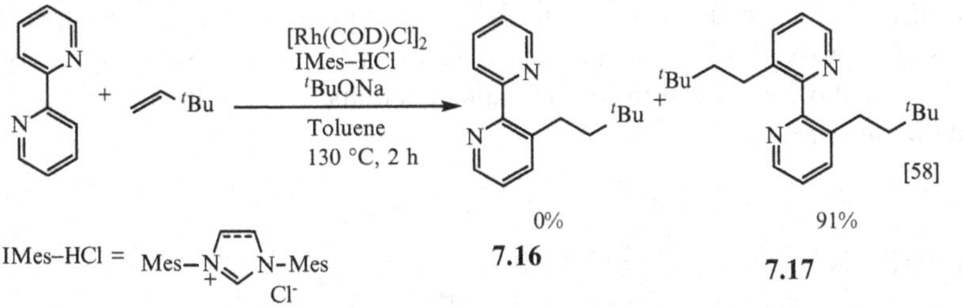

$$(7.13)$$

7.3 Applications for Organometallic Intramolecular-Coordination Five-Membered Ring Products

7.3.1 Introduction

From these two reviews [16, 18], this section shows alkenylations (reactions with alkenes), alkynylations (reactions with alkynes), acylations, carbonylations, isocyanations, and halogenations as applications for organometallic intramolecular-coordination five-membered products, as seen in Eqs. (7.14)–(7.44) and Schemes 7.3, 7.4, 7.5, 7.6, 7.7, 7.8, 7.9, 7.10, 7.11, and 7.12.

7.3.2 Alkenylations

Representative intramolecular five-membered ring compounds **7.18** are prepared through reactions of *N,N*-dimethylbenzylamines with metal compounds.

Alkenylations take place with these species, as shown in Eq. (7.14).

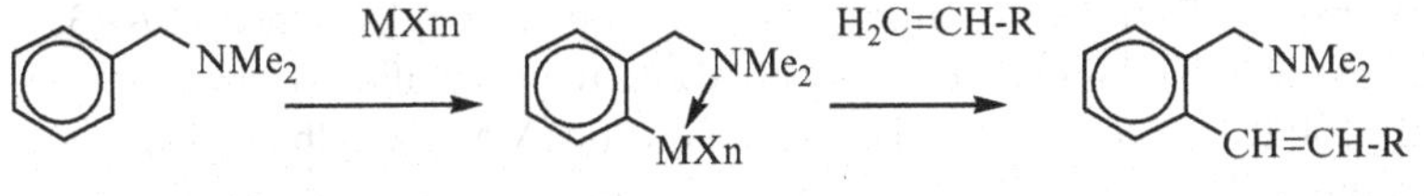

$$(7.14)$$

7.18

These alkenylations using five-membered ring compounds are widely employed in synthetic organic chemistry. The reaction of alkyl or aryl vinyl ketones with a cyclometalated palladium compound is shown as an example of alkenylation in Eq. (7.15) [72].

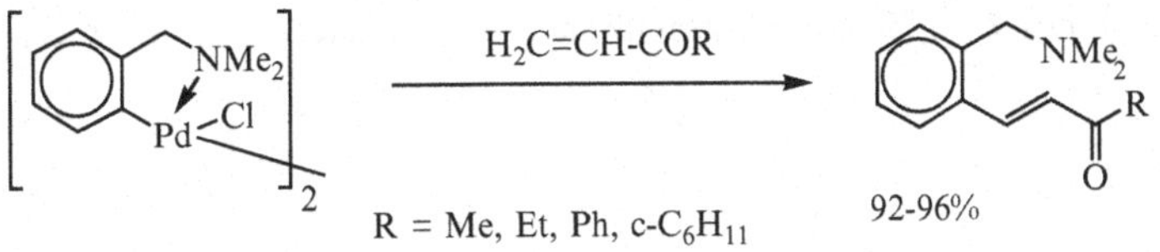

$$(7.15)$$

$R = Me, Et, Ph, c-C_6H_{11}$ 92-96%

Other alkenylations with dimethoxyphenyl, naphthyl, and ruthenocenyl metal derivatives also proceed as shown in Eqs. (7.16), (7.17), and (7.18), respectively [73–75]. Terminal olefins are used for these alkenylations, because the reactivity of branched olefins is low [76].

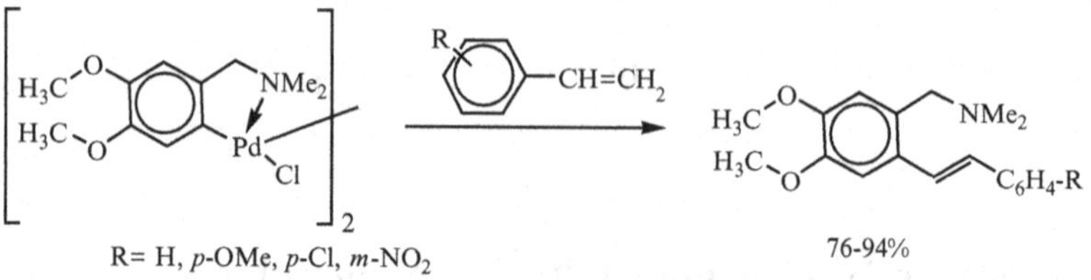

R= H, *p*-OMe, *p*-Cl, *m*-NO$_2$ 76-94%

$$(7.16)$$

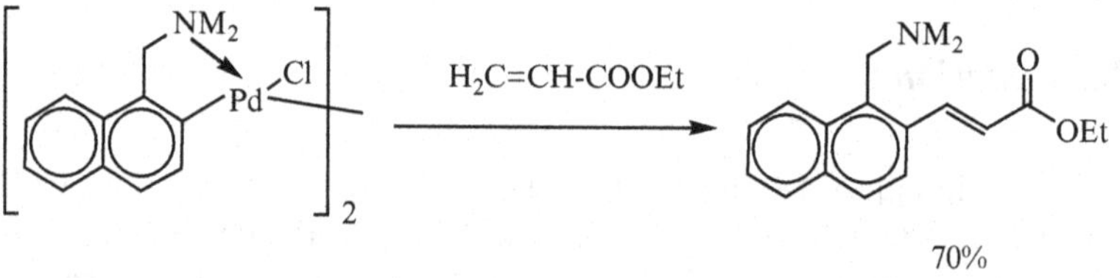

70%

$$(7.17)$$

R = COCH$_3$, COPh, Ph   89-94%

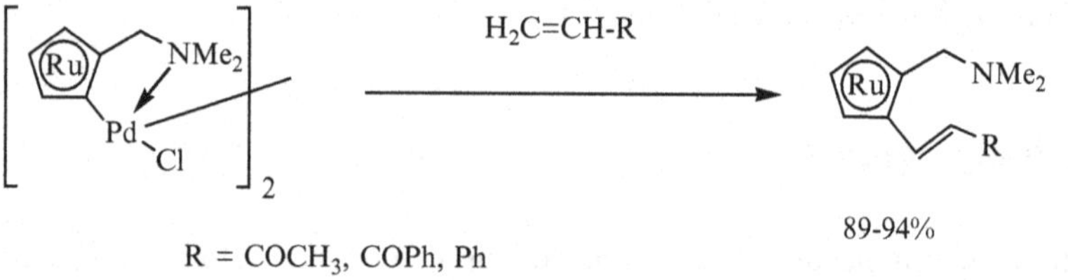

$$(7.18)$$

These alkenylation products are easily utilized for the formation of heterocyclic compounds by cyclization reactions. For example, ethyl *N*-methyl-*N*-(3,4-methylenedioxy)benzylglycinate is cyclopalladated regiospecifically at C(6) when treated with Li$_2$PdCl$_4$. The product, the di-μ-chloro-bis(*N*,*N*-dialkylbenzylamine-6-C,*N*)-dipalladium(II) complex **7.19**, undergoes a substitution reaction via the insertion of methyl vinyl ketone between the palladium metal and the phenyl carbon atom. The resultant β-aryl-α,β-unsaturated ketone **7.20** is cyclized using anhydrous potassium carbonate in ethanol to the corresponding ethyl *N*-methyl-1,2,3,4-tetrahydroisoquinolinium-3-carboxylate **7.21**, as shown in Eq. (7.19) [76, 77].

Cyclopalladated *t*-butylimine compounds are reacted with styrene and then treated with trifluoroacetic acid to give *o*-formylstilbenes in high yields. Their *N*-methylimine derivatives are converted to another heterocyclic compound, 3-aryl-N-methyl-isoquinolones **7.22**, by oxidation with mercuric acetate, as shown in Eq. (7.20) [78].

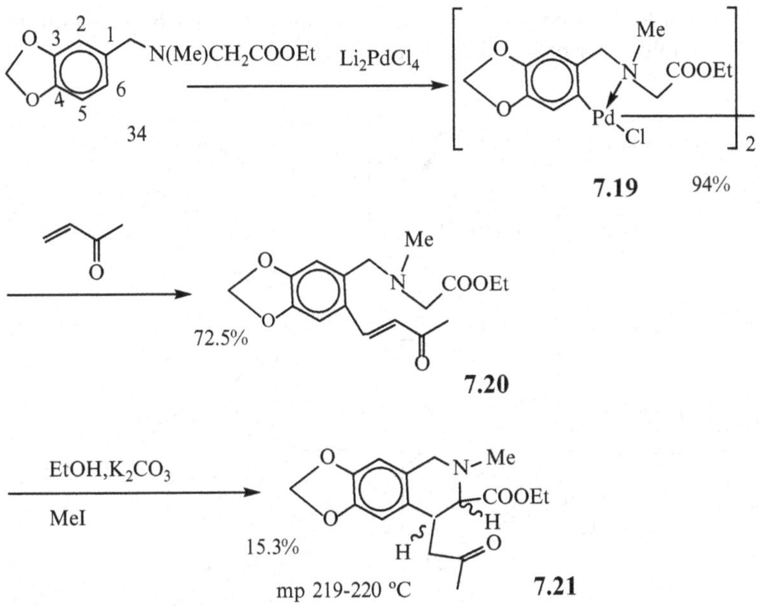

7.19   94%

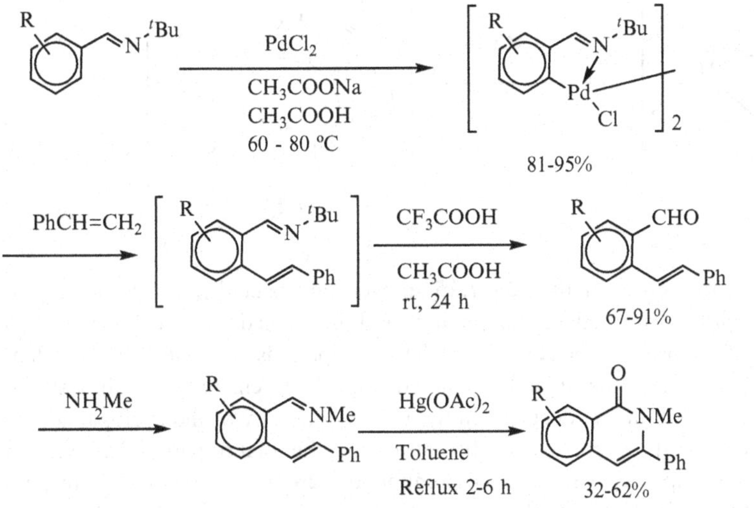

7.20

(7.19)

72.5%

EtOH, K$_2$CO$_3$

MeI

15.3%

mp 219-220 °C   **7.21**

PdCl$_2$

CH$_3$COONa
CH$_3$COOH
60 - 80 °C

81-95%

PhCH=CH$_2$ CF$_3$COOH

CH$_3$COOH
rt, 24 h

67-91%   (7.20)

NH$_2$Me   Hg(OAc)$_2$

Toluene
Reflux 2-6 h   32-62%

R = H, 5-Cl, 3-MeO, 4-MeO, 5-MeO, 4,5-(MeO)$_2$ **7.22**

7.3.3   Alkynylations

The reaction of hexafluorobut-2-yne with *N,N*-dimethylbenzylamine, 8-methylquinoline, or benzo[*h*]quinoline palladium dimer forms halide-bridged binuclear products with insertion of one hexafluorobut-2-yne **7.23–7.25** to the Pd–C bonds, as shown in Eqs. (7.21)–(7.23). These benzyl and quinolinyl dimeric complexes easily afford monomeric derivatives in high yield by bridge-splitting reactions with pyridine or triphenylphosphine [79–81].

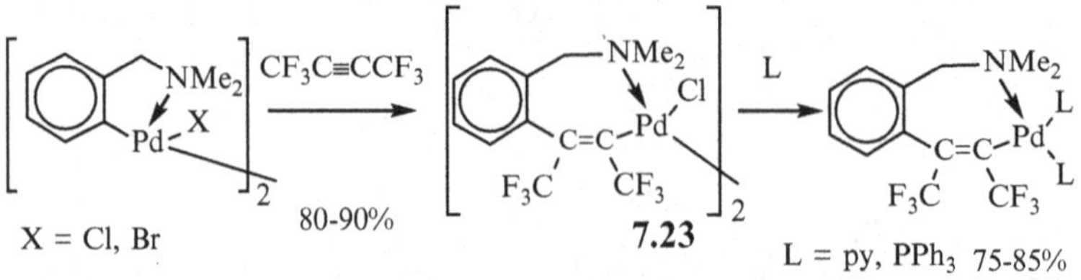

$$(7.21)$$

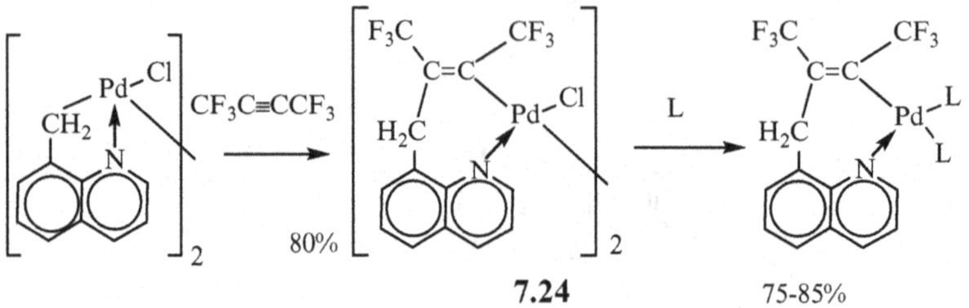

$$(7.22)$$

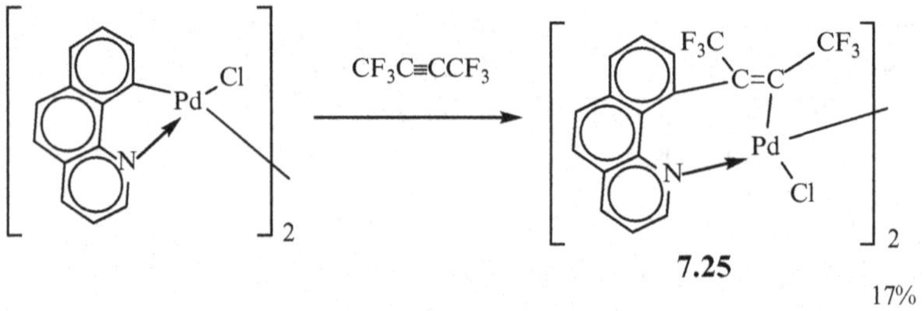

$$(7.23)$$

These inserted compounds **7.23–7.25** showed unexpected thermal stability so that no reaction performed to recover the modified palladium-free ligand led to clean products. However, the stability of these compounds is very much dependent upon the nature of the other ligands on the Pd atom. Thus, changing the chloride for an iodide led to a dramatic decrease in the thermal stability of the cyclopalladated compounds [82–84]. In reactions of this latter type, the total or partial dealkylation of the NMe$_2$ group occurs to produce compound **7.26** or compound **7.27**, respectively, as shown in Eq. (7.24). In a related reaction, it was possible to detect the presence of MeI together with an amount of CH$_4$ [82].

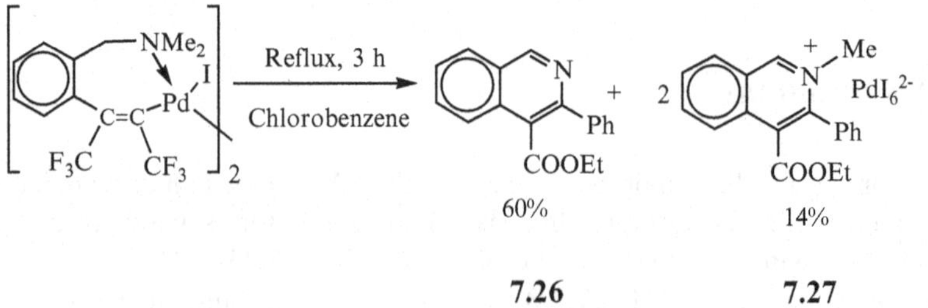

$$(7.24)$$

On the other hand, with diphenylacetylene or phenylmethylacetylene, cyclometalated compounds form nine-membered ring compounds **7.28** with bis-insertion of the alkynes, as shown in Eq. (7.25). When the R's are dissymmetric (R^1, R^3=Me, R^2, R^4=Ph; R^1, R^3=Ph, R^2, R^4=Me), head-to-tail and tail-to-tail arrangement isomers **7.28α** and **7.28β** are isolated; X-ray diffraction studies were made on a bromo-derivative **7.28α** and chloro-derivative **7.28β** [79].

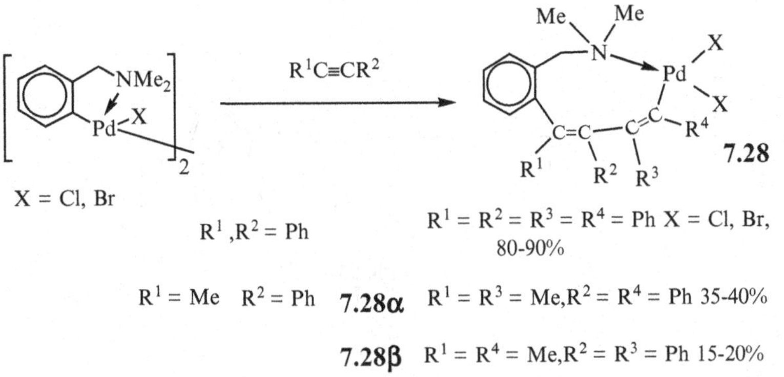

NMe2
Pd
X
2
X = Cl, Br
R1C≡CR2
R1,R2 = Ph
Me
Me
N
Pd
X
X
C=C
C=C
R1
R2
R3
R4
7.28
R1 = R2 = R3 = R4 = Ph X = Cl, Br,
80-90%
R1 = Me R2 = Ph
7.28α R1 = R3 = Me,R2 = R4 = Ph 35-40%
7.28β R1 = R4 = Me,R2 = R3 = Ph 15-20%

(7.25)

The insertion reactions of acetylene compounds with di-μ-iodo-bis(*N,N*-dimethyl-1-naphthylamine-*CN*)-dipalladium form substituted heterocyclic products **7.29** with formation of palladium metal and the loss of one *N*-methyl groups by refluxing in chlorobenzene (Eq. (7.26)) [79, 82, 84, 85]. The heterocyclic compounds are synthesized directly by the reaction of iododimethylaminonaphthalenes with acetylenes in the presence of a catalytic amount of the 1-dimethylaminonaphthalene palladacycle, as shown in Eq. (7.27) [85].

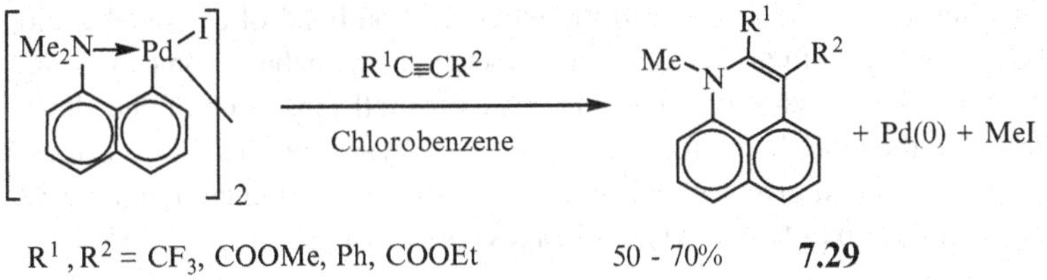

Me2N→Pd
I
2
R1C≡CR2
Chlorobenzene
Me
N
R1
R2
+ Pd(0) + MeI
R1, R2 = CF3, COOMe, Ph, COOEt
50 - 70%
7.29

(7.26)

Me2N
I
+
R1C≡CR2
Cat. 5%
Chlorobenzene
Reflux
Me
N
R1
R2
32 - 87%
7.30
Cat. =
Me2N→Pd
I
2
R1 = CF3, COOMe, Ph
R2 =4-NO2C6H4, COOEt
4-MeC6H4, COOMe
4-MeOC6H4, CHO
3-CF3C6H4, COMe,
CF3, Ph

(7.27)

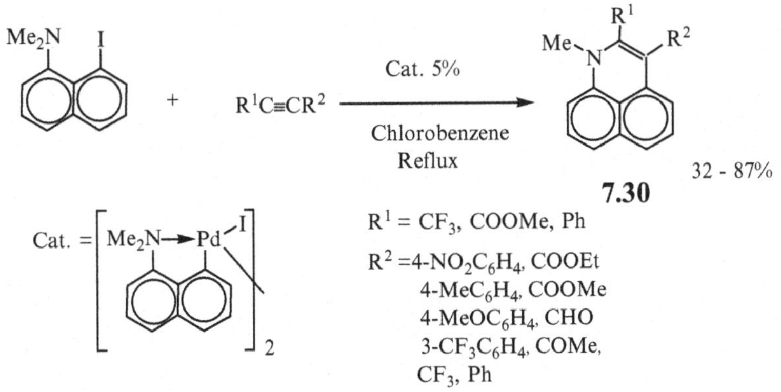

With dimethylaminomethylferrocene at room temperature, the insertion of two alkynes into the Pd–C bond of the cyclopalladated derivatives proceeds to a high yield, as shown in Eq. (7.28). When reaction with diphenylacetylene is performed at higher temperatures, depalladation occurs readily to give six- and seven-membered *ortho*-fused rings, that is, a bis-insertion product, through new annulation reactions of the phenyl groups, and the formation of one of these also involves cleavage of a C–N bond, as shown in Eq. (7.29) [79].

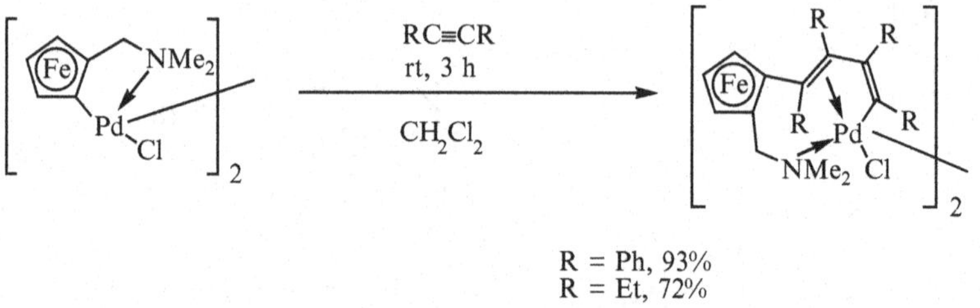

$$RC\equiv CR \quad rt, 3\ h \quad CH_2Cl_2$$

(7.28)

R = Ph, 93%
R = Et, 72%

PhC≡CPh

Chlorobenzene
Reflux 3 h

15%

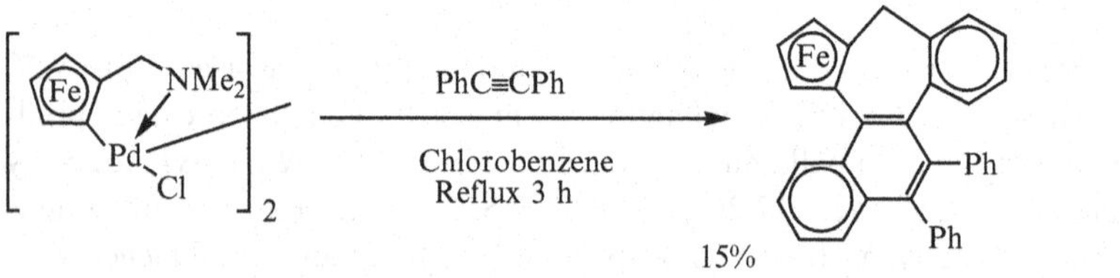

(7.29)

Dupont and Pfeffer reported one-pot synthesis of heterocyclic compounds through two types of insertions of an alkyne into the metal–carbon bond of activated cyclometalated benzyl methyl sulfide. The reaction of the alkyne with the cationic cyclopalladated compound gives a six-membered ring arising from the insertion of one alkyne molecule **7.31**. On the other hand, the reaction of the alkyne with the chloro-bridged cyclopalladated compound gives an eight-membered thiocyclic ring compound **7.32**, arising from the insertion of two alkyne molecules, as shown in Scheme 7.3 [59].

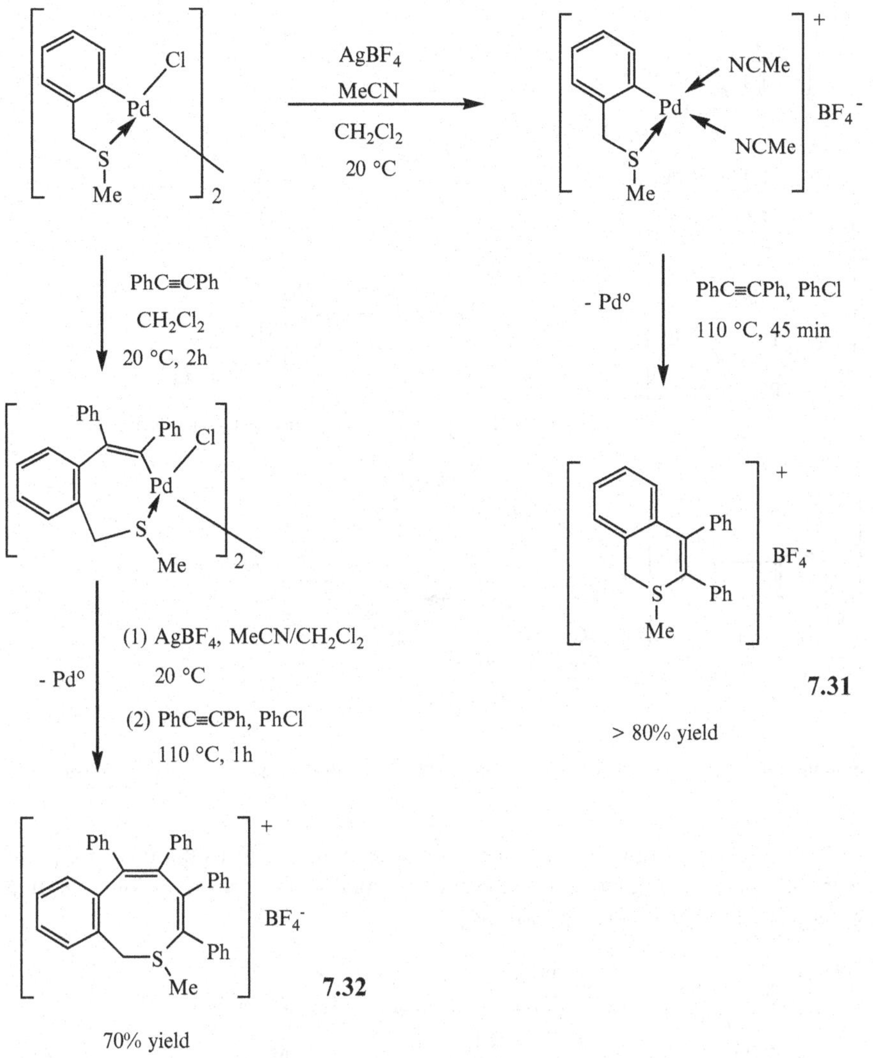

Scheme 7.3 One-pot synthesis of heterocyclic compounds through two types of insertions of an alkyne and two alkynes into the metal-carbon bond of activated cyclometalated benzyl methyl sulfide [67]

Alkynylation of cyclometalated compounds is utilized for the synthesis of indenols or indenones. Thermally promoted reactions of the alkynes with an *ortho*-manganated acetophenone **7.33** give 1*H*-indene-1-ols **7.34** in a high yield. However, the reaction of the corresponding *ortho*-manganated derivatives of methyl 3,4,5-trimethoxybenzoate gives indenone **7.35** in a high yield, as shown in Scheme 7.4 [60].

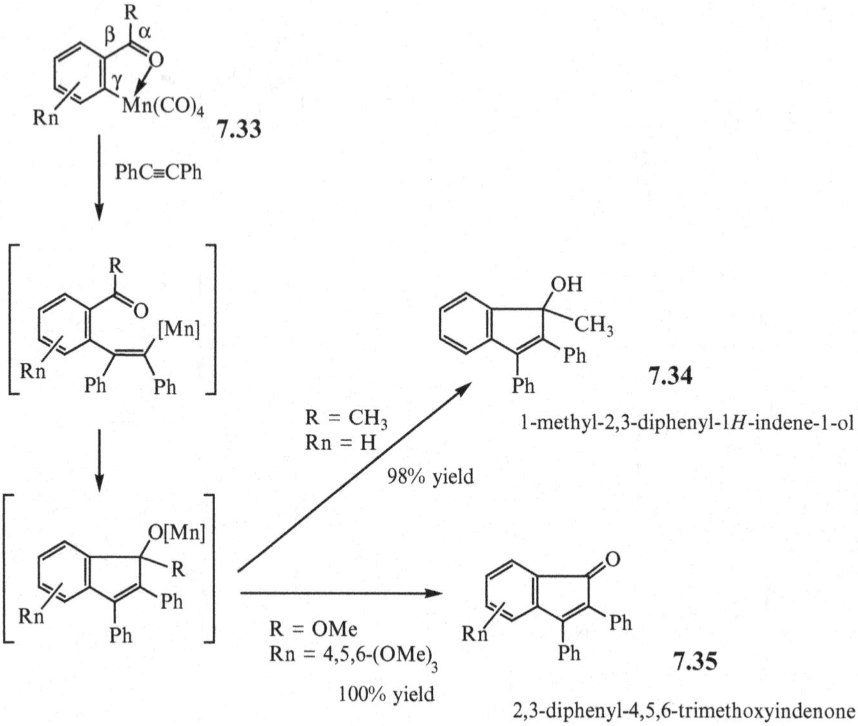

Scheme 7.4 Synthesis of indenols and indenones by alkynylation of cyclometalated compounds [68]

Further, alkynylation of cyclomanganated acetylindole gives the former type of thermally promoted mono-inserted product, 1-methyl-1-hydroxypyroloindole, in a good yield, as shown in Eq. (7.30) [86].

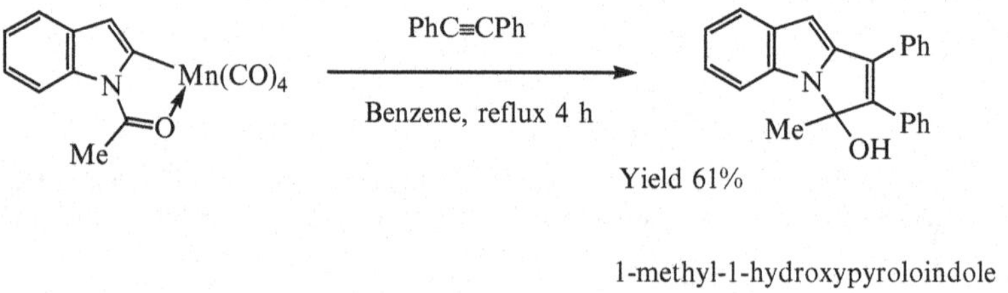

$$\text{Yield } 61\%$$

1-methyl-1-hydroxypyroloindole

7.3.4 Acylations

The regiospecific reaction of palladacycles with acyl halides gives a 2-acyl derivative **7.36** in a high yield (Eq. (7.31)) [87]. Cyclopalladated *N,N*-dimethylbenzylamine derivatives bearing methoxy groups on the ring behave similarly with respect to acetyl and benzoyl chlorides, also to give the corresponding ketones in high yield, as shown in Eq. (7.32). The heterocyclic compounds **7.38** are also synthesized by stirring these acyl compounds with a base such as sodium ethoxide in dry ethanol solution. The presence of an electron-withdrawing substituent appears to drastically decrease the rate; a complex di-μ-chloro-bis[N,N-dimethyl-5-chlorobenzylamino-$2C,N$]dipalladium(II) is recovered unchanged after 3 days under reflux under the same reaction conditions. These reactions are presumed to proceed through a two-step process in which the first step is the insertion of acyl halides and the second is the potassium cyanide-induced dissociation of the presumed intermediates **7.37** [76, 88].

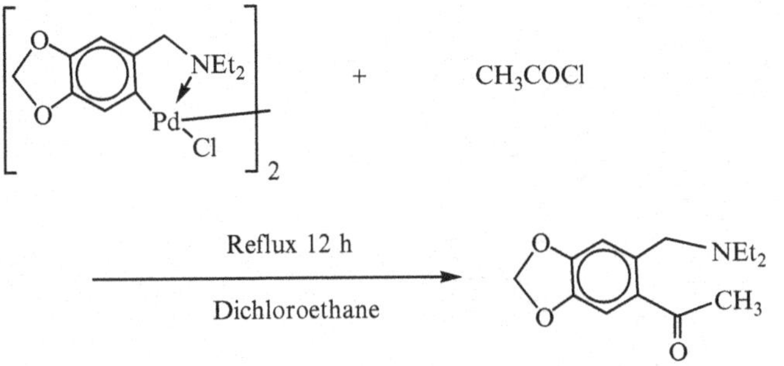

(7.31)

81% **7.36**

70 - 97% **7.37**

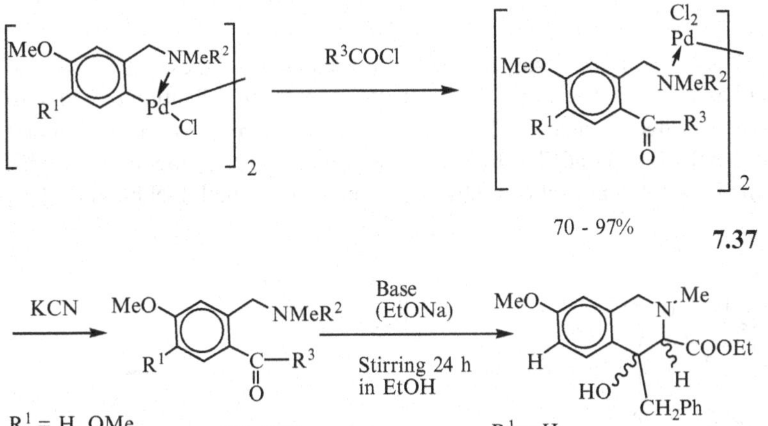

(7.32)

R^1 = H, OMe
R^2 = Me,-CH$_2$COOEt
R^3 = -CH$_2$Ph, -CH=CH-Ph,
   -CH(CH$_3$)-Ph, -CH$_2$CH$_2$Ph

77 - 98%

R^1 = H
R^2 = -CH$_2$COOEt
R^3 = -CH$_2$Ph,

12 - 25%

7.38

7.3.5 Carbonylations

Carbonylation with alkenylation is one of the most important reactions in the application of intramolecular five-membered ring compounds. Heterocyclic ketones are synthesized by the insertion of carbon monoxide into a metal–carbon bond followed by demetalation reactions. Because metal–carbon or metal–nitrogen σ-bonds are usually highly reactive to carbon monoxide, these bonds undergo carbon monoxide insertion and finally yield heterocyclic compounds through hydrolysis of the metal–nitrogen bond [61].

In 1967, Takahashi and Tsuji [61] reported that carbonylation of substituted azo-benzene in protic solvents such as alcohols or water proceeded smoothly under mild conditions to afford 2-aryl-3-indazolinones with separation of metallic palladium in a high yield, as shown in Eq. (7.33). Hence, the reactions are a very good synthetic method for the synthesis of substituted 2-aryl-3-indazolinones **7.39**.

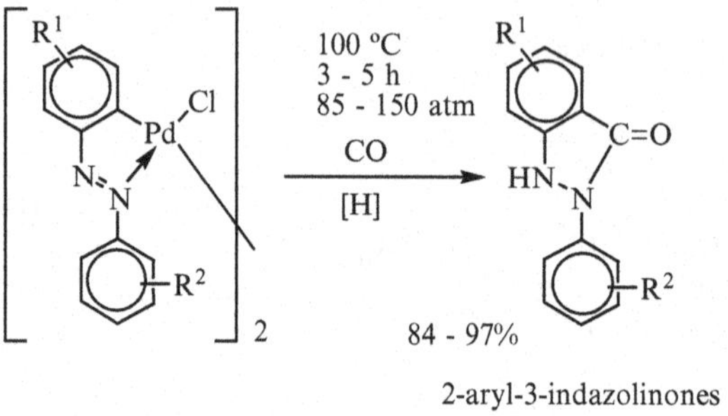

(7.33)

2-aryl-3-indazolinones

R^1, R^2= Me, Cl
R^1 = Me, MeO, Cl   **7.39**
R^2 = H

The reporters tentatively assumed that the mechanisms of these carbonylation reactions are as shown in Scheme 7.5: The first step is the coordination of CO with splitting of the bridged structure. The subsequent insertion of CO at the σ-bond gives an acyl–palladium bond. The final step seems to be the insertion of the –N=N– bond into the Pd–acyl bond, followed by hydrogenolysis of the Pd–N bond [61].

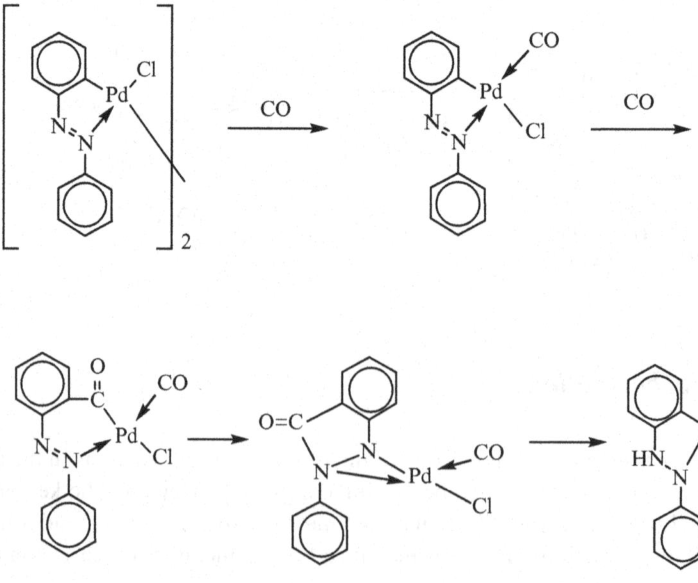

Scheme 7.5 Carbonylation of a diazobenzene palladium chloro complex [61]

Carbonylations yield various kinds of heterocyclic carbonyl compounds with organopalladium intramolecular five-membered ring compounds of *N,N*-dimethylbenzylamine, benzaldehyde–Schiff base compounds, and benzylamine–Schiff base compounds (e.g., forming 2-methylphthalimidine, 3-acetoxy-2-phthalimidine, and 2-benzylphthalimidine), as shown in Eqs. (7.34)–(7.36) [62]. Based on Eqs. (7.34) and (7.36), Thompson and Heck [62] assumed a similar mechanism to the Takahashi and Tsuji reactions [61]. Initially, carbonyl insertion into the palladium–carbon bond and breaking of a bridge bond formed a compound **7.40**, which may undergo internal addition of the acyl–palladium group to the nitrogen–carbon double bond to form compounds **7.41**. The simple reductive elimination of compounds **7.41** yields compounds **7.43** and **7.44**. Carbonylation of these in xylene or benzene forms various kinds of heterocyclic compounds, as shown in Eqs. (7.34)–(7.36) and Schemes 7.5 and 7.6.

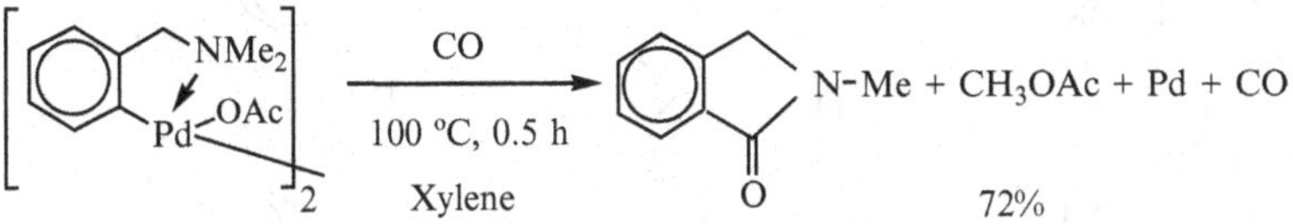

(7.34)

2-methylphthalimidine

(7.35)

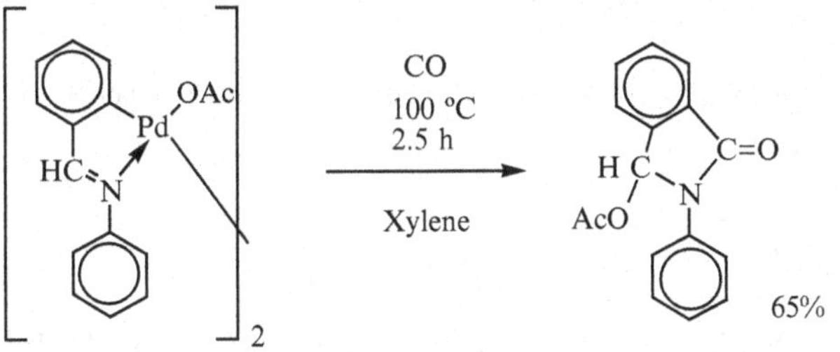

3-acetoxy-2-phthalimidine

(7.36)

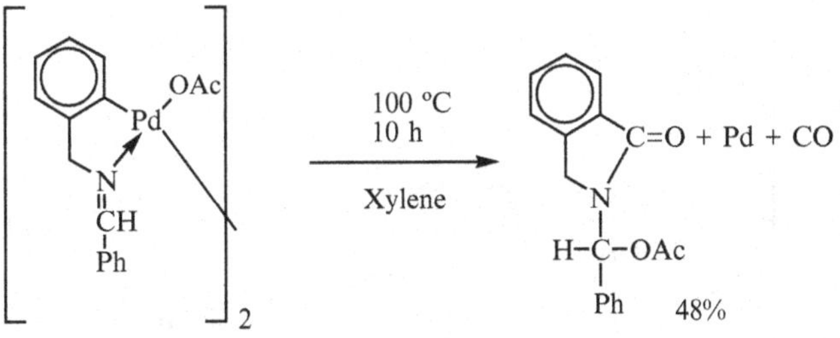

2-benzylphthalimidine

However, in alcoholic solvents, this carbonylation or carbonylation with ferrocenyl or pyrrole compounds also forms various kinds of uncyclized esters, as shown in Scheme 7.6 the compound **7.42** and in Eqs. (7.37)–(7.39) the compounds **7.45–7.47** [58, 68, 76, 89].

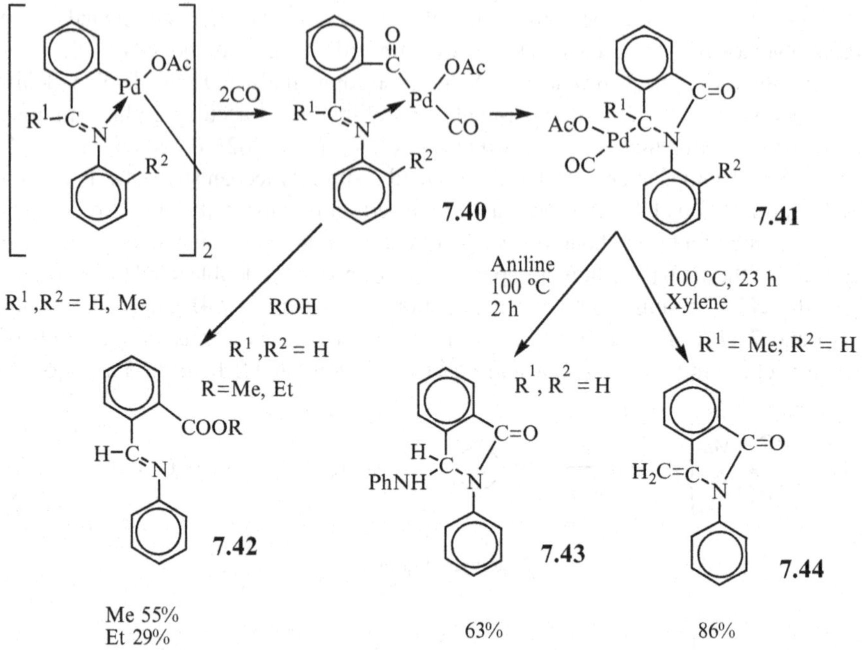

Scheme 7.6 Synthesis of various uncyclized esters and heterocyclic compounds via carbonylation with cyclometalated compounds [73]

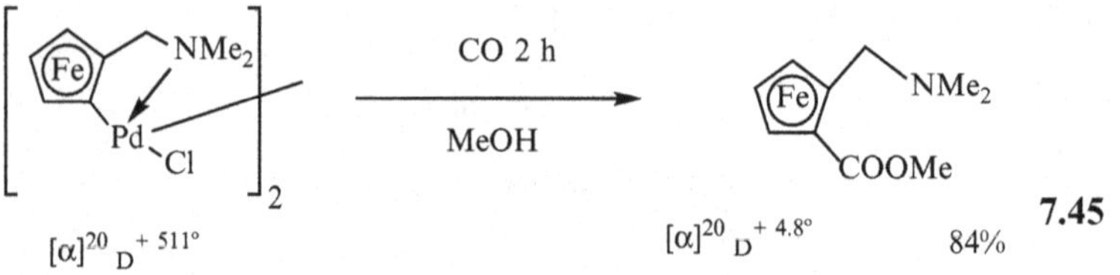

$$[\alpha]^{20}_{D} \, {}^{+511°} \qquad [\alpha]^{20}_{D} \, {}^{+4.8°} \qquad 84\% \qquad \textbf{7.45}$$

(7.37)

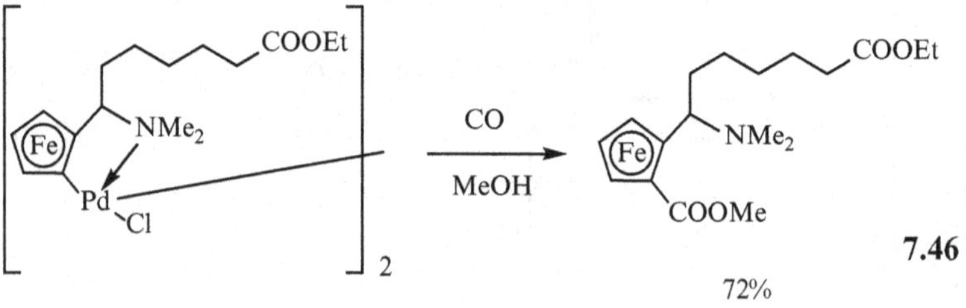

72%   **7.46**

(7.38)

69%   **7.47**

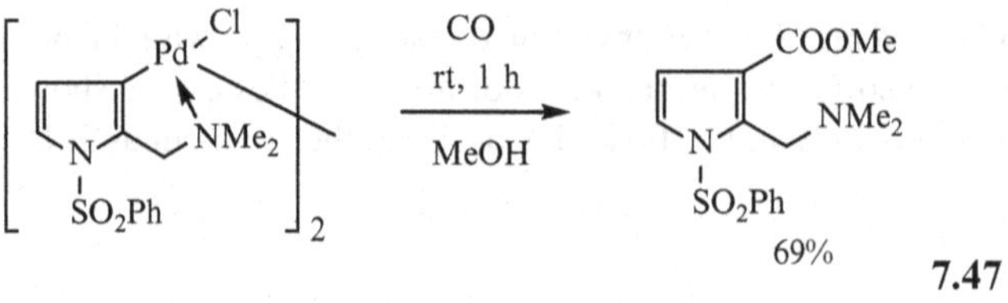

(7.39)

The carbonylation of *ortho*-palladated acetophenonephenylhydrazone yields an isoindolinone derivative **7.49**, which is of the same type as the methyleneisoindolinone **7.44** in Scheme 7.6 [62]. However, the addition of a stoichiometric amount of NaOMe to the carbonyl inserted product **7.48** in solution leads instantaneously to reductive elimination, yielding the indazole derivative 3-methyl-1-phenyl-indazole **7.50** in a high yield. This process is presumed to proceed by complete deprotonation of the carbonyl coordinated product **7.48**, followed by *E/Z* isomerization with reductive elimination, as shown in Scheme 7.7 [63].

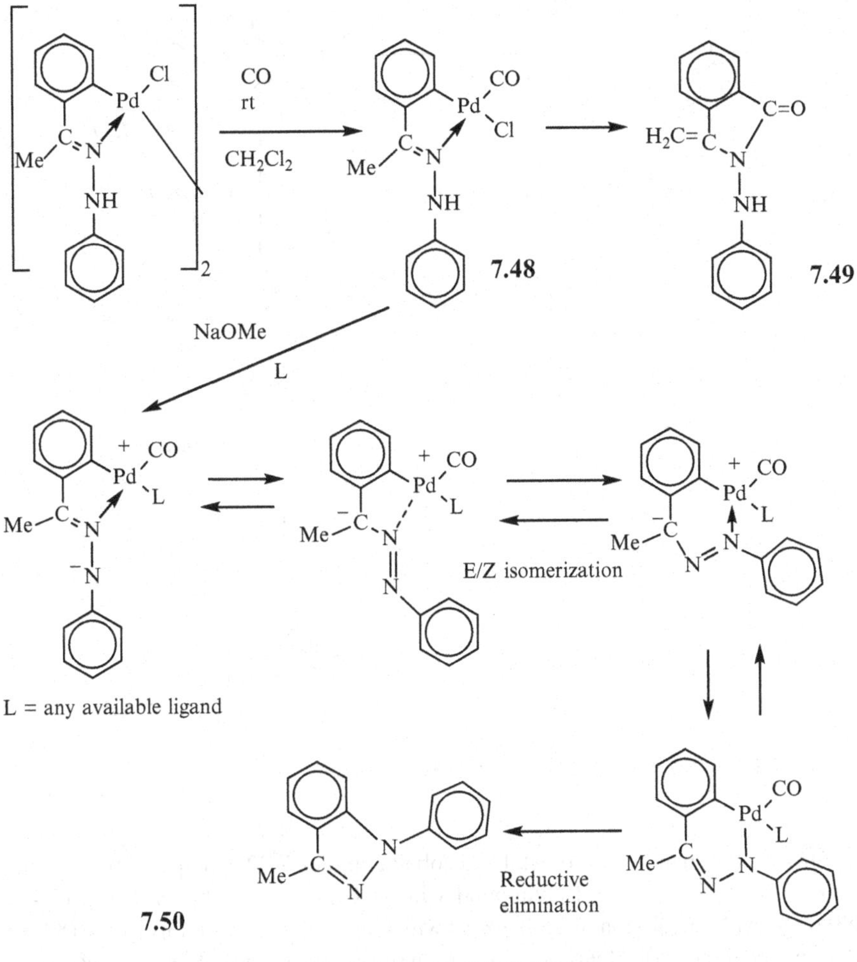

Scheme 7.7 Synthesis of 3-methyl-1-phenyl-indazole **7.50** [77]

In 1956, Murahashi and Horiie [64] reported the carbonylation of azobenzene with an octacarbonyldicobalt catalyst, as shown in Scheme 7.8. Monocarbonylation of azobenzene with a cobalt catalyst at 190 °C forms indazolones **7.39**, as shown in Scheme 7.8, which is also obtained by monocarbonylation of the cyclopalladium azobenzene complex as shown in Eq. (7.33) and Scheme 7.5. Dicarbonylation of azobenzenes at 230 °C forms 3-phenyl-2,4-dioxo-1,2,3,4-tetrahydroquinazolines **7.51** in a quantitative yield. The dicarbonylation product is assumed to occur through further carbonylation of the monocarbonylation product [64–67]. This process may be utilized for industrial preparation of anthranilic acid, because the hydrolysis of the quinazolines **7.51** gives anthranilic acid in a high yield [65].

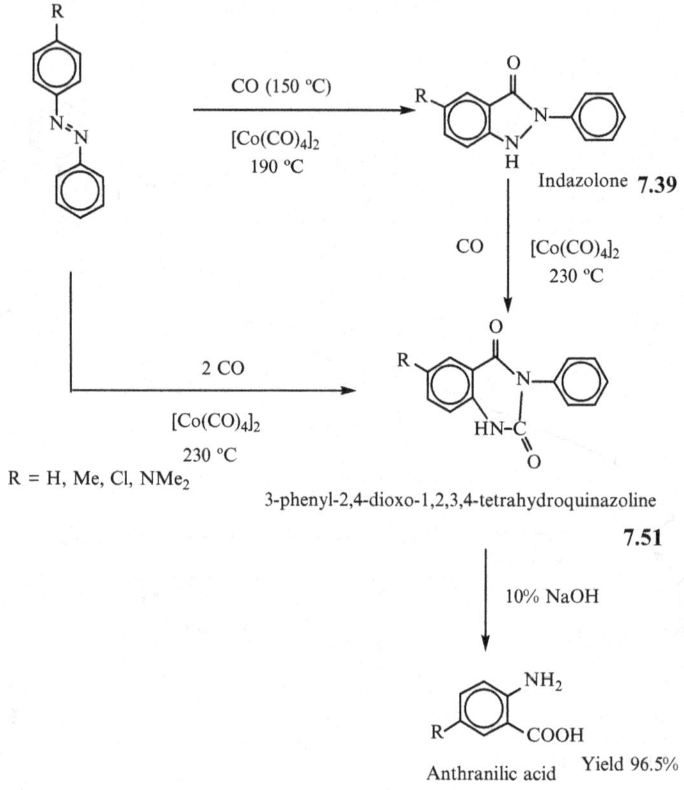

Scheme 7.8 Synthesis of anthranilic acid [78–81]

On the other hand, a cyclometalated cobalt complex **7.52** is prepared by the reaction of the cyclopalladated compound with $NaCo(CO)_4$, as shown in Eq. (7.40), although cyclometalation of azobenzene with $Co_2(CO)_8$ at a low temperature (80 °C) does not proceed [90]. Hence, the formation of isoindazolone **7.39** from azobenzene and CO in the presence of $Co_2(CO)_8$ at 190 °C is presumed to proceed via the cyclocobalt complex **7.52**.

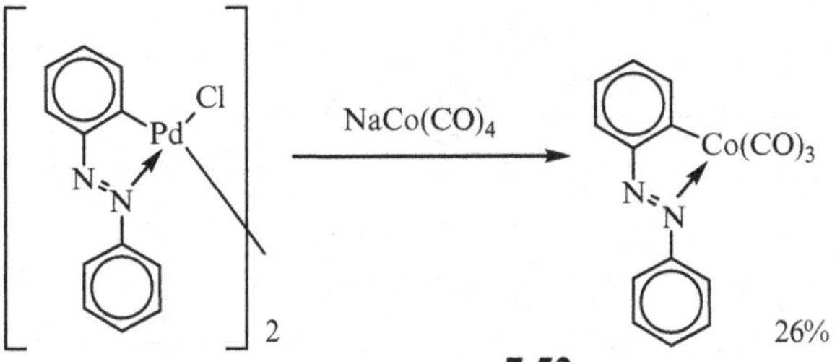

$$(7.40)$$

7.52

Considering the process of these reactions with a cobalt catalyst shown in Scheme 7.8, therefore, it can easily be assumed that carbonylation proceeds via the cyclometalation reaction of the cobalt compound. One year previously, moreover, in 1955, Murahashi also reported the same type of heterocyclic compound 2-phenylphthalimide **7.53** resulting from carbonylation of benzylideneaniline (Ph–CH=N–Ph) in the presence of $Co_2(CO)_8$ as a catalyst as shown in Fig. 7.2 [91, 92]. Hence, it is assumed that cyclometalation was actually performed in 1955, which was 8 years before the publication of Kleiman and Dubeck's report in 1963, although the product was not isolated as an intermediate. The first application of carbonylation was therefore performed before organotransition-metal intramolecular five-membered ring compounds were first reported [93].

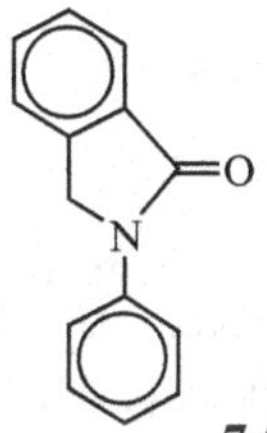

Fig. 7.2 Heterocyclic compound (2-phenylphthalimine **7.53**) resulting from carbonylation of benzylideneaniline in the presence of $CO_2(CO)_8$ as a catalyst [91, 92]

7.53

The various kinds of 2-ferrocene derivatives are synthesized by reaction with transition metal compounds in a manner similar to the orthometalation of phenyl compounds. The palladium compounds **7.54** of dimethylaminomethylferrocene react with carbon monoxide in methyl alcohol to give the 2-methoxycarbonyl compound **7.45**, and subsequent treatment of the product with methyl iodide, sodium amalgam, and phosphoric acid gives 1-methyl-2-carboxylic acid **7.55**. Reduction of the 2-methoxycarbonyl compound **7.45**, followed by treatment with NaOH, produces an alcohol **7.56**, and oxidation of the compound **7.56** yields an aldehyde **7.57**, as shown in Scheme 7.9. Since the starting material **7.54** is an optically active compound, as shown in Scheme 7.9, all of its derivatives are also optically active 1,2-ferrocene derivatives [68].

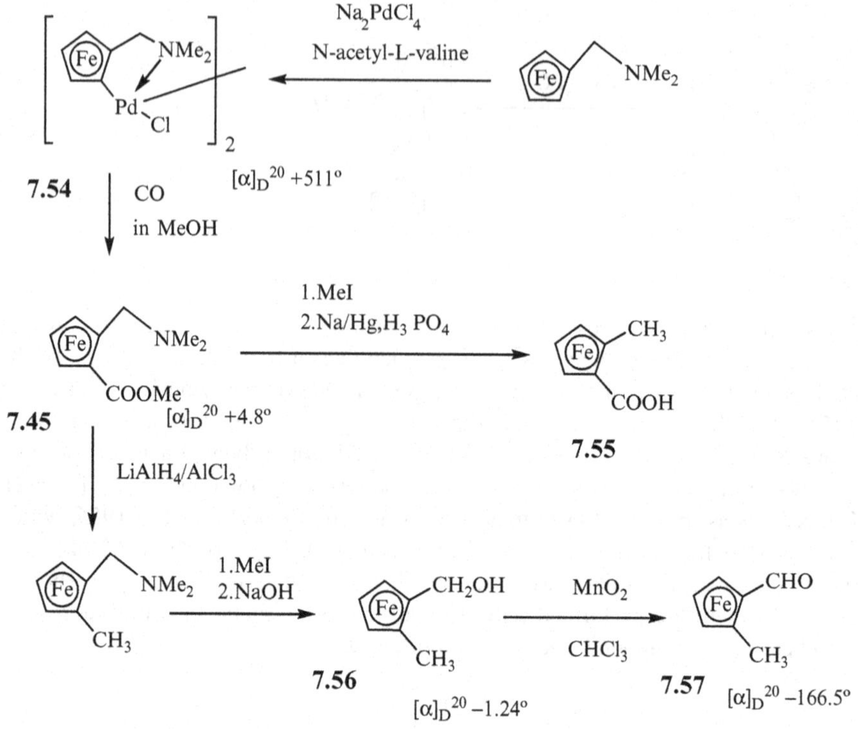

Scheme 7.9 Synthesis of various 1,2-ferrocene derivatives [74]

Asymmetric cyclopalladation of dimethylaminomethylferrocene in the presence of *N*-acetyl-(*R*)- or (*S*)-leucine gives an enantiomerically enriched palladacycle, (*S*)- or (*R*)-[Pd{C$_5$H$_3$(CH$_2$NMe$_2$)FeC$_5$H$_5$}(μ–Cl)]$_2$ **7.58**, **7.59**, respectively. Carbonylation of each enantiomer followed by iodomethylation and reduction by sodium amalgam gives (*S*)- or (*R*)-2-methylferrocene carboxylic acid **7.60**, **7.61** with an optical purity of 80 or 90 %, respectively, as shown in Scheme 7.10 [69].

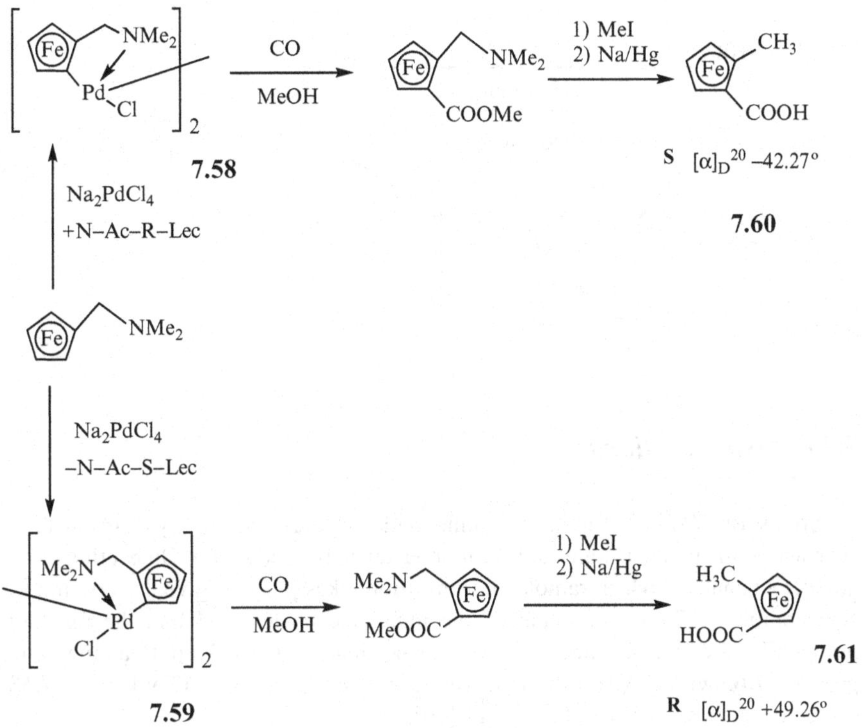

Scheme 7.10 Synthesis of (*S*) and (*R*)-2-methylferrocene carboxylic acids [86]

These reactions are utilized for the syntheses of pharmaceutical intermediates, e.g., diastereomeric 1,2-disubstituted ferrocene, by using an enantiomeric ferrocene **7.46**, as shown in Eq. (7.38) [76, 89].

A chloro-bridged cyclopalladated pyridine compound is also carbonylated by the reaction of carbon monoxide with bubbling through a reaction mixture at room temperature in the presence of NEt$_3$ as a proton scavenger. Metallic palladium is separated out, while ethyl ester **7.62** is recovered in a 73 % yield, as shown in Eq. (7.41) [94].

On the other hand, palladium-catalyzed carbonylations with benzylamines yield benzolactams **7.63** in high yield as shown Eq. (7.42) [94].

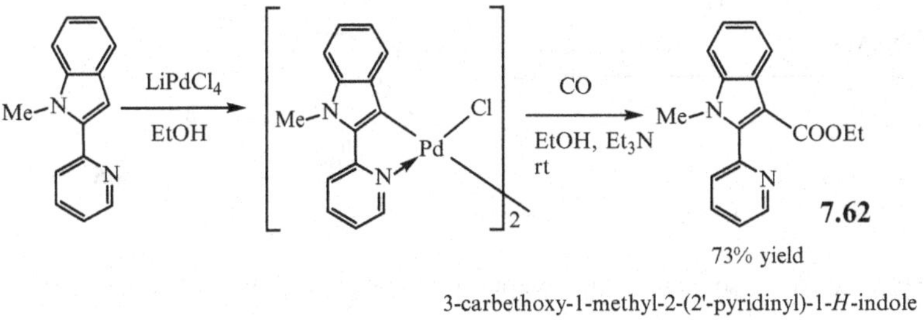

(7.41)

73% yield

3-carbethoxy-1-methyl-2-(2'-pyridinyl)-1-*H*-indole

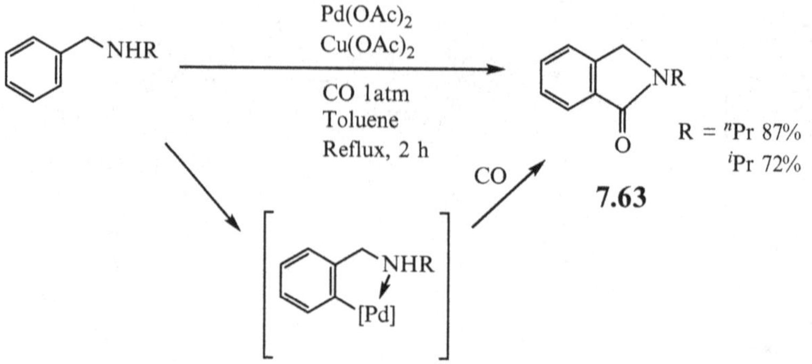

$$(7.42)$$

7.3.6   Isocyanations

Reactions of *N,N*-dimethylbenzylamine palladacycles with isocyanates lead to cleavage of the halide bridges to give monocyanate products **7.64**. On heating of the monoisocyanates **7.64**, intramolecular insertion takes place to give dimeric imino-acyl complexes **7.65**. Treatment of the monoisocyanates **7.64** and the insertion products **7.65** with isocyanide produces diisocyanate products **7.66**. Reaction of the products **7.65** with LiAlH$_4$ or Grignard reagent gives diamines **7.67** or ketones **7.68**, respectively, as shown in Scheme 7.11 [70].

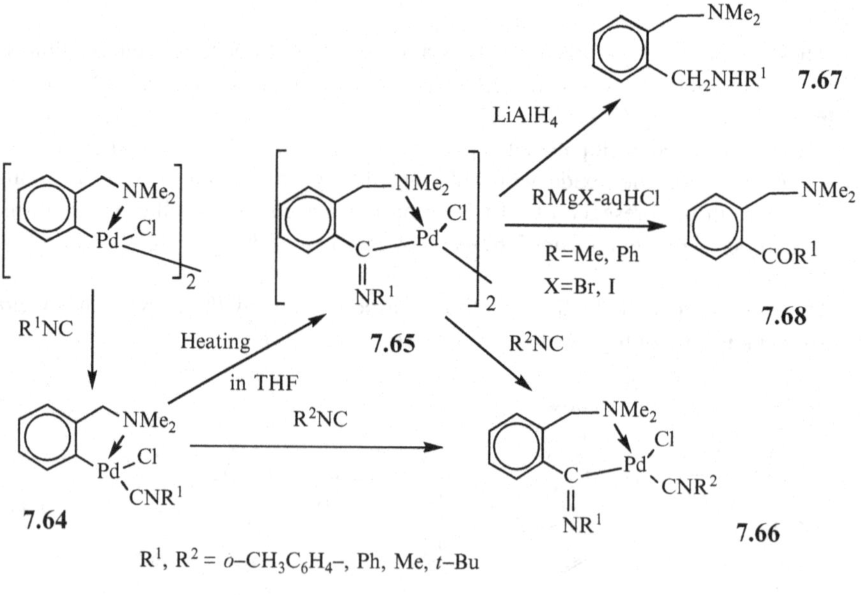

R^1, R^2 = *o*–CH$_3$C$_6$H$_4$–, Ph, Me, *t*–Bu

Scheme 7.11 Synthesis of isocyanation products and their diamine and ketone derivatives [89]

Some heterocyclic compounds are easily synthesized simply by thermal degradation of the cyclometalated compounds. Reactions of azobenzene cyclopalladated products with isocyanides proceed, for example, with cleavage of the chloride bridges to yield the highly stable yellow–orange complexes **7.69**. Thermal degradation of the cleaved products proceeds smoothly at 100–130 °C in toluene to give 3-imino-2-phenylindazolines **7.70** with separation of metallic palladium [71]. These indazolines **7.70** may also be obtained from reaction of the azobenzene cyclopalladation products in toluene at 120 °C with isocyanides in a 1:2 molar ratio, respectively, as shown in Scheme 7.12 [71].

The 2-aryl-3-indazolines **7.39** that are synthesized by reaction of the azobenzene cyclopalladated complexes with carbon monoxide are also obtained from the cleavage products **7.69** by reaction with monoisocyanides and carbon monoxide in methanol at 40 °C under a pressure of 35 kPa/cm^2 [71].

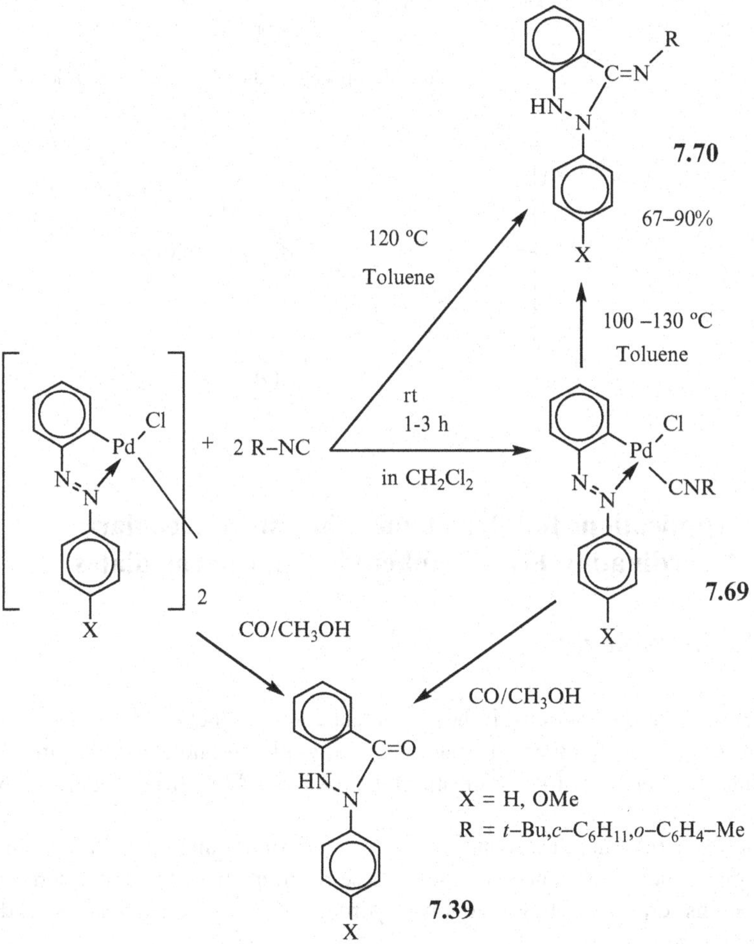

Scheme 7.12 Synthesis of 2-aryl-3-indazolines **7.39** and 3-imino-2-phenylindazolines **7.70** [90]

7.3.7 Halogenations

It is well documented [95] that direct halogenation of arenes, bromination in particular, is one of the most selective electrophilic reactions yielding almost exclusively *para*-substituted products. Evidently, however, the use of cyclometalated compounds might drastically change the selectivity in favor of *ortho*-halogenated compounds, as shown in Eq. (7.43) [76, 96]. This halogenation is also widely applied as a regioselective reaction method, as shown below in Eq. (7.44) [97].

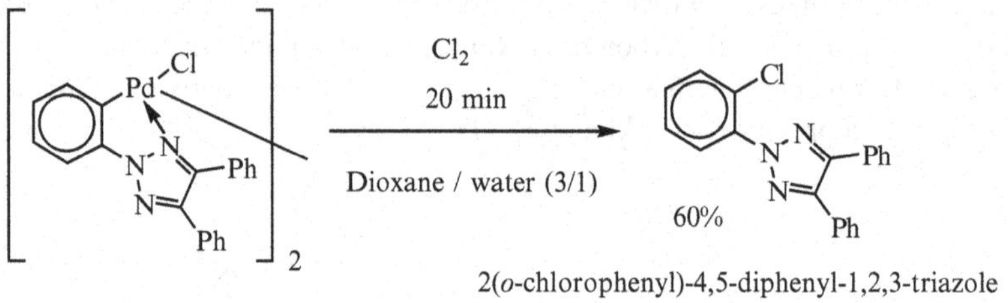

$$(7.43)$$

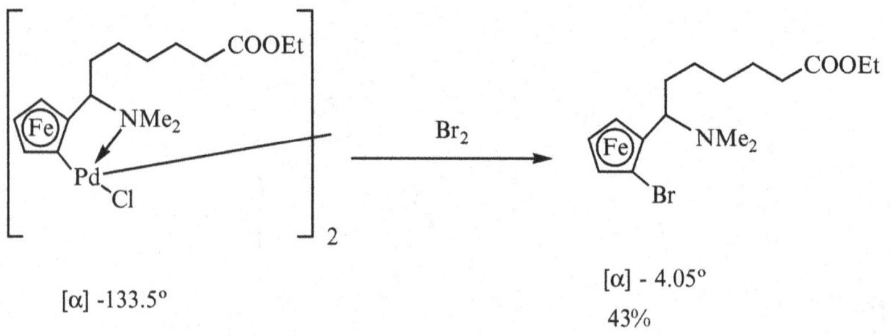

2(*o*-chlorophenyl)-4,5-diphenyl-1,2,3-triazole

$$(7.44)$$

7.4 Applications for Organometallic Intramolecular-Coordination Five-Membered Ring Intermediates

7.4.1 Introduction

In cyclometalation reactions, if the cyclometalation products **7.71** are highly labile, they react easily with other substrates **ZXm** through intramolecular five-membered ring intermediates **7.71** to give substitution products **7.72** with a metal moiety **MXn**, as shown in Eq. (7.45).

These reactions are the second type of reactions of intramolecular five-membered ring compounds in organic syntheses. These reactions have been reported for carbonylations, cross-coupling reactions, hydroacylations, ring expansions, carbocyclization reactions, etc.

$$\gamma\ \beta\ \alpha \atop C\text{-}Q\text{-}G\text{-}Y \quad + \quad MXm \quad \longrightarrow \quad \left[\begin{array}{c} Y \longrightarrow MXn \\ \quad | \\ \alpha\ G \diagdown \underset{\beta}{Q} \diagup C\ \gamma \end{array} \right] \quad \overset{ZXm}{\underset{-\,MXn}{\longrightarrow}} \quad \begin{array}{c} Y \qquad ZXn \\ | \qquad | \\ \alpha\ G \diagdown \underset{\beta}{Q} \diagup C\ \gamma \end{array}$$

$$\qquad\qquad\qquad\qquad\qquad \textbf{7.71} \qquad\qquad\qquad\qquad \textbf{7.72}$$

Five-membered ring intermediate

$$(7.45)$$

MXm, MXn = metal compounds
M = transition metals and main group metals,
    69 kinds of metals
Y = N, P, O, S, etc.
G, Q = C, N, P, O, S, etc.
ZXm, ZXn = subtrates

7.4.2 Carbonylations

In the previous section, the synthesis of 2-methylphthalimidine **7.73** by the carbonylation of a cyclopalladation product of *N,N*-dimethylbenzylamine at 100 °C and under 150 atm of carbon monoxide was shown in Eq. (7.46) [61]. These reactions represent the first type of reaction of intramolecular five-membered ring compounds in organic syntheses. However, 2-aryl-3-indazolinone is also synthesized by the second type of carbonylation. This is because 2-phenyl-3-indazolinone is directly synthesized by carbonylation of diazobenzene in the presence of $Co_2(CO)_8$ at a higher temperature (190 °C) and under the same high pressure with carbon monoxide, as shown in Eq. (7.47) [64], and the reaction is assumed to proceed via the intramolecular five-membered ring intermediate. In fact, a similar cyclometalated organocobalt compound was prepared by the reaction of azobenzene with methyltetra(trimethylphosphine)cobalt, as shown in Eq. (7.48) [98].

7.73
2-methylphthalimidine

$$(7.46)$$

Intramolecular five-membered
ring intermediate

2-phenyl-3-indazolinone

55%

$$(7.47)$$

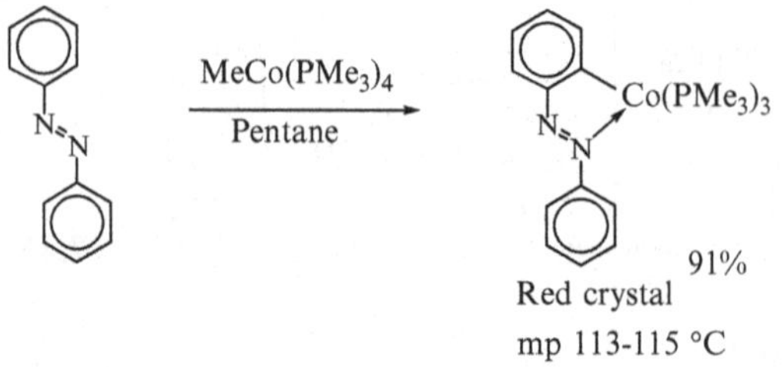

(7.48)

91%
Red crystal
mp 113-115 °C

In the reaction at 230 °C, on the other hand, 2-phenyl-3-indazolinone is further carbonylated to give 3-phenyl-2,4-dioxo-1,2,3,4-tetrahydroquinazoline **7.51**. Anthranilic acid is easily prepared with a high yield by hydrolysis of a quinazoline derivative **7.51**. Hence, double carbonylation of diazobenzene at 230 °C is a good method for the preparation of anthranilic acid, as shown in Scheme 7.8 in the previous section [64, 65, 67].

Catalytic carbonylative [4+1]cycloaddition is also considered to be a type of carbonylation with ruthenium catalysts via the intramolecular five-membered ring structure. The substrates are α,β-unsaturated imines. A β,γ-unsaturated γ-lactam is prepared in a high yield by reaction with carbon monoxide at 180 °C for 20 h in the presence of $Ru_3(CO)_{12}$ [99]. Representative reactions and their reaction mechanisms are shown in Eq. (7.49) and Scheme 7.13, respectively [99].

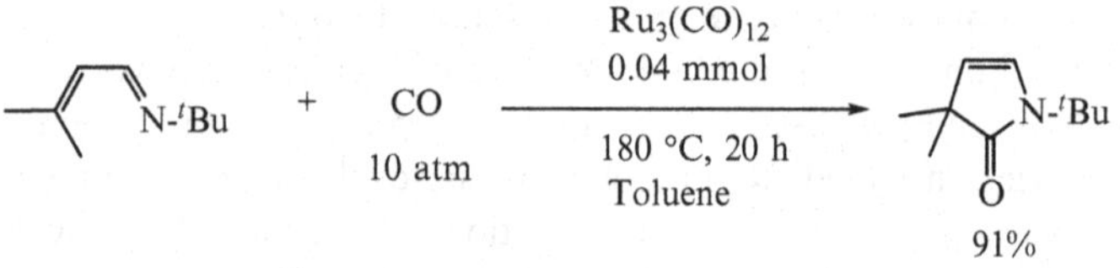

(7.49)

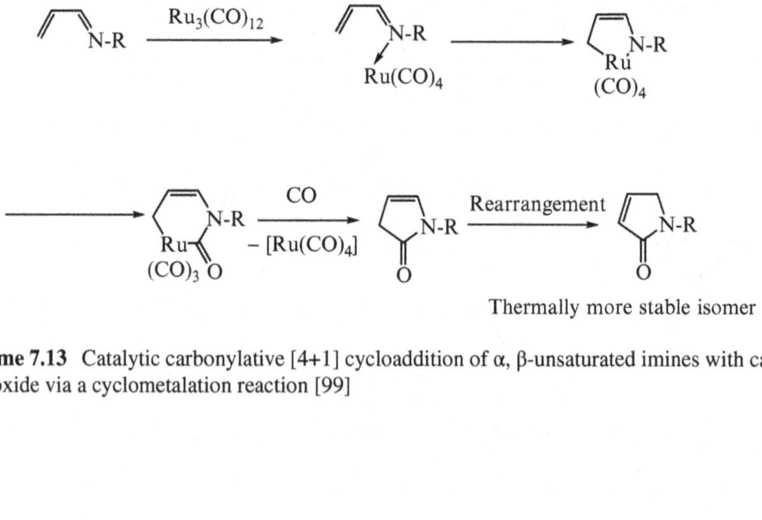

Thermally more stable isomer

Scheme 7.13 Catalytic carbonylative [4+1] cycloaddition of α, β-unsaturated imines with carbon monoxide via a cyclometalation reaction [99]

7.4.3 Cross-Coupling Reactions

Murai et al. [100–111] reported on many cross-coupling reactions regarding representative substrates such as methyl phenyl ketone **5.7** and phenylimine **5.6**, as shown in Table 5.6, in the presence of metal catalysts such as ruthenium triphenylphosphines. A representative example of a C–H/olefin coupling reaction is presented in Eq. (7.50) [103].

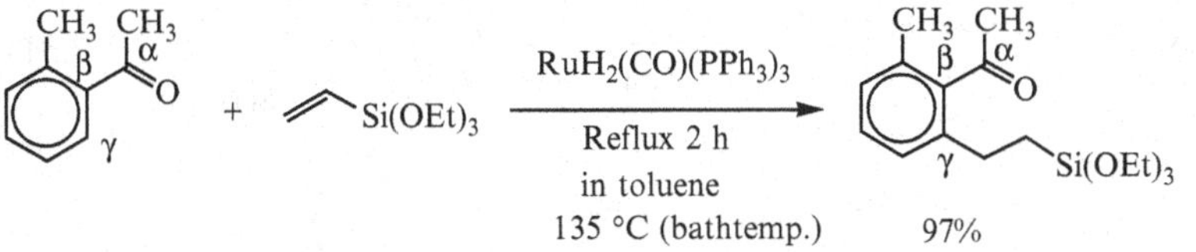

$$(7.50)$$

Ruthenium complexes with three triphenylphosphine ligands such as $RuH_2(CO)$ $(PPh_3)_3$ or $Ru(CO)_2(PPh_3)_3$ are excellent catalysts [103]. They are assumed to act via orthometalation of the triphenylphosphine ruthenium catalyst and reductive elimination of its metal element via the five-membered ring structure of the ruthenium catalyst **7.74**, as shown in Scheme 7.14 [103].

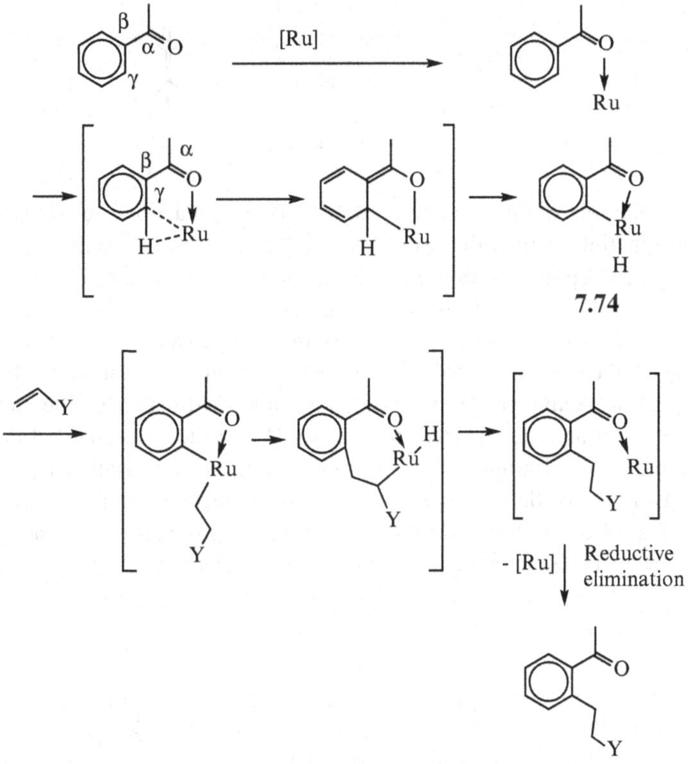

Scheme 7.14 Cross-coupling reactions via cyclometalation reactions [103]

As shown in Scheme 7.14, these metal compounds act as catalysts because these cyclometalation products are highly reactive and react easily with other reagents. The substrates of these reactions are not only aromatic compounds, such as methyl phenyl ketone (Eq. (7.50)) [100, 101, 103] and phenylimine (Eq. (7.51)) [100, 103], but also nonbenzenoid compounds, such as thiophenyl methyl ketone (Eq. (7.52)) [100, 101] and cyclohexenyl methyl ketone (Eq. (7.53)) [103].

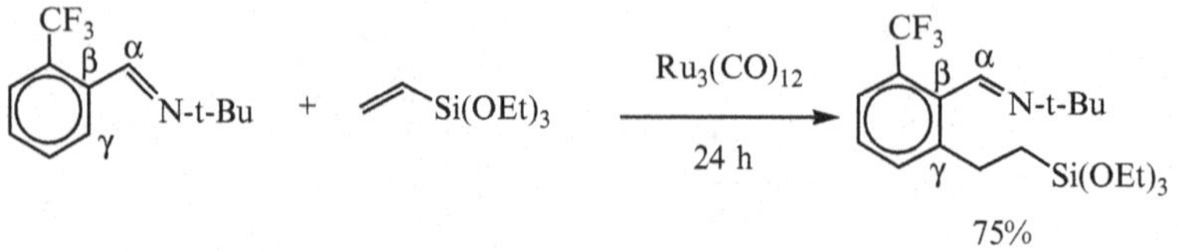

$$(7.51)$$

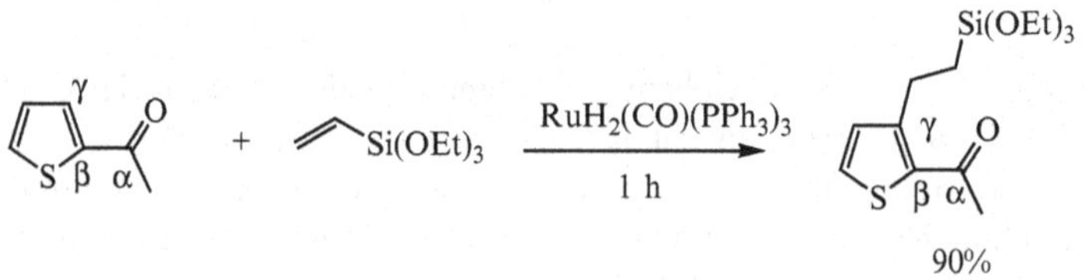

$$(7.52)$$

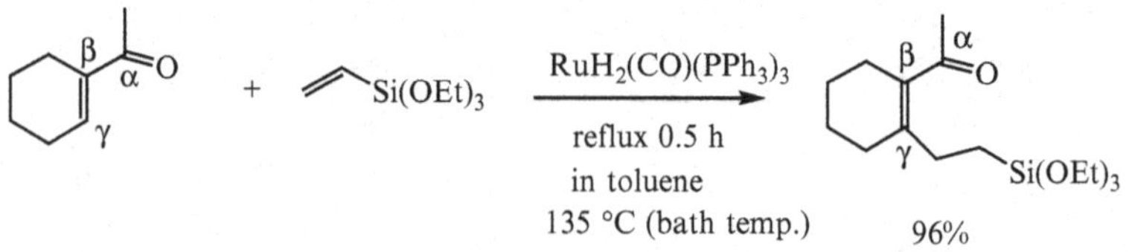

$$(7.53)$$

In these reactions of imines or ketones, which have a reaction site at the γ-position to the coordinating atom with vinylsilanes in the presence of a ruthenium catalyst, the C–H/olefin coupling reactions proceed as shown in Eqs. (7.51)–(7.53). However, as shown in Eq. (7.54), C–H/SiR$_3$ coupling reactions proceed only in the presence of highly reactive trialkylhydrosilanes. First, the nitrogen atom of the oxazoline ring coordinates to the ruthenium atom, and the activated ruthenium atom bonds with the *ortho*-carbon of phenyl ring (γ-position) and trialkylhydrosilanes. The olefins then coordinate to Ru and are inserted into an Ru–H bond to give an alkyl derivative. After the reductive elimination of a corresponding alkane, trialkylsilylruthenium compounds are formed. From this intermediate, reductive elimination produces a C–Si bond and leads to the corresponding silylation products, and an active ruthenium(0) species is regenerated, as shown in Scheme 7.15 [111].

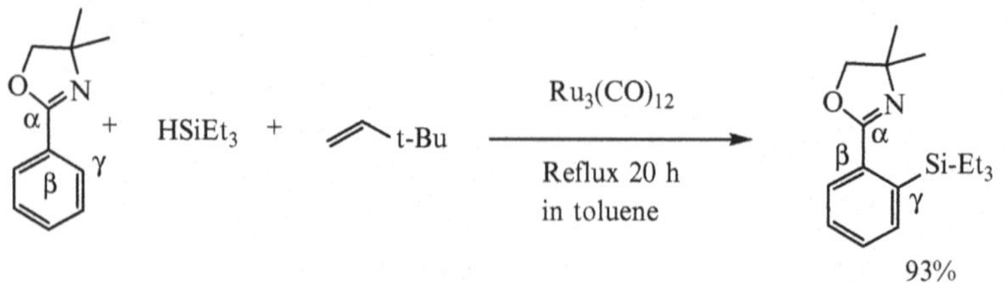

$$(7.54)$$

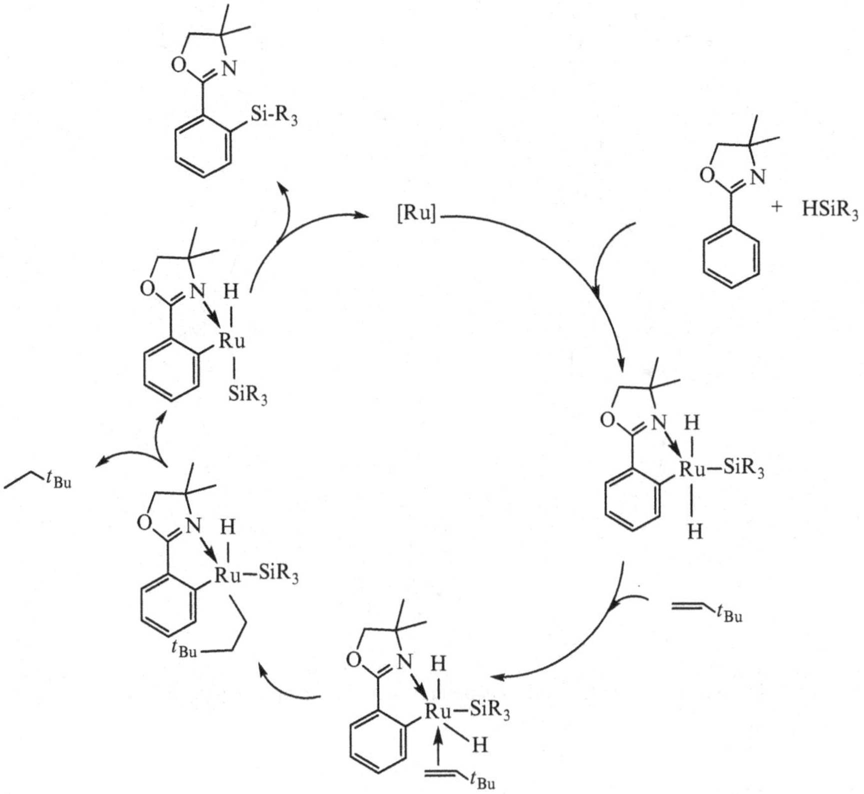

Scheme 7.15 CH/SiR$_3$ coupling reactions via cyclometalation reactions [111]

Intramolecular C–H/olefin coupling proceeds through the reaction of 1-(2-pyridyl)-1,5-dienyl compounds in the presence of rhodium compounds, as shown in Eq. (7.55) [99, 106, 107]. First, the metal **7.75** is activated by the coordinating atom, and the activated metal **7.75** then bonds with the γ-carbon atom through the insertion of a vinylic C–H bond to form a five-membered ring and metal hydrogen bond **7.76**. The intramolecular insertion of an olefin into the metal–hydrogen bond gives a tricyclic intermediate **7.77** and reductive elimination of the metal atom to form a cyclic product **7.78**, as shown in Scheme 7.16 [105].

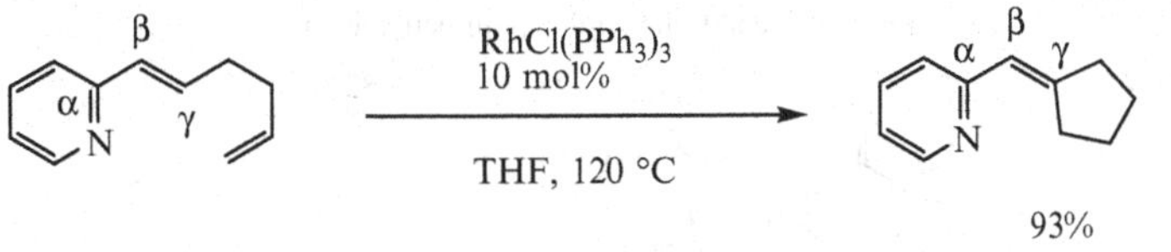

(7.55)

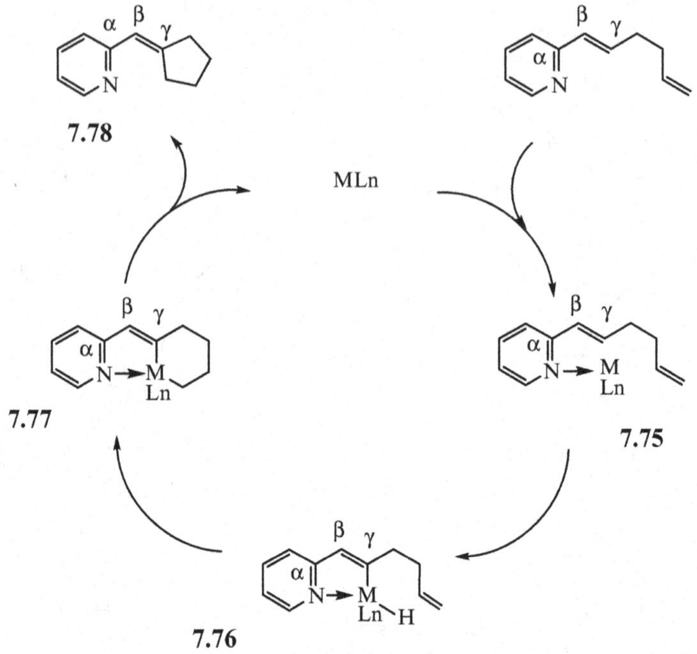

Scheme 7.16 Intramolecular C-H/olefin coupling reactions via cyclometalation reactions [105]

7.4.4 Hydroacylations

In the reactions of sulfanyl aldehydes with alkenes in the presence of a rhodium catalyst, hydroacylation proceeds via an intramolecular five-membered ring intermediate with the insertion of an alkenyl moiety between the rhodium and carbonyl carbon at the γ-position to the coordinating atom [112–114]. For example, β-methylsulfanyl aldehyde reacts with an amide alkene to give an intermolecular hydroacylation product with the insertion of an amide alkenyl moiety at a high yield, as shown in Eq. (7.56) [112].

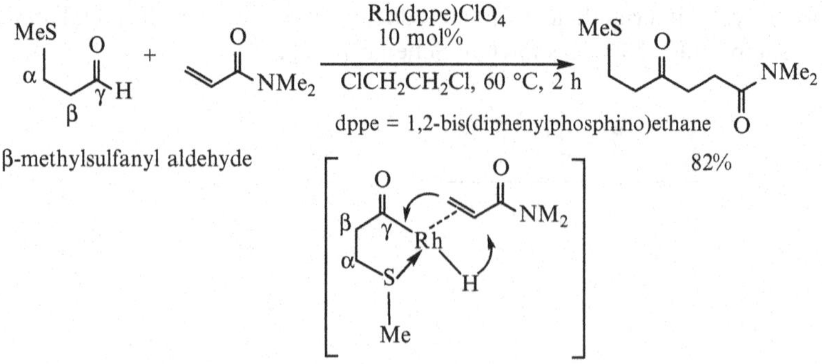

β-methylsulfanyl aldehyde

dppe = 1,2-bis(diphenylphosphino)ethane

82%

(7.56)

On the other hand, in the reaction of sulfanyl aldehydes containing a carbon–carbon double bond at the terminal position or a carbon–carbon triple bond, metal-catalyzed cyclizations by intramolecular hydroacylation proceed via the intramolecular five-membered ring intermediate, as shown in Eqs. (7.57) and (7.58), respectively [114].

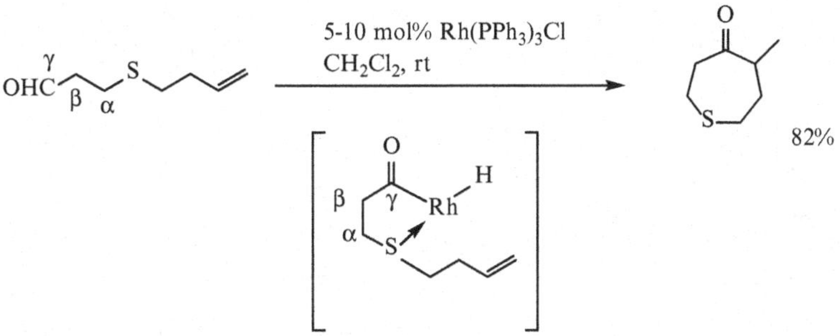

$$(7.57)$$

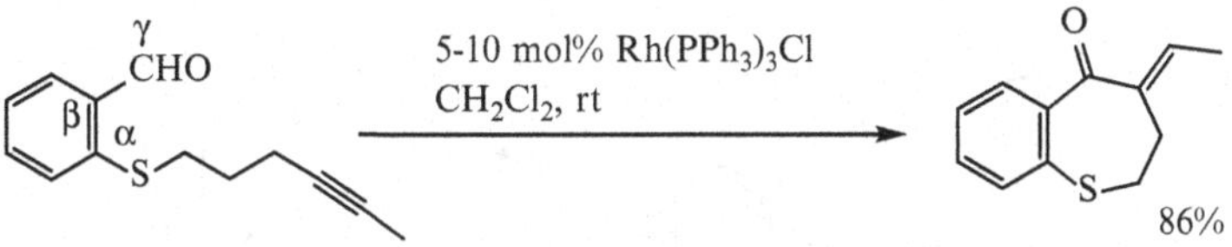

$$(7.58)$$

7.4.5 Ring Expansion Reactions

As three-membered ring compounds generally exhibit ring strain, ring expansion reactions often proceed via intramolecular five-membered ring intermediates. For example, cationic Zr alkyl complexes react with 2-Me-pyridine (α-picoline) to give a zirconium pyridine three-membered ring compound **7.79**. The reaction with propene produces a ring expansion compound containing a five-membered ring compound **7.80**, as shown in Eq. (7.59) [115]. The coupling of propene and α-picoline is indeed catalyzed by the zirconium five-membered ring compound **7.80** in the presence of H_2, as shown in Eq. (7.60). The mechanism of coupling reactions via ring expansion with a zirconium compound and hydrogenolysis is considered to be as shown in Scheme 7.17 [116].

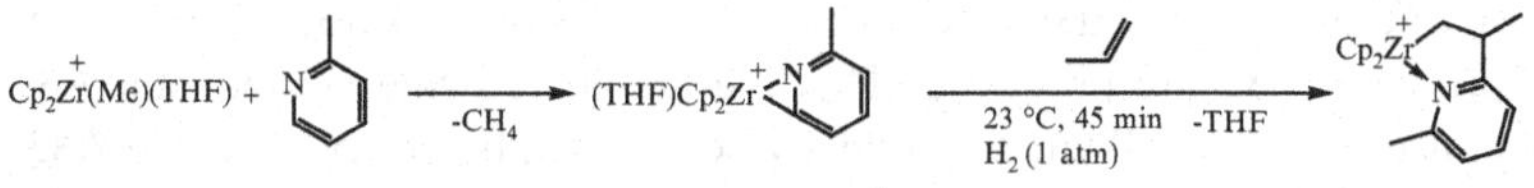

$$(7.59)$$

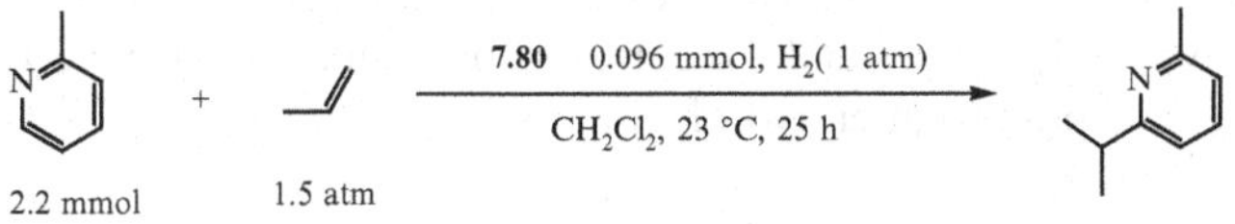

$$(7.60)$$

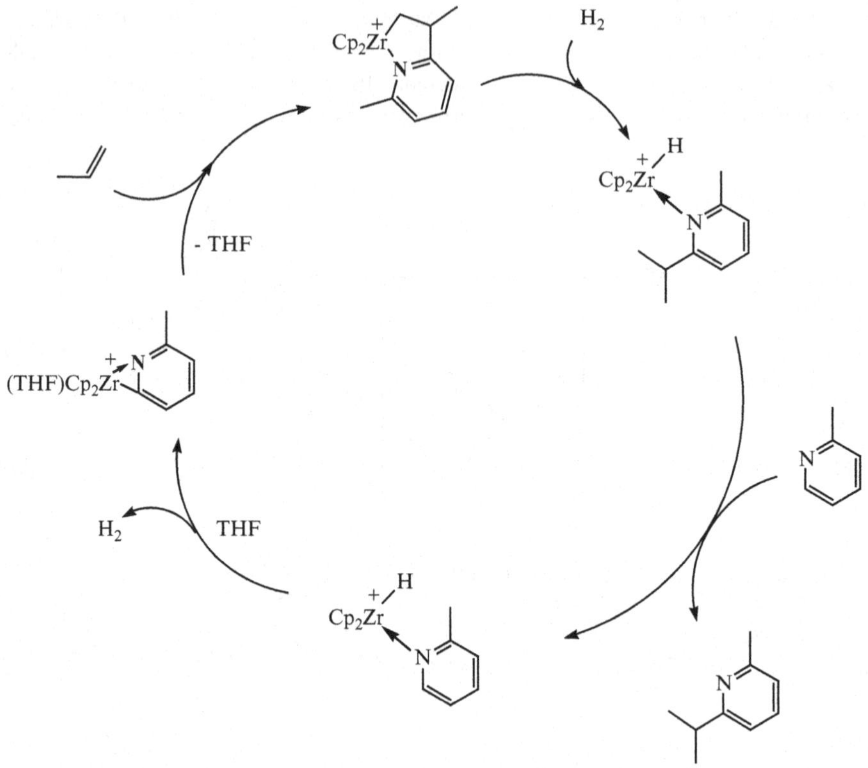

Scheme 7.17   A ring expansion reaction from a three-membered ring to a five-membered ring via cyclometalation reactions [117]

7.4.6   Carbocyclizations and Other Reactions

o-Iodophenyl ketones or aldehydes react with alkynes to give regioselectively carbocyclic compounds **7.81** in high yields in the presence of cobalt catalysts via the five-membered phenyl carbonyl compounds **5.7** in Table 5.5, as shown in Eq. (7.61) (see Compound **7.86** in Scheme 7.18) [117, 118].

The reaction mechanism is considered to be as follows: The reduction of Co(II) to Co(I) by zinc dust initiates catalysis. The cobalt catalyst is then activated by the coordination of the oxygen in the *o*-iodephenylaldehydes or the ketones, and cyclometalated products **7.82** are produced. The five-membered ring compounds **7.82** undergo the insertion of an alkyne to produce a seven-membered cobaltacycle **7.83**. The intramolecular nucleophilic addition of the cobalt–carbon bond in **7.83** to the carbonyl group leads to the formation of a cobalt alkoxide **7.84**. The reduction of the latter by zinc powder affords a Co(I) alkoxide **7.85**. The transmetalation of **7.85** with ZnI$_2$ gives an active Co(I) species and the corresponding zinc alkoxide **7.86**, which is converted to a final product after hydrolysis.

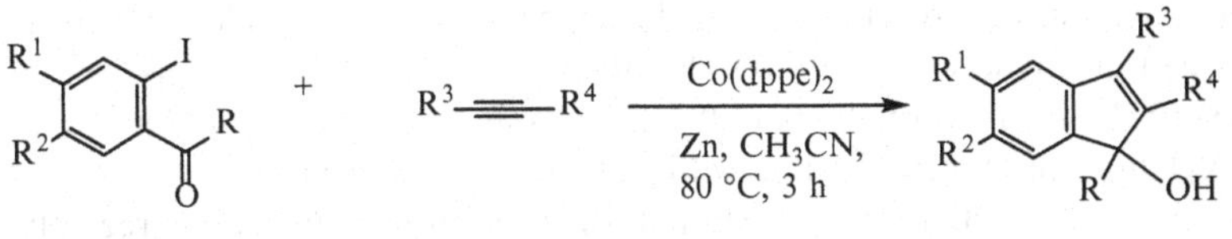

(7.61)

dppe = bis(diphenylphosphino)ethane

7.81

R = H, Me, n-Bu, Ph

54 -99%

R^1-R^2 = H, t-Bu,OMe, -OCH$_2$O-, Ph, SiMe$_3$

R^3-R^4 = Me, Et, n-Pr, n-Bu, n-hexl, Ph, COOMe, COOEt

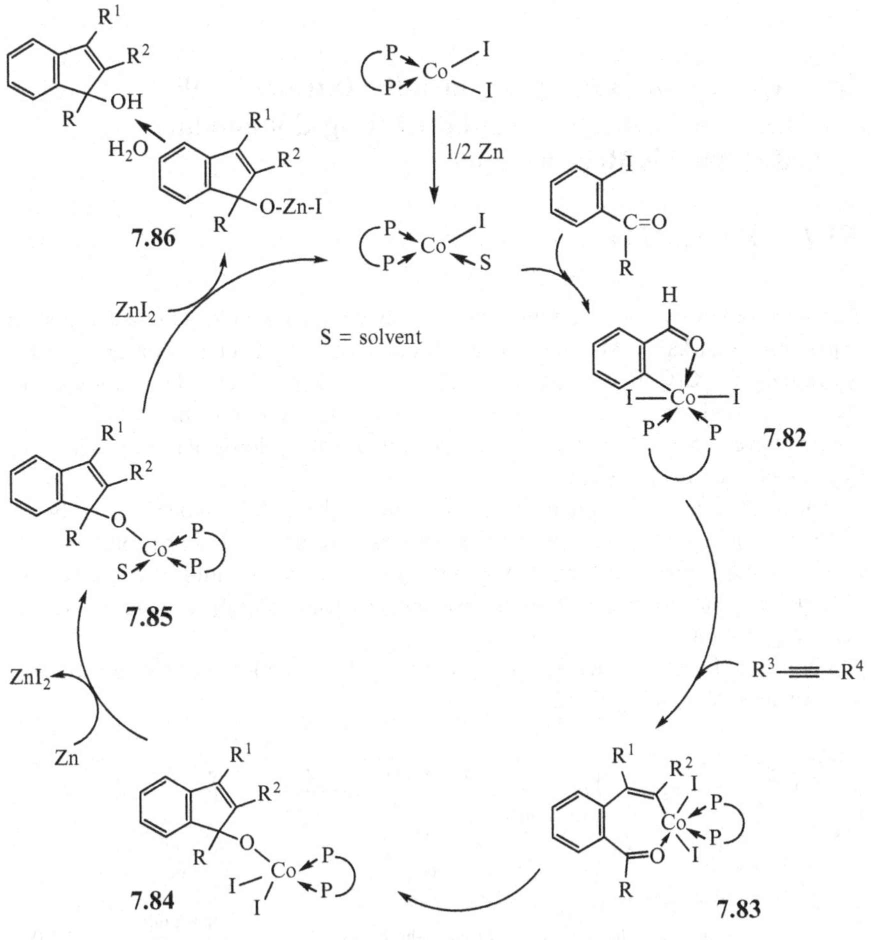

Scheme 7.18 Carbocyclization of o-iodophenyl ketones or aldehyde with alkynes via cyclometalation reactions [118]

The other reactions occurring via intramolecular five-membered ring intermediates are decarbonylative cleavages with organoruthenium catalysts [119], asymmetric hydrogenations with rhodium catalysts [120], reductive eliminations with rhodium catalysts [121], one-pot preparations of chiral homoallylic alcohol or amine derivatives through reactions of vinylic copper reagents, bis(iodomethyl)zinc carbenoid and aldehyde or N-sulfonyl aldimines[122], CO/olefins copolymerizations [123–127], lactone formations [128], hydrolysis of methyl parathion [129] , and acetoxylations of O-methyloxime with palladium catalysts [130].

A huge number of articles have been published on these applications recently. The cross-coupling reactions shown in Eqs. (7.50)–(7.53), in particular, obviously proceed via cyclometalation reactions with compounds **5.7**, **5.6**, and **5.19**, respectively, in Fig 5.11 as their substrates. These reactions are understandable from their reaction mechanisms [103], as shown in Scheme 7.14, in which they proceed very easily via cyclometalation reactions.

7.5 Applications for Organometallic Intramolecular-Coordination Five-Membered Ring Compounds Reported in Recent Articles

7.5.1 Introduction

Recently, organosynthetic applications for cyclometalation reactions have been expanding remarkably. Many recent articles include agostic interactions, C–H bond activations, C–X (C–F, C–Cl, C–Br, C–I, C–C, C–O, C–Si, etc.) bond activations, chelate-assisted reactions, and C–H bond functionalizations in their titles. These articles have reported on syntheses of derivatives of cyclometalation products or cyclometalation intermediates.

Their substrates are 2-phenylpyridines, imines, benzo[h]quinones, phosphines, oxygen compounds, sulfur compounds, etc., as described in the previous sections or in Fig. 5.11, Tables 5.5 and 5.6. These reactions are arylations, alkenylations, alkylations, acylations, cyclizations, hydrogenations, oxidations, hydrosilylations, dehydrogenations, etc.

For example, such reactions proceed via cyclometalation and the substitution reaction as shown in Eq. (7.62).

$$
\begin{array}{c}
\underset{\substack{| \\ H(X)}}{\overset{\gamma\ \beta\ \alpha}{C\text{-}Q\text{-}G\text{-}Y}} \xrightarrow{MXm}
\left[\underset{\substack{| \\ H(X)\ \ MXm}}{\overset{\gamma\ \beta\ \alpha}{C\text{-}Q\text{-}G\text{-}Y}} \right]
\xrightarrow{\text{Cyclization}}
\underset{\substack{\beta \\ }}{\overset{Y}{\underset{\alpha}{G}}}\!\!-\!\!MXn \xrightarrow{ZEm}_{\text{Substitution}}
\underset{\beta}{\overset{Y}{G}}\,\overset{ZEn}{\underset{\gamma}{C}}
\end{array}
$$

Metal activation	Chelate effect	Substitution
Coordination by a hetero atom	Five-membered ring intermediates	products

(7.62)

MXm, MXn = metal compounds
M = transition metals and main group metals:
    (69 kinds of metals)
Xm, Xn = Lingands (Cp,Cp*, CO, phosphine, hetero atom, halogen, etc.)
Y = N, P, O, S, etc.
G, Q = C or Y

Some of these reviews have also reported on cyclometalation reactions [131], C–H bond activation [132–134], and C–H bond functionalizations [135–137].

7.5.2 Arylations

Many articles on arylations have reported on synthetic applications of cyclometalation reactions, especially. For example, Song and Ackermann reported that the cobalt acac complex reacts with 2-phenylpyridine first to give a cobalt intermediate bonded with the γ-carbon atom **7.87**, and the intermediate then reacts with an aryl sulfamate to give a final substitution product in a high yield, as shown in Eq. (7.63) [133, 134, 138, 139].

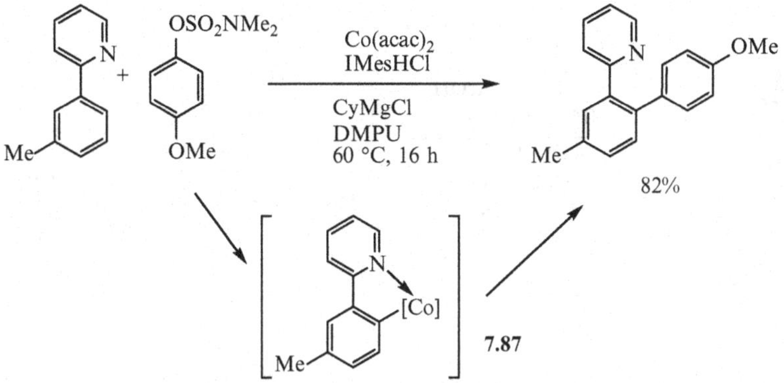

$$(7.63)$$

IMesH = N,N -bis(mesityl)imidazolium
DMPU = 1,3-dimethyl-3,4,5,6-tetrahydro-2-pyrimidinone

The arylations shown in Eq. (7.63) also proceed easily with the following many other substrates to give arylation products in high yields, as shown in Eqs. (7.64)–(7.66).

The substrates are 2-indolepyridines (Eq. (7.64)) [138], benzo[h]quinolines (Eq. (7.65)) [133, 134, 138, 140], 8-methylquinolines (Eq. (7.66)) [141, 142], pyrrolidinones (Eq. (7.67)) [140], oxazolidinones (Eq. (7.68)) [141], 6-phenylpurines (Eq. (7.69)) [143, 144], benzoic acids (Eq. (7.70)) [145], 2-vinylpyridines (Eq. (7.71)) [146], and benzamides (Eq. (7.72)) [147]. The arylation products are prepared by substitution reactions with the five-membered products in Eq. (7.62), such as cyclometalation intermediates **7.87–7.90** in the cyclometalation reactions.

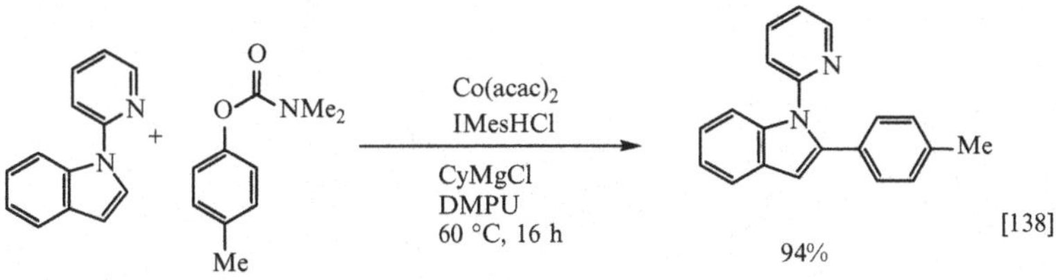

$$(7.64)$$

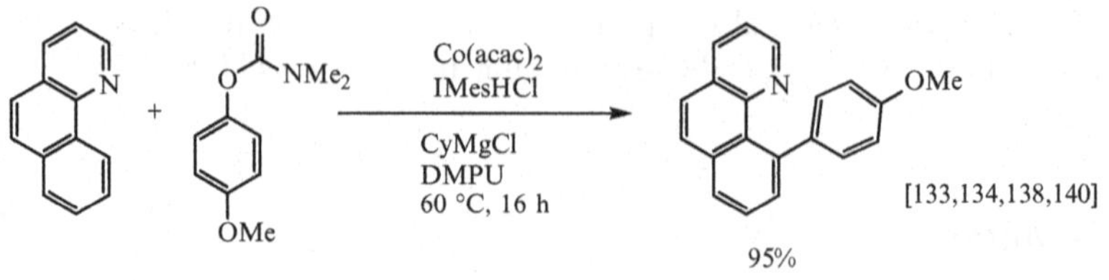

$$Co(acac)_2 \\ IMesHCl \\ \overline{\hspace{2cm}} \\ CyMgCl \\ DMPU \\ 60\ °C,\ 16\ h$$

[133,134,138,140]

95%

(7.65)

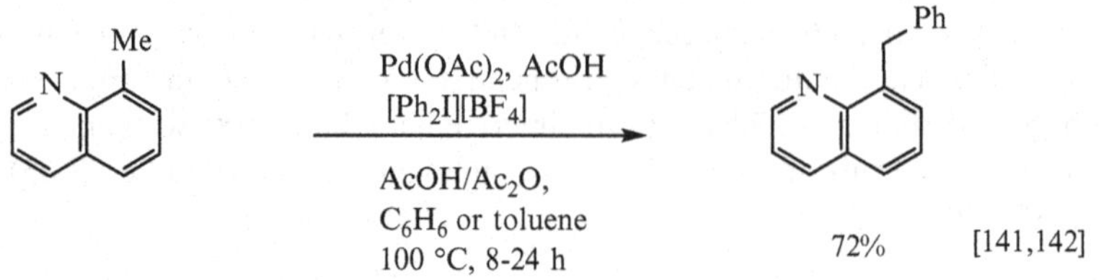

$$Pd(OAc)_2,\ AcOH \\ [Ph_2I][BF_4] \\ \overline{\hspace{2cm}} \\ AcOH/Ac_2O, \\ C_6H_6\ or\ toluene \\ 100\ °C,\ 8\text{-}24\ h$$

72%   [141,142]

(7.66)

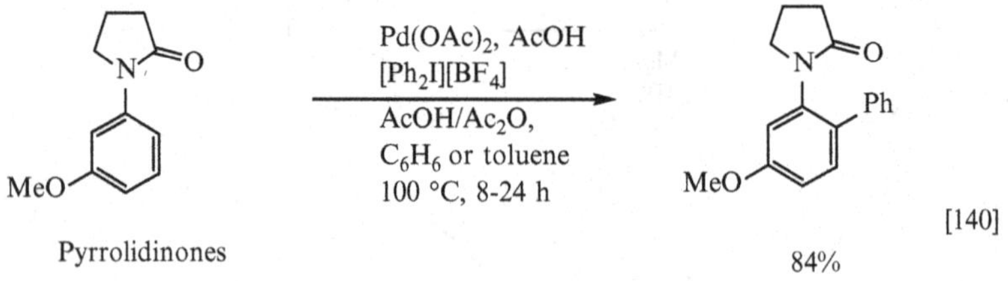

Pyrrolidinones

$$Pd(OAc)_2,\ AcOH \\ [Ph_2I][BF_4] \\ \overline{\hspace{2cm}} \\ AcOH/Ac_2O, \\ C_6H_6\ or\ toluene \\ 100\ °C,\ 8\text{-}24\ h$$

[140]

84%

(7.67)

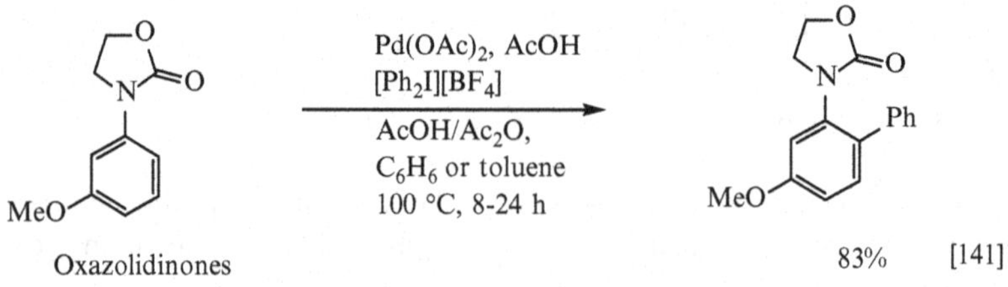

Oxazolidinones

$$Pd(OAc)_2,\ AcOH \\ [Ph_2I][BF_4] \\ \overline{\hspace{2cm}} \\ AcOH/Ac_2O, \\ C_6H_6\ or\ toluene \\ 100\ °C,\ 8\text{-}24\ h$$

83%   [141]

(7.68)

+ PhI

30 equiv

$$Pd(OAc)_2 \\ AgOAc \\ \overline{\hspace{2cm}} \\ HOAc \\ N_2 \\ 120\ °C,\ 60\ h$$

[143,144]

82%

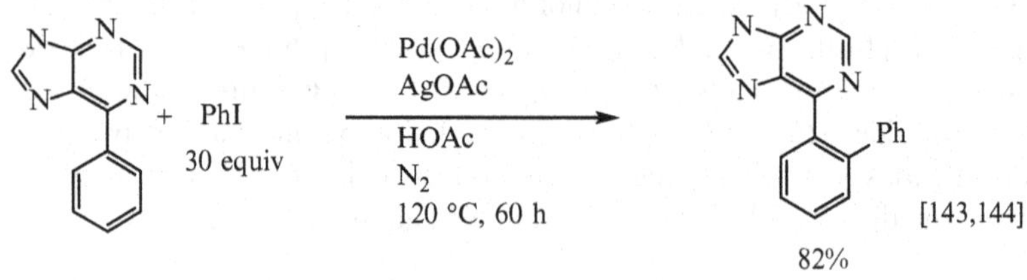

(7.69)

$$(7.70)$$

$$(7.71)$$

$$(7.72)$$

7.5.3 Alkenylations

The alkenylation products are also prepared by reactions similar to the previous arylations with similar substrates, such as arylamides and 1-phenylpyrazoles.

For example, N-methoxybenzamides react with alkenes in the presence of rhodium complex to afford oxidative *ortho*-derivatives in high yields, as shown in Eq. (7.73) [148].

$$
\begin{array}{c}
\text{[structure with NHOMe]} + \text{[alkene } C_6H_4\text{-}R^1\text{]} \xrightarrow[\substack{\text{MeOH} \\ 60\,°\text{C, 3-16 h}}]{\substack{\text{[Cp*RhCl}_2]_2 \\ \text{CsOAc}}} \text{[product with NHOMe, } C_6H_4\text{-}R^1\text{]}
\end{array}
\tag{7.73}
$$

R^1= H, Cl, Br, Me, ^{t}Bu, OMe

Yield 52 - 90%

Similar alkenylation reactions with phenyl esters as substrates also proceed via cyclometalation reaction intermediate **7.91** to give an alkenyl derivative in a high yield, as shown in Eq. (7.74) [149].

$$
\begin{array}{c}
\text{[Me, OEt structure with =O]} + \text{[alkene COOEt]} \xrightarrow[\substack{\text{ClCH}_2\text{CH}_2\text{Cl} \\ 110\,°\text{C, 12 h}}]{\substack{\text{[Cp*RhCl}_2]_2 \\ \text{AgSbF}_6 \\ \text{Cu(OAc)}_2}} \text{[product Me, OEt, COOEt]}
\end{array}
\tag{7.74}
$$

80%

[intermediate Me, OEt, [Rh] ---- COOEt]   **7.91**

With 1-phenylpyrazoles [150], 2-phenyloxazolines [151], phenyltriazines [152], and N-(1-naphthyl)sulfonamides [153] as substrates, their alkenylations also proceed in the presence of ruthenium or rhodium compounds, as shown in Eqs. (7.75)–(7.78).

$$
\begin{array}{c}
\text{[1-phenylpyrazole]} + \text{[alkene COO}^n\text{Bu]} \xrightarrow[\substack{\text{DMF, N}_2 \\ 100\,°\text{C, 4 h}}]{\substack{\text{[Ru(}p\text{-cymene)Cl}_2]_2 \\ \text{Cu(OAc)}_2\cdot\text{H}_2\text{O}}} \text{[product pyrazole, COO}^n\text{Bu]}
\end{array}
\tag{7.75}
$$

67%   [150]

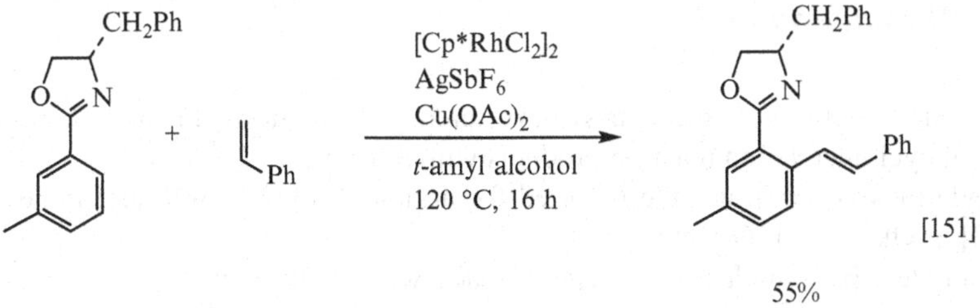
CH2Ph
O
N
+
Ph
[Cp*RhCl2]2
AgSbF6
Cu(OAc)2
t-amyl alcohol
120 °C, 16 h
CH2Ph
O
N
Ph
[151]
55%

(7.76)

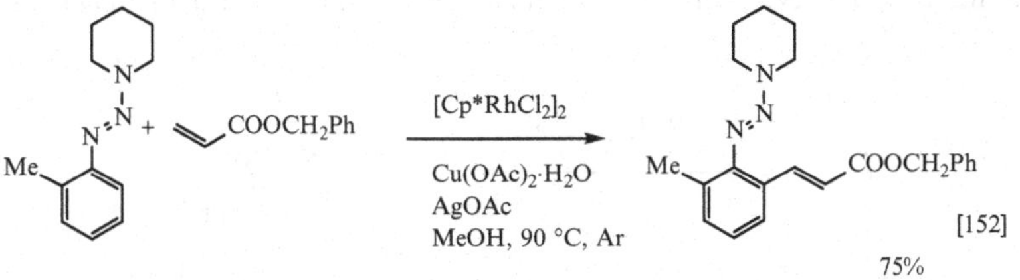
N
N
N
+
COOCH2Ph
Me
[Cp*RhCl2]2
Cu(OAc)2·H2O
AgOAc
MeOH, 90 °C, Ar
N
N
N
Me
COOCH2Ph
[152]
75%

(7.77)

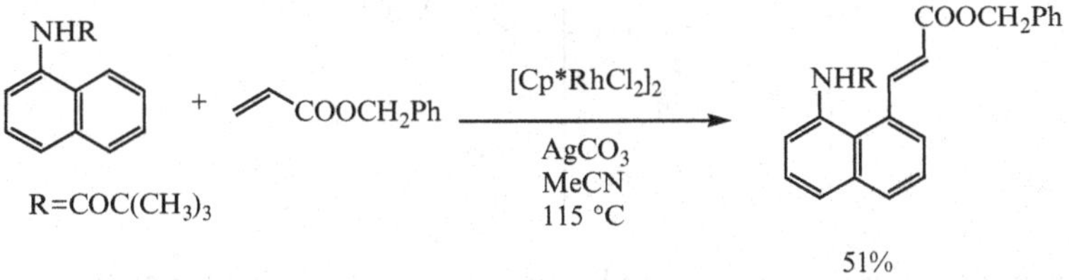
NHR
+
COOCH2Ph
R=COC(CH3)3
[Cp*RhCl2]2
AgCO3
MeCN
115 °C
COOCH2Ph
NHR
51%

(7.78)

As shown in Eq. (7.73), alkenylation reactions with benzamide as the substrate give *ortho*-alkenylation derivatives as the cyclometalation derivatives. However, an alkenylation product can undergo a further Michael reaction to give a γ-lactam in the case of electron-withdrawing alkenes in a high yield, as shown in Eq. (7.79) [154, 155].

Furthermore, in alkenylation reactions with phenyl carboxylic acid as the substrate, a similar annulation reaction proceeds to give a lactone by cyclization reaction via the alkenylation of the cyclometalation product, as shown in Eq. (7.80) [156].

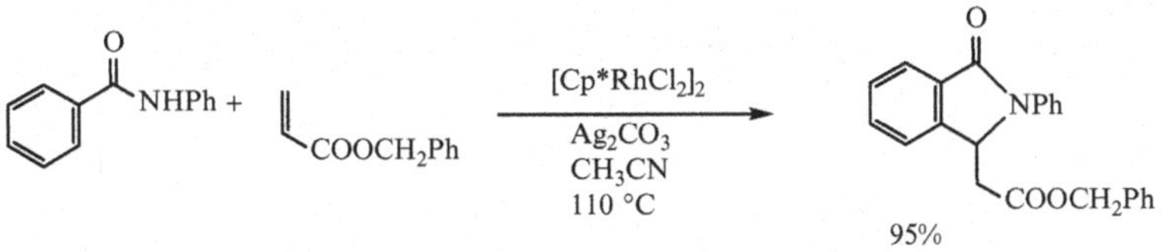
O
NHPh +
COOCH2Ph
[Cp*RhCl2]2
Ag2CO3
CH3CN
110 °C
O
NPh
COOCH2Ph
95%

(7.79)

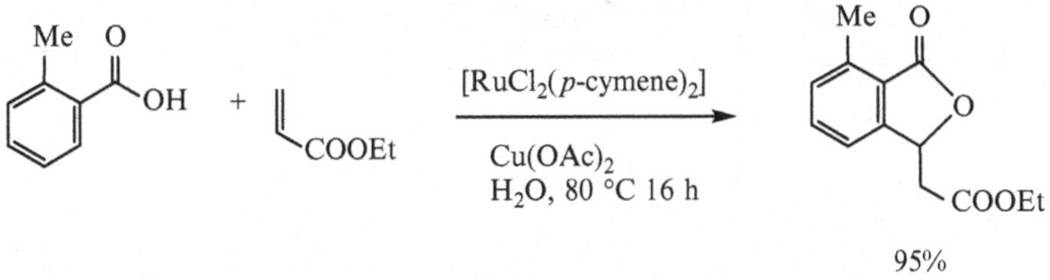
Me O
OH
+
COOEt
[RuCl2(p-cymene)2]
Cu(OAc)2
H2O, 80 °C 16 h
Me O
O
COOEt
95%

(7.80)

7.5.4   Alkynylations

Alkynylations (reactions with alkynes) have already been shown in Sect. 7.3 as reactions of cyclometalation reaction products with alkynes.

Alkynylations as substrates are reported for arylamides [157–160], arylimines [161–163], arylketones [164], etc. [165].

For example, a benzamide as an arylamide reacts with an alkyne in the presence of a rhodium compound to give the cyclization product isoquinolone **7.93** via a cyclometalation intermediate **7.92** coordinated by an alkyne, as shown in Eq. (7.81) [157].

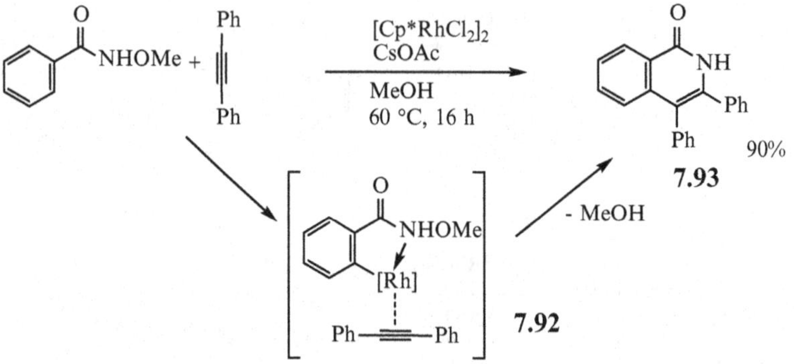

$$\text{(7.81)}$$

Other examples of arylimines and arylketones as substrates in cyclometalation reactions are shown in Eqs. (7.82) and (7.83), respectively. These products are also prepared via cyclometalation intermediates **7.94** and **7.95**, followed by insertion of an alkyne moiety between the metal and γ-carbon atom, that is, at the *ortho*-carbon [161, 164].

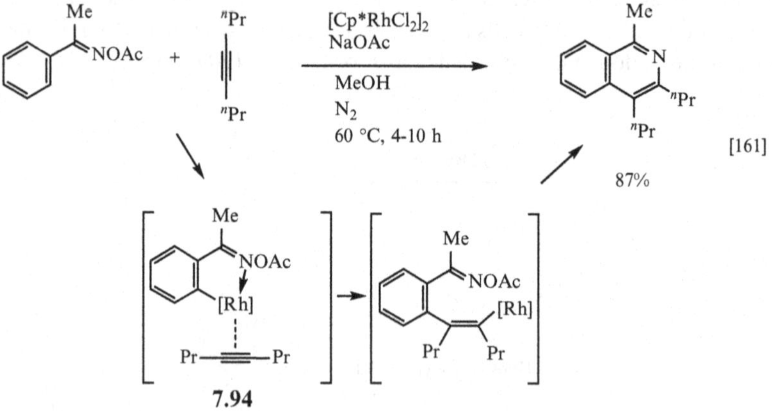

$$\text{(7.82)}$$

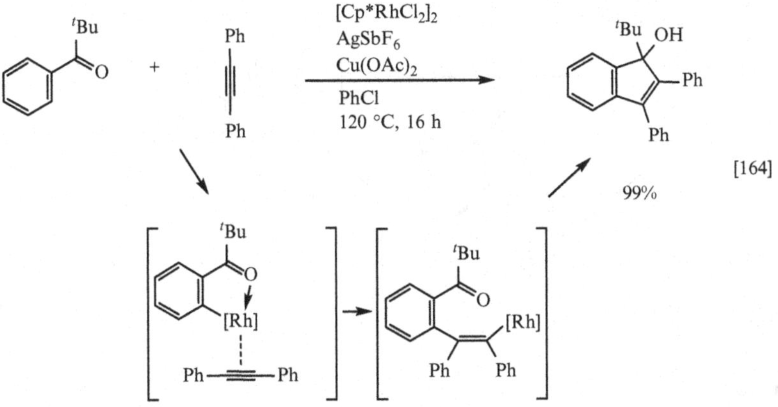

(7.83)

7.95

7.5.5 Alkylations

Alkylation products are also easily prepared in good or excellent yields by cyclo-metalation reactions of substrates such as benzaldehydes with alkyl halides, 1-alkenes, or 2-alkenes [166–168].

For example, benzamide reacts with an alkyl chloride in the presence of cobalt compounds at room temperature to give an alkyl derivative at the *ortho*-position in a high yield, as shown in Eq. (7.84) [166].

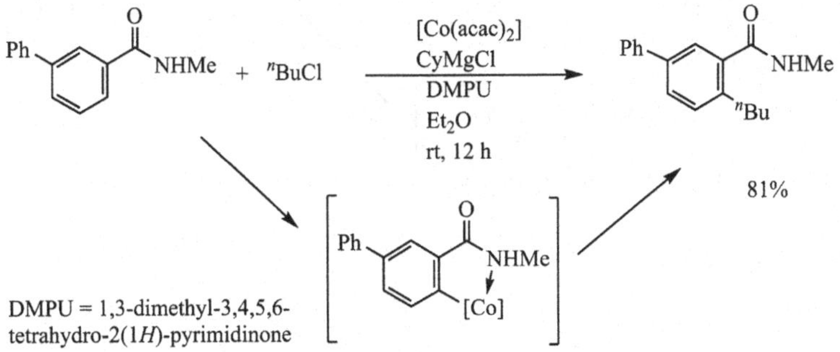

(7.84)

DMPU = 1,3-dimethyl-3,4,5,6-tetrahydro-2(1*H*)-pyrimidinone

7.5.6 Other Reactions

Other reactions include acylation [169–171], amination and imination [172–176], halogenation [177–179], silylation [180, 181], and carbonylation [182], as shown in Eqs. (7.85)–(7.89), respectively.

Acylation

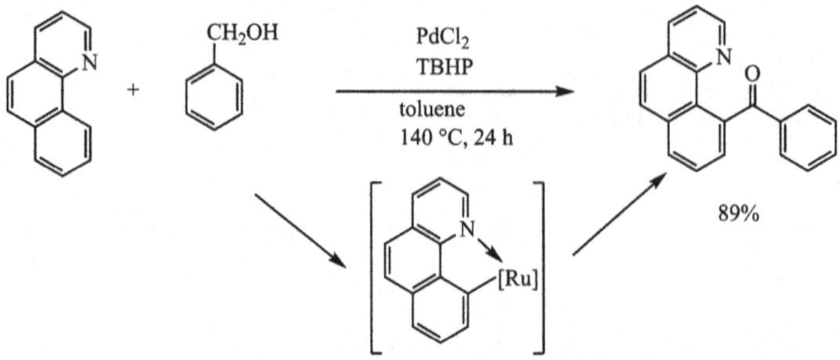

(7.85)

TBHP = *t*-butyl hydroperoxide

Amination

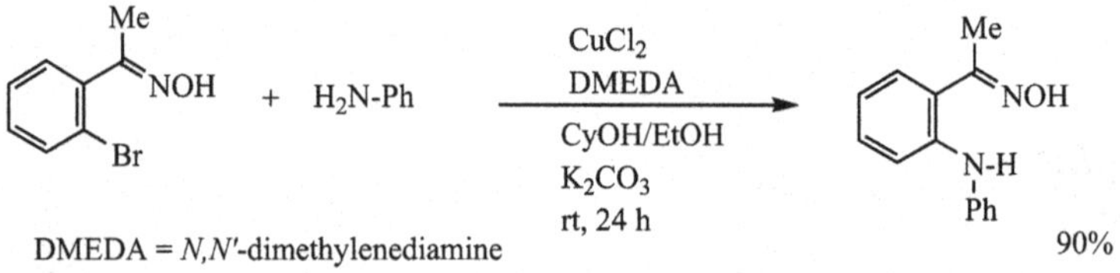

(7.86)

DMEDA = *N,N'*-dimethylenediamine

Halogenation

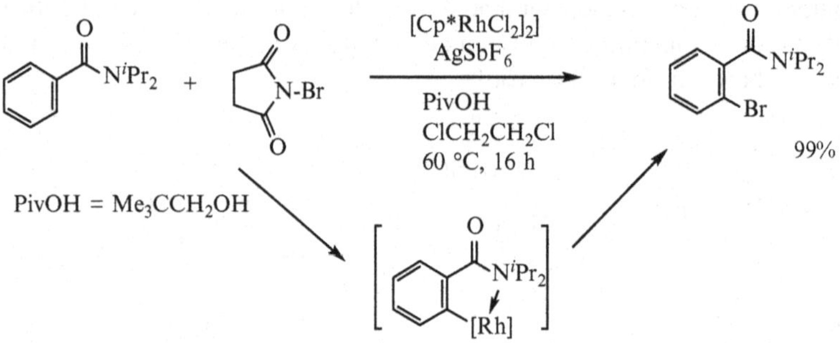

(7.87)

PivOH = Me$_3$CCH$_2$OH

Silylation

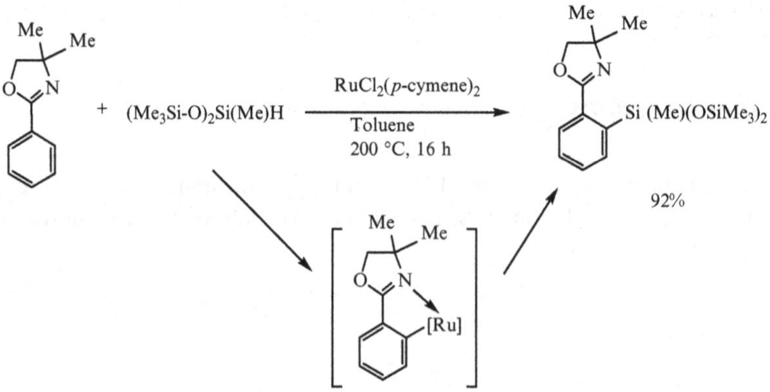

(7.88)

Carbonylation

(7.89)

Other reactions include cyanations [183] and benzoxylations [184–191].

References

1. Omae I (1972) Rev Silicon Germanium Tin Lead Comp 1:59
2. Omae I (1979) Chem Rev 79:287
3. Omae I (1979) Coord Chem Rev 28:97
4. Omae I (1979) Kagaku No Ryoiki 33:767
5. Omae I (1980) Coord Chem Rev 32:235
6. Omae I (1982) Coord Chem Rev 42:31
7. Omae I (1982) Angew Chem Int Ed 21:889; (1982) Angew Chem 94:902
8. Omae I (1982) Yuki Gosei Kagaku Kyokaishi 40:147
9. Omae I (1982) Coord Chem Rev 42:245
10. Omae I (1982) Kagaku Kogyo 33:989
11. Omae I (1983) Coord Chem Rev 51:1
12. Omae I (1984) Coord Chem Rev 53:261
13. Omae I (1988) Coord Chem Rev 83:137
14. Omae I (1998) Kagaku Kogyo 49:303
15. Omae I, Aoki A, Horiguchi K (1998) Kagaku Kogyo 49:469
16. Omae I (2004) Coord Chem Rev 248:995
17. Omae I (2004) Phosphorus Sulfur Silicon 179:891
18. Omae I (2007) J Organomet Chem 692:2608
19. Omae I (2010) Appl Organometal Chem 24:347
20. Omae I (2011) J Organomet Chem 696:1128
21. Omae I (1986) Organometallic intramolecular-coordination compounds, J Organomet Chem Library 18. Elsevier, Amsterdam
22. Omae I (1989) Organotin chemistry, J Organomet Chem Library 21. Elsevier, Amsterdam
23. Omae I (1998) Applications of organometallic compounds. Wiley, NewYork
24. Jun C-H, Lee H, Hong J-B (1997) J Org Chem 62:1200
25. Lee D-Y, Moon CW, Jun C-H (2002) J Org Chem 67:3945
26. Chang D-H, Lee D-Y, Hong B-S, Choi J-H, Jun C-H (2004) J Am Chem Soc 126:424
27. Kim D-W, Lim S-G, Jun C-H (2006) Org Lett 8:2937
28. Park J-W, Park J-H, Jun C-H (2008) J Org Chem 73:5598
29. Padilla R, Salazar V, Paneque M, Alvarado-Rodríguez JG, Tamariz J, Pacheco-Cuvas H, Vattier F (2010) Organometallics 29:2835

30. Basolo F, Johnson R (1964) Coordination chemistry. The Benjamin/Cummings Publishing Company, London
31. Zucca A, Cordeschi D, Stoccoro S, Cinellu MA, Minghetti G, Chelucci G, Manassero M (2011) Organometallics 30:3064
32. Zhao S-B, Wang R-Y, Wang S (2007) J Am Chem Soc 129:3092
33. Crosby SH, Clarkson GJ, Rourke JP (2009) J Am Chem Soc 131:14142
34. Crosby SH, Clarkson GJ, Deeth RJ, Rourke JP (2010) Organometallics 29:1966
35. Crosby SH, Clarkson GJ, Rourke JP (2011) Organometallics 30:3603
36. Thomas HR, Deeth RJ, Clarkson GJ, Rourke JP (2011) Organometallics 30:5641
37. Canty AJ, Honeyman RT (1990) J Organomet Chem 387:247
38. Maidich L, Zuri G, Stoccoro S, Cinellu MA, Masia M, Zucca A (2013) Organometallics 32:438
39. Petretto GL, Rourke JP, Maidich L, Stoccoro S, Cinellu MA, Minghetti G, Clarkson GJ, Zucca A (2012) Organometallics 31:2971
40. Zucca A, Doppiu A, Cinellu MA, Stoccoro S, Minghetti G, Manassero M (2002) Organometallics 21:783
41. Minghetti G, Stoccoro S, Cinellu MA, Soro B, Zucca A (2003) Organometallics 22:4770
42. Zucca A, Petretto GL, Stoccoro S, Cinellu MA, Minghetti G, Manassero M, Manassero C, Male L, Albinati A (2006) Organometallics 25:2253
43. Minghetti G, Stoccoro S, Cinellu MA, Petretto L, Zucca A (2008) Organometallics 27:3415
44. Butschke B, Schröder D, Schwarz H (2009) Organometallics 28:4340
45. Butschke B, Schwarz H (2010) Organometallics 29:6002
46. Butschke B, Schlangen M, Schröder D, Schwarz H (2008) Chem Eur J 14:11050
47. Butschke B, Tabrizi SG, Schwarz H (2010) Chem Eur J 16:3962
48. Zucca A, Petretto GL, Stoccoro S, Cinellu MA, Manassero M, Manassero C, Minghetti G (2009) Organometallics 28:2150
49. Britovsek GJP, Taylor RA, Sunley GJ, Law DJ, White AJP (2006) Organometallics 25:2074
50. Zucca A, Cinellu MA, Pinna MV, Stoccoro S, Minghetti G (2000) Organometallics 19:4295
51. Shibata T, Takayasu S, Yuzawa S, Otani T (2012) Org Lett 14:5106
52. Kwak J, Ohk Y, Jung Y, Chang S (2012) J Am Chem Soc 134:17778
53. Young KJH, Yousufuddin M, Ess DH, Periana RA (2009) Organometallics 28:3395
54. Young KJH, Mironov OA, Periana RA (2007) Organometallics 26:2137
55. Tyo EC, Castleman AW Jr, Schröderk D, Milko P, Roithova J, Ortega JM, Cinellu MA, Cocco F, Minghetti G (2009) J Am Chem Soc 131:13009
56. Cocco F, Cinellu MA, Minghetti G, Zucca A, Stoccoro S, Maiore L, Manassero M (2010) Organometallics 29:1064
57. Zhao J-W, Wang C-M, Zhang J, Zheng S-T, Yang G-Y (2008) Chem Eur J 14:9223
58. Cartoon MEK, Cheeseman GWH (1982) J Organomet Chem 234:123
59. Dupont J, Pfeffer M (1987) J Organomet Chem 321:C13
60. Robinson NP, Depree GJ, de Wit RW, Main L, Nicholson BK (2005) J Organomet Chem 690:3827
61. Takahashi H, Tsuji J (1967) J Organomet Chem 10:511
62. Thompson JM, Heck RF (1975) J Org Chem 40:2667
63. Carbayo A, Cuevas JV, García-Herbosa G (2002) J Organomet Chem 658:15
64. Murahashi S, Horiie S (1956) J Am Chem Soc 78:4816
65. Horiie S (1958) Nippon Kagaku Zasshi 79:499
66. Horiie S (1960) Chem Abstr 54:4607
67. Horiie S, Murahashi S (1960) Bull Chem Soc Jpn 33:88
68. Sokolov VI, Troitskaya LL, Reutov OA (1979) J Organomet Chem 182:537
69. Ryabov AD, Firsova YN, Goral VN, Ryabova ES, Shevelkova AN, Troitskaya LL, Demeschik TV, Sokolov VI (1998) Chem Eur J 4:806
70. Yamamoto Y, Yamazaki H (1980) Inorg Chim Acta 41:229
71. Yamamoto Y, Yamazaki H (1976) Synthesis 750
72. Holton RA (1977) Tetrahedron Lett 355
73. Brisdon BJ, Nair P, Dyke SF (1981) Tetrahedron 37:173

74. Julia M, Duteil M, Lallemand JY (1975) J Organomet Chem 102:239
75. Kamiyama S, Kimura T, Kasahara A, Izumi T, Maemura M (1979) Bull Chem Soc Jpn 52:142
76. Ryabov AD (1985) Synthesis 233
77. Barr N, Dyke SF, Quessy SN (1983) J Organomet Chem 253:391
78. Girling IR, Widdowson DA (1982) Tetrahedron Lett 23:1957
79. Bahsoun A, Dehand J, Pfeffer M, Zinsius M (1979) J Chem Soc Dalton Trans 547
80. Dehand J, Mutet C, Pfeffer M (1981) J Organomet Chem 209:255
81. Mutet C, Pfeffer M (1979) J Organomet Chem 171:C34
82. Maassarani F, Pfeffer M, Borgne GL (1987) Organometallics 6:2029
83. Beydoun N, Pfeffer M, Cian AD, Fischer J (1991) Organometallics 10:3693
84. Pfeffer M (1992) Pure Appl Chem 64:335; Beydoun N, Pfeffer M (1990) Synthesis 729; Ossorf H, Pfeffer M, Jastrzebski JTBH, Stam CH (1987) Inorg Chem 26:1169
85. Pfeffer P, Rotteveel MA, Sutter J-P, Cian AD, Fischer J (1989) J Organomet Chem 371:C21
86. Depree GJ, Main L, Nicholson BK, Robinson NP, Jameson GB (2006) J Organomet Chem 691:667
87. Holton RA, Natalie KJ Jr (1981) Tetrahedron Lett 22:267
88. Clark PW, Dyke HJ, Dyke SF, Perry G (1983) J Organomet Chem 253:399
89. Troitskaya LL, Khrushcheva NS, Sokolov VL, Reutov OA (1982) Zh Org Khim 18:2606; (1983) Chem Abstr 98:126327d
90. Heck RF (1968) J Am Chem Soc 90:313
91. Murahashi S (1955) J Am Chem Soc 77:6403
92. Horiie S, Murahashi S (1960) Bull Chem Soc Jpn 33:247
93. Kleiman JP, Dubeck M (1963) J Am Chem Soc 85:1544
94. Cravotto G, Demartin F, Palmisano G, Penoni A, Radice T, Tollari S (2005) J Organomet Chem 690:2017
95. Stock LM, Brown HC (1963) Adv Phys Org Chem 1:35
96. Allison JAC, El Khadem HS, Wilson CAJ (1975) Heterocyclic Chem 12:1275
97. Sokolov VI, Troitskaya LL, Khrushchova NS (1983) J Organomet Chem 250:439
98. Klein H-F, Helwig M, Koch U, Flörke U, Haupt H-J (1993) Z Naturforsch 48b:778
99. Morimoto T, Chatani N, Murai S (1999) J Am Chem Soc 121:1758
100. Murai S (1994) J Synth Org Chem Jpn 52:992
101. Borman S (1993) Chem Eng News 6, Dec 13
102. Murai S, Kakiuchi F, Sekine S, Tanaka Y, Kamatani A, Sonoda M, Chatani N (1993) Nature 366:529
103. Murai S, Kakiuchi F, Sekine S, Tanaka Y, Kamatani A, Sonoda M, Chatani N (1994) Pure Appl Chem 66:1527
104. Kakiuchi F, Sekine S, Tanaka Y, Kamatani A, Sonoda M, Chatani N, Murai S (1995) Bull Chem Soc Jpn 68:62
105. Fujii N, Kakiuchi F, Chatani N, Murai S (1996) Chem Lett 939
106. Fujii N, Kakiuchi F, Yamada A, Chatani N, Murai S (1997) Chem Lett 425
107. Fujii N, Kakiuchi F, Yamada A, Chatani N, Murai S (1998) Bull Chem Soc Jpn 71:285
108. Kakiuchi F, Gendre PL, Yamada A, Ohtaki H, Murai S (2000) Tetrahedron Asymmetry 11:2647
109. Kakiuchi F, Tsujimoto T, Sonoda M, Chatani N, Murai S (2001) Synlett 948
110. Kakiuchi F, Murai S (2002) Acc Chem Res 35:826
111. Kakiuchi F, Matsumoto M, Tsuchiya K, Igi K, Hayamizu T, Chatani N, Murai S (2003) J Organomet Chem 686:134
112. Willis MC, McNally SJ, Beswick PJ (2004) Angew Chem Int Ed 43:340
113. Tanaka M, Imai M, Yamamoto Y, Tanaka K, Shimowatari M, Nagumo S, Kawahara N, Suemune H (2003) Org Lett 5:1365
114. Bendorf HD, Colella CM, Dixon EC, Marchetti M, Matukonis AN, Musselman JD, Tiley TA (2002) Tetrahedron Lett 43:7031
115. Jordan RF, Taylor DF (1989) J Am Chem Soc 111:778

116. Chang K-J, Rayabarapu DK, Cheng C-H (2003) Org Lett 5:3963
117. Chang K-J, Rayabarapu DK, Cheng C-H (2004) J Org Chem 69:4781
118. Chatani N, Ie Y, Kakiuchi F, Murai S (1999) J Am Chem Soc 121:8645
119. Knowles WS (1983) Acc Chem Res 16:106
120. Halpern J (1982) Acc Chem Res 15:332
121. Chinkov N, Sklute G, Chechik H, Abramovitch A, Amsallem D, Varghese J, Majumdar S, Marek I (2004) Pure Appl Chem 76:517
122. Dremt E, Budzelaar PHM (1996) Chem Rev 96:663
123. Sen A (1986) Adv Polym Sci 73(74):125
124. Sen A (1993) Acc Chem Res 26:303
125. Ash CE (1994) J Mater Educ 16:1
126. Liu J, Heaton BT, Iggo JA, Whyman R (2004) Angew Chem Int Ed 43:90
127. Murakami M, Tsuruta T, Ito Y (2000) Angew Chem Int Ed 39:2484
128. Kim M, Liu Q, Gabbaï FP (2004) Organometallics 23:5560
129. Matsuda T (2006) Yuki Gosei Kagaku Kyokaishi 64:780
130. Herrmann WA, Böhm VPW, Reisinger C-P (1999) J Organomet Chem 576:23
131. Albrecht M (2010) Chem Rev 110:576
132. Balcells D, Clot E, Eisenstein O (2010) Chem Rev 110:749
133. Kuhl N, Hopkinson MN, Wencel-Delord J, Glorius F (2012) Angew Chem Int Ed 51:10236
134. Niu J-L, Hao X-Q, Gong J-F, Song M-P (2011) Dalton Trans 40:5135
135. Lyons TW, Sanford MS (2010) Chem Rev 110:1147
136. Neufeldt SR, Sanford MS (2012) Acc Chem Res 45:936
137. Engle KM, Mei T-S, Wasa M, Yu J-Q (2012) Acc Chem Res 45:788
138. Song W, Ackermann L (2012) Angew Chem Int Ed 51:8251
139. Li H, Wei W, Xu Y, Zhang C, Wan X (2011) Chem Commun 47:1497
140. Lyons TW, Hull KL, Sanford MS (2011) J Am Chem Soc 133:4455
141. Kalyani D, Deprez NR, Desai LV, Sanford MS (2005) J Am Chem Soc 127:7330
142. Baudoin O (2011) Chem Soc Rev 40:4902
143. Guo H-M, Jiang L-L, Niu H-Y, Rao W-H, Liang L, Mao R-Z, Li D-Y, Qu G-R (2011) Org Lett 13:2008
144. Lakshman MK, Deb AC, Chamala RR, Pradhan P, Protap R (2011) Angew Chem Int Ed 50:11400
145. Cornella J, Righi M, Lassosa I (2011) Angew Chem Int Ed 50:9429
146. Ackermann L, Vicente R, Potukuchi HK, Pirovano V (2010) Org Lett 12:5032
147. Karthikeyan J, Cheng C-H (2011) Angew Chem Int Ed 50:9880
148. Rakshit S, Grohamann C, Besset T, Glorius F (2011) J Am Chem Soc 133:2350
149. Park SH, Kim JY, Chang S (2011) Org Lett 13:2372
150. Hashimoto Y, Ueyama T, Fukutani T, Hirano K, Satoh T, Miura M (2011) Chem Lett 40:1165
151. Schröder N, Besset T, Glorius F (2012) Adv Synth Catal 354:579
152. Wang C, Chen H, Wang Z, Chen J, Huang Y (2012) Angew Chem Int Ed 51:7242
153. Li X, Gong X, Zhao M, Song G, Deng J, Li X (2011) Org Lett 13:5808
154. Wang F, Song G, Li X (2010) Org Lett 12:5430
155. Zhu C, Falck JR (2011) Org Lett 13:1214
156. Ackermann L, Pospech J (2011) Org Lett 13:4153
157. Guimond N, Gouliaras C, Fagnou K (2010) J Am Chem Soc 132:6908
158. Ackermann L, Fenner S (2011) Org Lett 13:6548
159. Ackermann L, Lygin AV, Hofmann N (2011) Angew Chem Int Ed 50:6379
160. Su Y, Zhao M, Han K, Song G, Li X (2010) Org Lett 12:5462
161. Too PC, Wang Y-F, Chiba S (2010) Org Lett 12:5688
162. Yoshida Y, Kurahashi T, Matsubara S (2011) Chem Lett 40:1140
163. Kuninobu Y, Takai K (2012) Bull Chem Soc Jpn 85:656
164. Patureau FW, Besset T, Kuhl N, Glorius F (2011) J Am Chem Soc 133:2154
165. Aguilar D, Bielsa R, Soler T, Urriolabeitia EP (2011) Organometallics 30:642

166. Chen Q, Ilies L, Nakamura E (2011) J Am Chem Soc 133:428
167. Zhao Y, Chen G (2011) Org Lett 13:4850
168. Ilies L, Chen Q, Zeng X, Nakamura E (2011) J Am Chem Soc 133:5221
169. Xiao F, Shuai Q, Zhao F, Baslé O, Deng G, Li C-J (2011) Org Lett 13:1614
170. Shen Z-L, Goh KKK, Cheong H-L, Wong CHA, Lai Y-C, Yang Y-S, Loh T-P (2010) J Am Chem Soc 132:15852
171. Sharma S, Park E, Park J, Kim IS (2012) Org Lett 14:906
172. Xie R, Fu H, Ling Y (2011) Chem Commun 47:8976
173. Xu W, Jin Y, Liu H, Jiang Y, Fu H (2011) Org Lett 13:1274
174. Yoo EJ, Ma S, Mei T-S, Chan KSL, Yu J-Q (2011) J Am Chem Soc 133:7652
175. Xu S, Lu J, Fu H (2011) Chem Commun 47:5596
176. Ryu J, Shin K, Park SH, Kim JY, Chang S (2012) Angew Chem Int Ed 51:9904
177. Schröder N, Wencel-Delord J, Glorius F (2012) J Am Chem Soc 134:8298
178. Chan KSL, Wasa M, Wang X, Yu J-Q (2011) Angew Chem Int Ed 50:9081
179. Niu L, Yang H, Yang D, Fu H (2012) Adv Synth Catal 354:2211
180. Sakurai T, Matsuoka Y, Hanataka T, Fukuyama N, Namikoshi T, Watanabe S, Murata M (2012) Chem Lett 41:374
181. Simmons EM, Hartwig JF (2010) J Am Chem Soc 132:17092
182. Du Y, Hyster TK, Rovis T (2011) Chem Commun 47:12074
183. Kim J, Chang S (2010) J Am Chem Soc 132:10272
184. Li L, Yu P, Cheng J, Chen F, Pan C (2012) Chem Lett 41:600
185. Wang W, Pan C, Chen F, Cheng J (2011) Chem Commun 47:3978
186. Ackermann L, Kozhushkov SI, Yufit DS (2012) Chem Eur J 18:12068
187. Tsai AS, Tauchert ME, Bergman RG, Ellman JA (2011) J Am Chem Soc 133:1248
188. Li Y, Zhang X-S, Chen K, He K-H, Pan F, Li B-J, Shi Z-J (2012) Org Lett 14:636
189. Peng X, Zhu Y, Ramirez TA, Zhao B, Shi Y (2011) Org Lett 13:5244
190. Kitahara M, Umeda N, Hirano K, Satoh T, Miura M (2011) J Am Chem Soc 133:2160
191. Enthaler S, Company A (2011) Chem Soc Rev 40:4912

Chapter 8
Applications of Five-Membered Ring Products as Catalysts in Cyclometalation Reactions

Abstract Applications of five-membered ring products as catalysts in cyclometalation reactions include chiral reactions, metathesis reactions, cross-coupling reactions, and polymerization reactions. Other reactions include reductions, Michael addition reactions, dehydrogenations, Diels–Alder reactions, and hydrogenations.

Keywords Catalysts • Chiral • Cross-coupling • Cyclometalation • Dehydrogenation • Metathesis • Polymerization

8.1 Introduction

As concerns the applications for organometallic intramolecular-coordination five-membered ring compounds, several catalysts employed for the three most representative recent synthetic organic reactions have already been described in Chap. 1. In previous reports entitled "Intermolecular Five-Membered Ring Compounds and Their Applications [1]" and "Three Types of Reactions with Intramolecular Five-Membered Ring Compounds in Organic Synthesis [2]" published in 2004 and 2007, respectively, the catalytic activities of organometallic intramolecular-coordination five-membered ring compounds were reported in detail for the common bidentate organometallic intramolecular-coordination five-membered ring compounds and tridentate complexes, or pincer complexes.

In this section, many reactions, including these three reactions from the earlier reviews and recent articles, are described as applications for the organometallic intramolecular-coordination five-membered ring compounds as the catalysts.

I. Omae, *Cyclometalation Reactions: Five-Membered Ring Products as Universal Reagents*, DOI 10.1007/978-4-431-54604-7_8, © Springer Japan 2014

8.2 Chiral Reactions

8.2.1 Introduction

Most of the reported compounds among organometallic intramolecular-coordination five-membered ring compounds are palladacycles. Many reviews on palladacycles have consequently been published, as described in Chap. 1. Palladacycles are also reported to serve as chiral catalysts. The following reports have been published on these chiral catalysts:

In 1997, "Resolutions of Tertiary Phosphines and Arsines with Orthometallated Palladium(II)-Amine Complexes," by Wild [3]

In 2006, "Synthesis and Applications of Optically Active Metallacycles Derived from Amines," by Albert et al. [4]

In 2008, "Palladacycles: Synthesis, Characterization and Applications," Dupont and Pfeffer, eds [5, 6]

In 2009, "The Asymmetric Aza-Claisen Arrangement: Development of Widely Applicable Pentaphenylferrocenyl Palladacycle Catalysts," by Fischer et al. [7]

In 2011, "Cyclopalladated Complexes in Enantioselective Catalysis," by Dunina [8, 9]

Other reviews [10–15] have also been published on catalysts using such other metal compounds as Rh, Ir, Pt, and Cu compounds, in addition to those on palladium compounds:

In 1998, "Bioorganometallic Chemistry-Transition Metal Complexes with α-Amino Acids and Peptides," by Severin et al. [14]

In 2006, "α-Chelation-Directed C-H Functionalizations Using Pd(II) and Cu(II) Catalysts: Regioselectivity, Stereoselectivity and Catalytic Turnover," by Yu et al. [13]

In 2007, "Synthesis and Use of Bisoxazolinyl-Phenyl Pincers," by Nishiyama [12]

In 2008, "Non-racemic (Scalemic) Planar-Chiral Five-Membered Metallacycles: Routes, Means, and Pitfalls in Their Synthesis and Characterization," by Djukic et al. [11]

In 2010, "Bis(oxazolinyl)phenyl Transition-Metal Complexes: Asymmetric Catalysis and Some Reactions of the Metal," Nishijima and Ito [9]

Some of the compounds used as catalysts for chiral reactions are shown in Fig. 8.1.

Chiral catalysts are usually bulky chiral moieties, such as pincer-type compounds, naphthalenes, biaryl compounds such as 2-phenylpyridines, pentaphenylferrocenes, and bicyclic compounds, or oxazolines.

Rearrangements, Diels–Alder reactions, and Michael addition reactions are described below as chiral reactions of chiral organometallic intramolecular-coordination five-membered ring compounds.

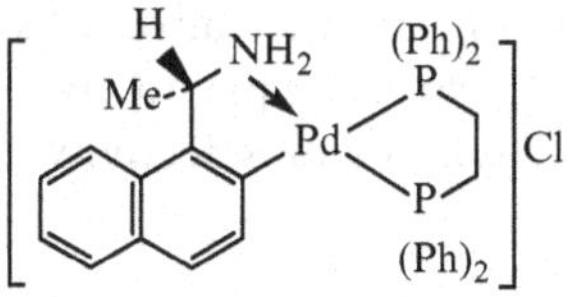

8.1 [4]

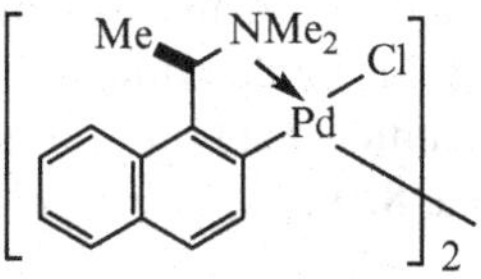

8.2 [5]

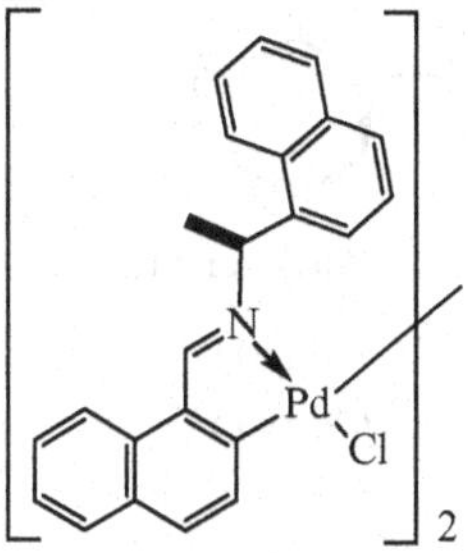

8.3 [5]

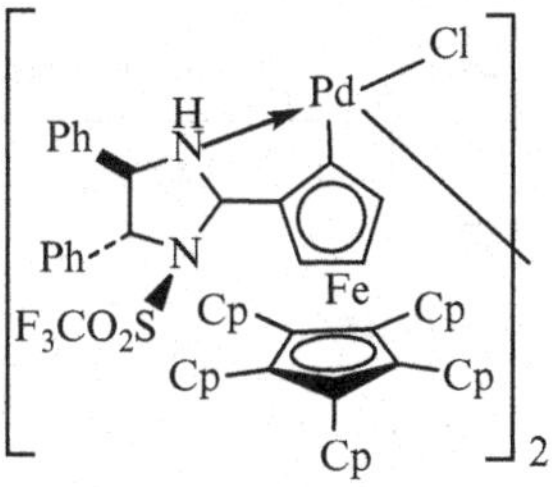

8.4 [7]

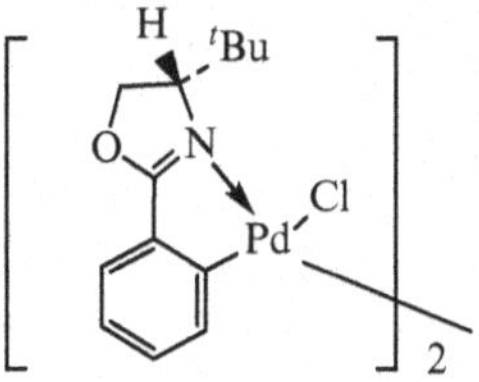

8.5 [8]

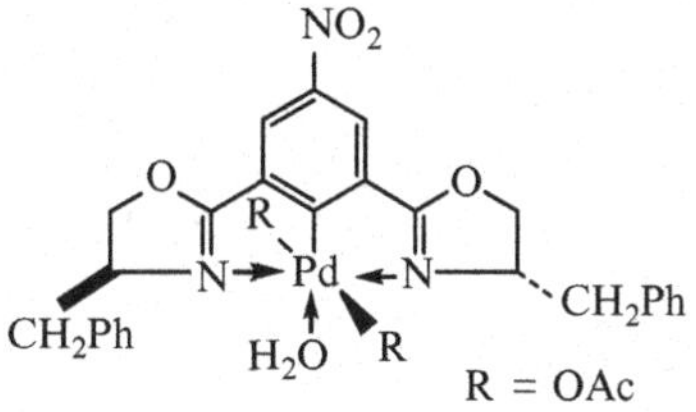

8.6 [15]

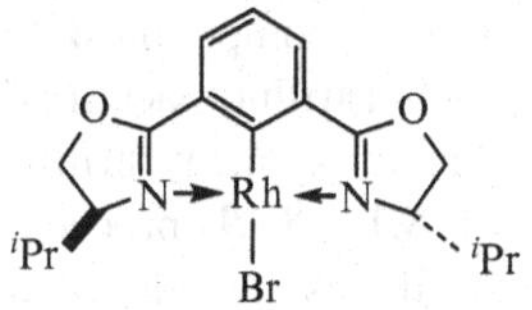

8.7 [12]

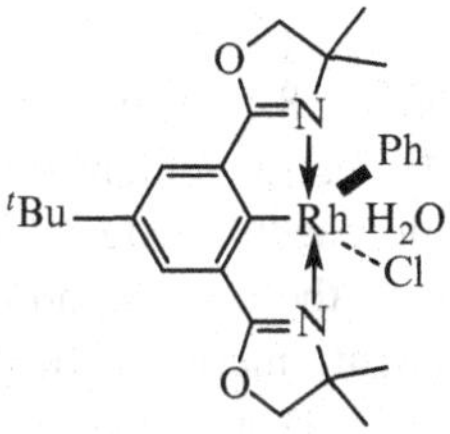

8.8 [10]

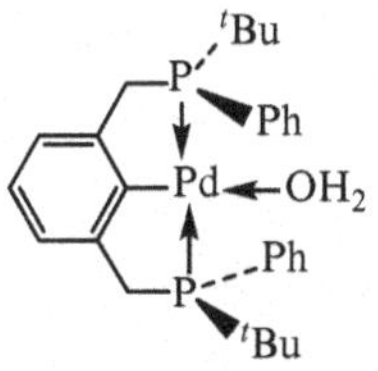

8.9 [9]

Fig. 8.1 Some palladacycles and the other cyclometalated compounds employed for chiral reactions

8.2.2 *Rearrangements*

With chiral cyclopalladated ferrocenyl compounds, allylic imidates were rearranged to allylic amides. Their enantioselectivities were dramatically improved by using catalysts containing planar chiral elements such as Pd, O, N, C, and Si.

A cyclopalladated ferrocenyl oxazoline 3-methoxy-3-pentyl derivative **8.12** was activated in CH_2Cl_2 by deiodination with CF_3COOAg, for example, to give rearranged allylic amide **8.11** in a high yield and a high enantioselectivity from *E*-allylic imidate **8.10**, as shown in Eq. (8.1) [16].

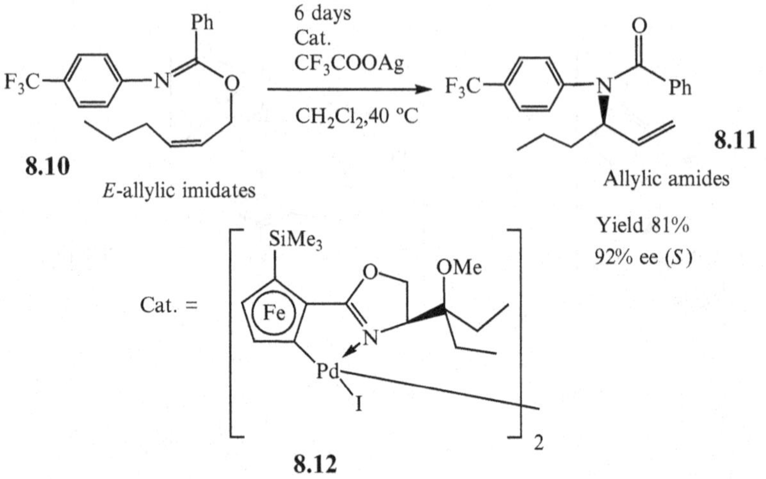

(8.1)

With a cyclopalladated (η^4-tetraphenylcyclobutadiene)cobalt oxazoline propyl chloro-bridged compound **8.17** and a trifluoroacetate-bridged compound **8.20** as catalysts, moreover, the rearrangement of *N*-(4-methoxyphenyl)trifluoroacetimidate **8.13** [17], allylic trichloroacetimidate **8.15** [18], and *N*-(4-methoxyphenyl)trifluoroacetimidate **8.18** [19] to the corresponding amides (**8.14, 8.16, 8.19**) proceeds in high yields and high enantiomeric purities without use of a silver salt as an activator, as shown in Eqs. (8.2), (8.3), and (8.4), respectively.

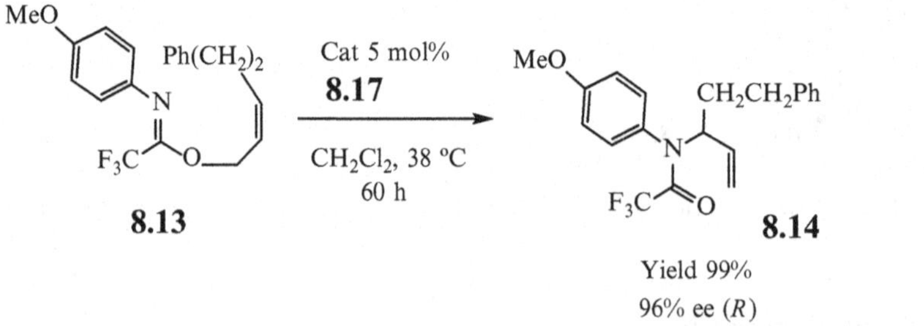

(8.2)

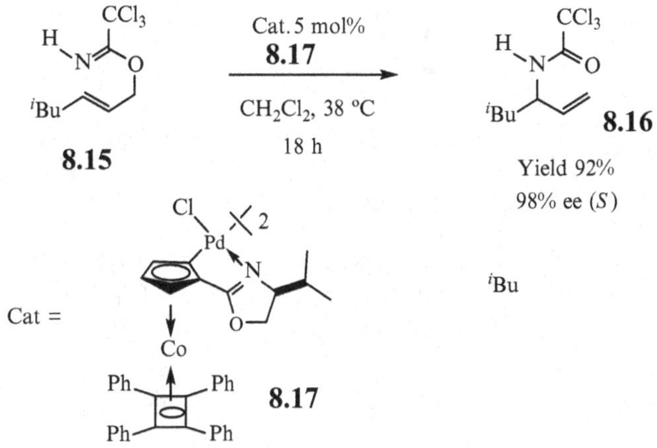
CCl3
H
N
O
iBu
8.15
Cat. 5 mol%
8.17
CH2Cl2, 38 °C
18 h
8.16
Yield 92%
98% ee (S)
Cat =
Cl
Pd
2
N
O
Co
Ph
Ph
Ph
Ph
8.17

(8.3)

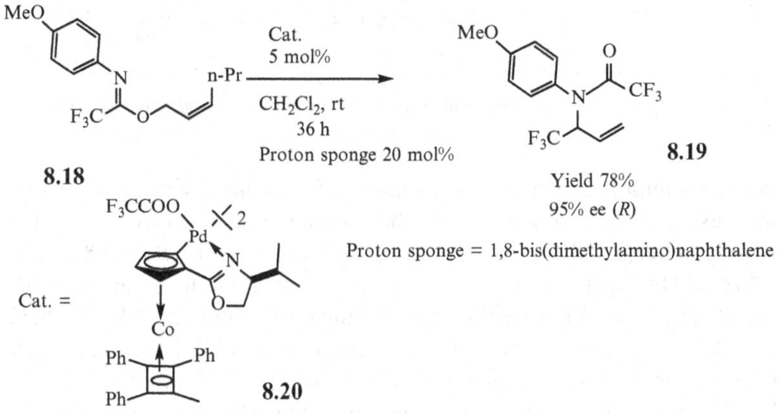
MeO
N
n-Pr
F3C
O
8.18
Cat.
5 mol%
CH2Cl2, rt
36 h
Proton sponge 20 mol%
O
CF3
F3C
8.19
Yield 78%
95% ee (R)
Proton sponge = 1,8-bis(dimethylamino)naphthalene
Cat. =
F3CCOO
Pd
2
N
O
Co
Ph
Ph
Ph
8.20

(8.4)

8.2.3 *Reactions of Chiral Compounds with Amino Acids and Enantioselective Rearrangements*

Optically pure α-amino acids such as L-leucine are used as chiral auxiliaries for the optical resolution of cyclopalladated ferrocenylimines, as shown in Scheme 8.1 [20, 21]. At the same time, optically pure cyclometalated ferrocenylimines were found to be equally useful in the resolution of racemic α-amino acids.

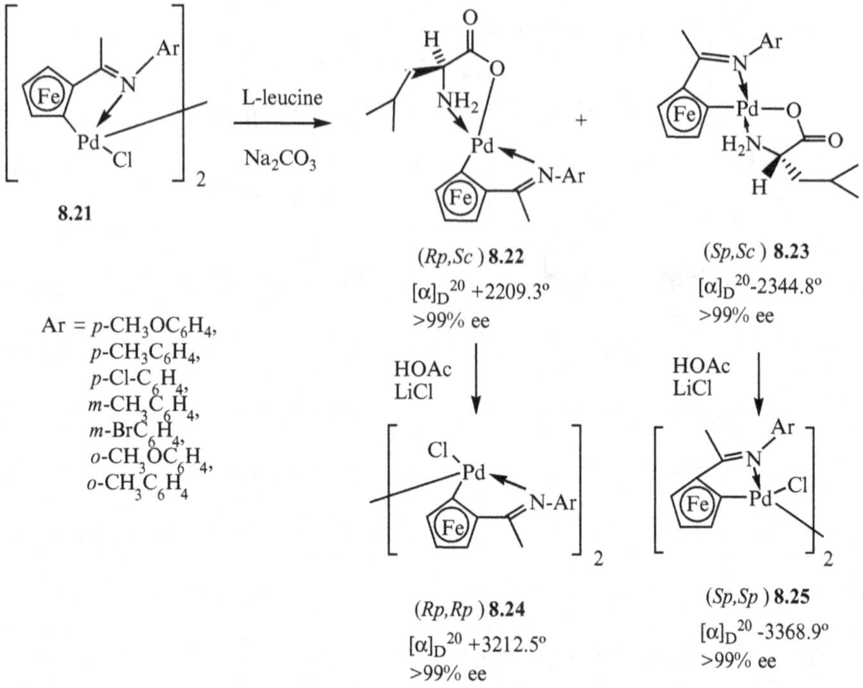

Scheme 8.1 Optical resolution of cyclopalladated ferrocenylimines with optically pure α-amino acids [20, 21]

The optical resolution of a racemic α-amino acid was successfully carried out, for example, using a chiral dimer (*Rp,Rp*) **8.24** (R-configuration) as a resolving agent, as shown in Scheme 8.2 [20]. The mixture of diastereomers (*Rp,Sc*) **8.22** and (*Rp,Rc*) **8.26** could be separated by chromatography. Optically active α-amino acids (*Sc* **8.27** and *Rc* **8.28**) were obtained by the treatment of compounds (*Rp,Sc*) **8.22** and (*Rp,Rc*) **8.26**, respectively, with glacial acetic acid and LiCl, and the dimer (*Rp,Rp*) **8.24** was recovered with no loss of optical activity [20–22].

Chiral cyclopalladated ferrocenyl compounds rearrange allylic imidates to allylic amides. It is apparent that allylic *N*-arylimidates are excellent substrates for providing allylic amides at good rates and good yields with moderate enantioselectivities. The variations in the electron-donating ability of the *N*-aryl group had little effect on the outcome of the reactions (Nos. 1–3), whereas the reaction rate decreased dramatically as the size of the substituent on the alkane increased (No. 3–6). The corresponding Z-allylic imidate (No. 7) provided an allylic amide product exhibiting an opposite absolute configuration with a decrease in the reaction rate, as shown in Eq. (8.5) and in Table 8.1 [16].

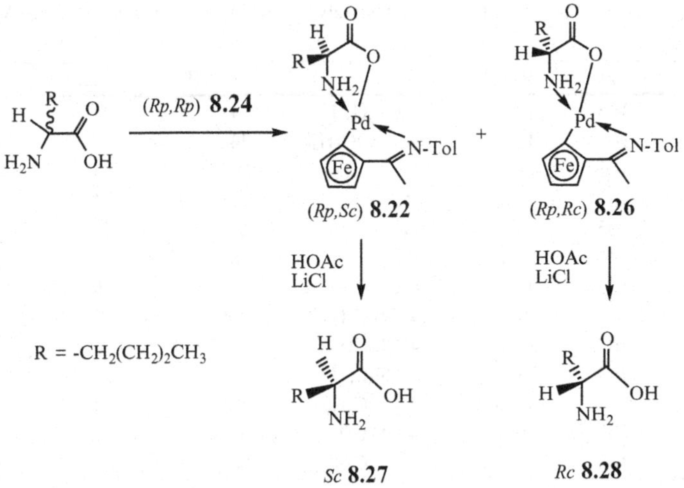

Scheme 8.2 Optical resolution of a racemic α-amino acid with a chiral dimer *(Rp, Rp)* (R-configuration) as a resolving agent [20]

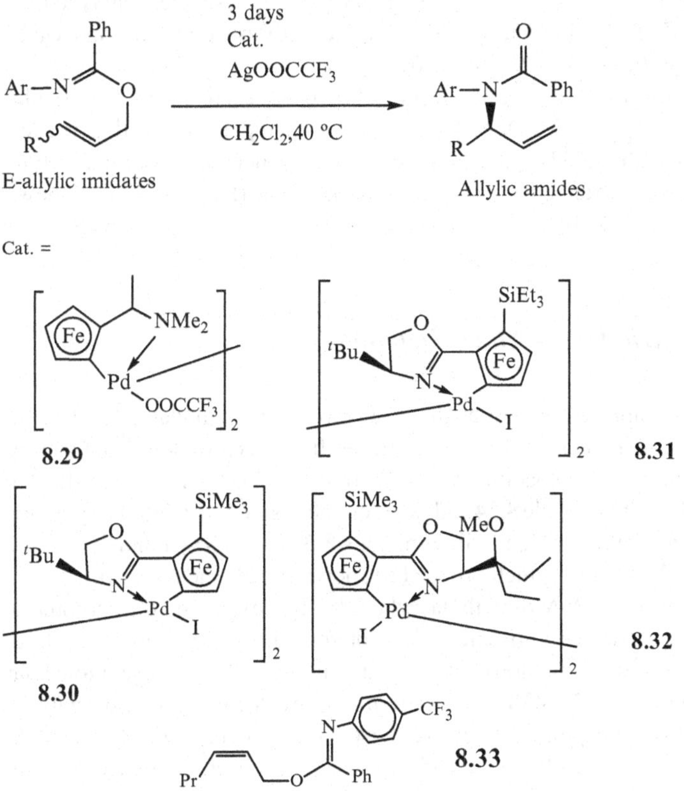

$$(8.5)$$

Table 8.1 Rearrangement of allylic imidate catalyzed by the palladacycle catalysts/AgOOCCF$_3$ in CH$_2$H$_2$

Numbers	Cat.	R	Ar	Time (h)	Yield (%)	% ee
1	105	Pr	4-CF$_3$C$_6$H$_4$	24	97	57
2	105	Pr	4-MeOC$_6$H$_4$	24	69	52
3	105	Pr	Ph	38	84	61
4	105	Me	Ph	16	94	54
5	105	Ph	Ph	27	47	47
6	105	t-Bu	Ph	48		
7	105	Pr	4-CF$_3$C$_6$H$_4$	6(d)	76	46
8	106	Pr	4-CF$_3$C$_6$H$_4$	2(d)	57	79(S)
9	107	Pr	4-CF$_3$C$_6$H$_4$	2.5(d)	76	76(S)
10	108	Pr	4-CF$_3$C$_6$H4	3(d)	95	72(R)
11	106	Pr (Z)	4-CF$_3$C$_6$H$_4$	3(d)	67	91(R)
12	107	Pr (Z)	4-CF$_3$CH$_4$	6(d)	89	90(R)
13	108	Pr (Z)	4-CF$_3$C$_6$H$_4$	6(d)	81	92(S)

The size of the oxazoline substituent in the catalysts (**8.30–8.32**) was increased to 3-methoxy-3-pentyl (**8.32**), and an iodide-bridged dimer was activated in CH$_2$Cl$_2$ by deiodination with Ag(OOCCF$_3$). These in *situ*-generated species catalyzed the rearrangement of *E*-allylic imidate to allylic amide with moderate to good yields with enantioselectivities of 72–79 % *ee* (Nos. 8–10).

Enantioselectivity was dramatically improved by using catalysts containing planar chiral elements, such as Pd, O, N, C, and Si. For example, cyclopalladated ferrocenyloxazolines (**8.30–8.32**) catalyzed the rearrangement of several *Z*-allylic imidates **8.33** to give rearranged allylic amides in >90 % *ee* (Nos. 11–13), as shown in Eq. (8.5) and Table 8.1 [16].

8.2.4   Asymmetric Diels–Alder Reactions

Organopalladium complexes containing the (*S*)-form of *ortho*-palladated (1-(dimethylamino)ethyl)-naphthalene have been used successfully as chiral templates to promote asymmetric cycloaddition reactions between coordinated 3,4-dimethyl-1-phenylphosphole **8.34** and two dienophiles (*N,N*-dimethylacrylamide and styrene) via two pathways, endo (compounds **8.35**) with X=Cl and exo (compounds **8.36**) with X=OClO$_3$, which proceed as shown in Scheme 8.3 [23].

In the reactions with *N,N*-dimethylacrylamide, for example, diastereomeric cycloadducts were separated into pale yellow prisms and yellow blocks by fractional crystallization with dichloromethane and diethyl ether. These molecular structures (compounds **8.37**, **8.38**) were determined by X-ray structural analysis. The optically active phosphinoamides were isolated as both pure *R*-endo **8.39** ([α]$_{365}$=+5.9) and *S*-endo **8.40** ([α]$_{365}$=−5.7) forms, moreover, by treatment with aqueous potassium cyanide, as shown in Scheme 8.4 [23].

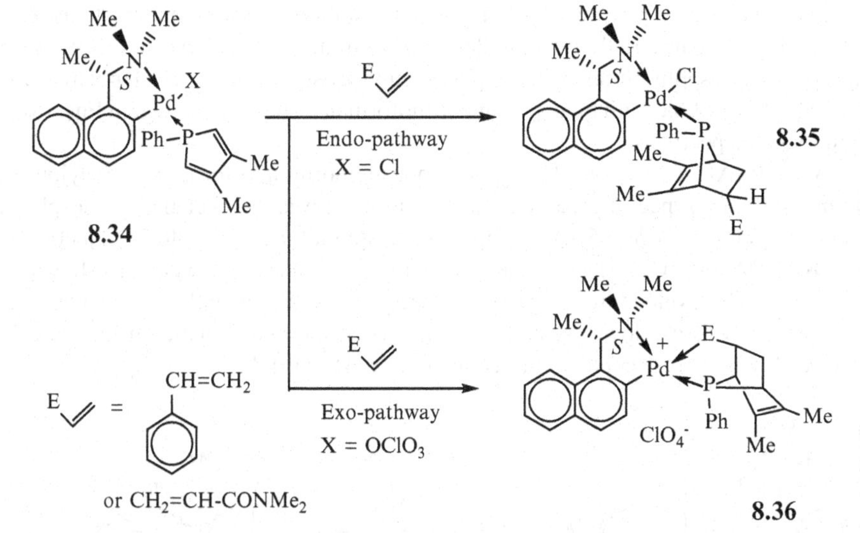

Scheme 8.3 Asymmetric Diels-Alder reactions of the (*S*)-form of *ortho*-palladated (1-(dimethylamino)ethyl)-naphthalene with two dienophiles [23]

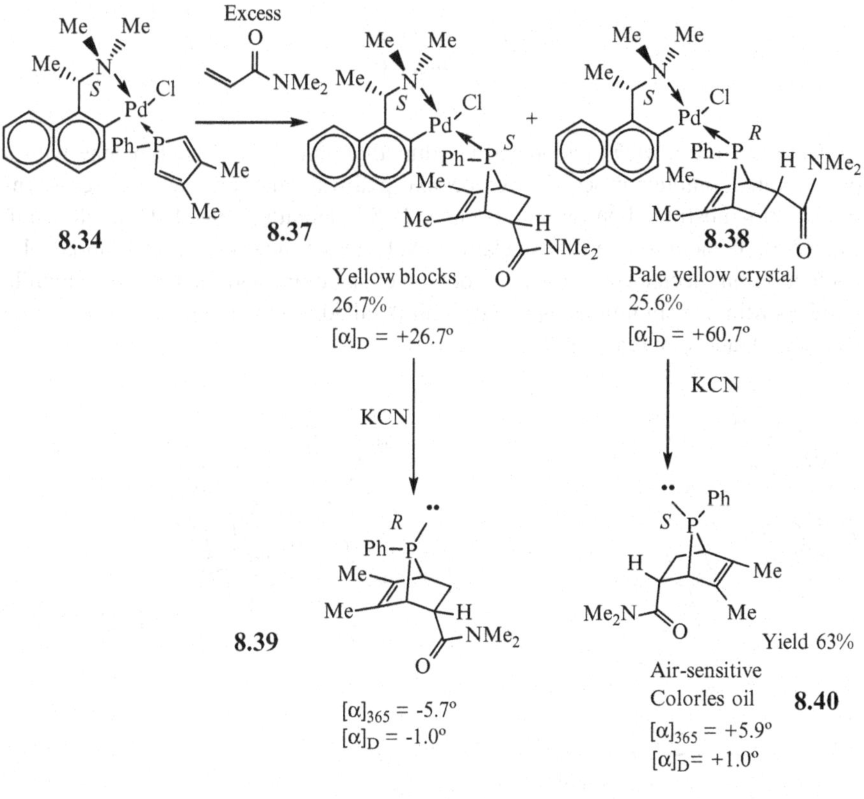

Scheme 8.4 Separation of diastereoisomeric cycloadducts by fractional crylatallization and isolations of pure *R*-endo and *S*-endo forms of optical active phosphinoamides by treatment with KCN [23]

Leug et al. [24, 25] reported the effects of palladium or platinum metal templates of (1-(dimethylamino)ethyl)naphthalene on asymmetric Diels–Alder reactions of 2-diphenylphosphinofuran with diphenylvinylphosphine and of 1-phenyl-3,4-dimethlylphosphole with methyl cyanodithioformate. These are shown in Eqs. (8.6) and (8.7), respectively.

A Diels–Alder reaction of 2-diphenylphosphinofuran with diphenylvinylphosphine in the presence of organoplatinum complex gives the chelating diphosphine exo-cycloadduct, 4(R), 5(R)-bis(diphenylphosphine)-7-oxabicyclo[2.2.1]hepta-2-ene **8.41** (M=Pt) in a 70 % isolated yield with many diastereoisomers, as shown in Eq. (8.6). The cycloaddition reaction proceeds at a significantly slower rate and exhibits a markedly lower stereoselectivity when the chiral platinum template is replaced with its organopalladium counterpart **8.41** (M=Pd) [24].

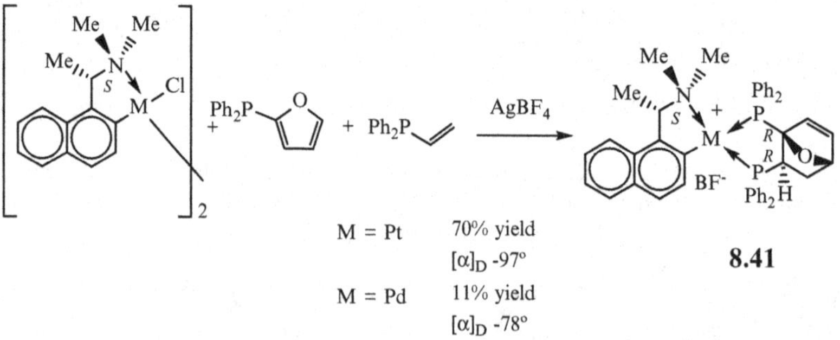

$$\text{(8.6)}$$

M = Pt   70% yield
           $[\alpha]_D$ -97°

                                8.41

M = Pd   11% yield
           $[\alpha]_D$ -78°

The reaction of *ortho*-palladated (1-(dimethylamino)ethyl)-naphthalene with a perchlorate complex proceeds by the exo-pathway and produces (+)-exo-syn-methylthio-substituted phosphanorbornene P–S bidentate chelate **8.42**, as shown in Eq. (8.7). Generation of the chelating cycloadduct involved an intramolecular cycloaddition mechanism, in which both the cyclic diene and the hetero dienophile were coordinated simultaneously to the chiral palladium template during the course of the cycloaddition reaction [25].

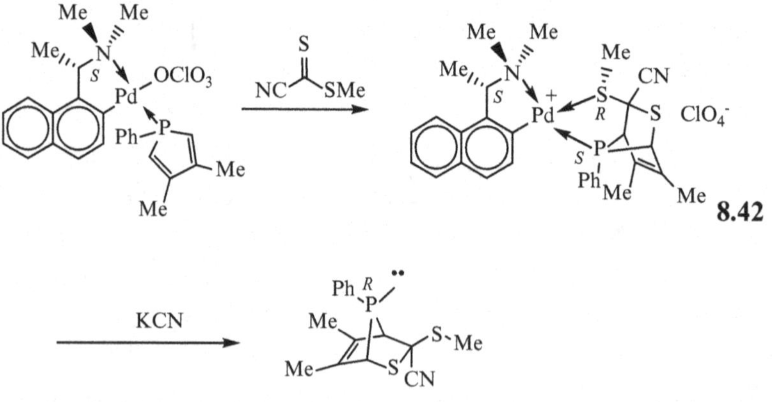

$$\text{(8.7)}$$

8.42

8.2.5 Other Diels–Alder Reactions

Tridentate bis(oxazolinyl)pyridinyl rhodium and ruthenium pincer complexes are useful as catalysts for hydrosilylations and cyclopropanations. These NNN-type inorganic pincer complexes are not as stable, however, as phosphine or salen-type pincer complexes. On the other hand, an organometallic tridentate bis(oxazolinyl) phenyl NCN-type complex is stable. These optically active NCN-type pincer complexes act as efficient catalysts for enantioselective hetero Diels–Alder reactions of Danishefsky's diene with glyoxylates [26].

Asymmetric hetero Diels–Alder reactions of 1-methoxy-3-[(t-butyldimethylsilyl) oxy]-1,3-butadiene (Danishefsky's diene) with n-butyl glyoxylate are catalyzed highly enantioselectively with cis(endo)-diastereoselectivity, for example, by chiral bis(oxazolinyl)phenylrhodium aqua dichloride **8.43**, as shown in Eq. (8.8) [27].

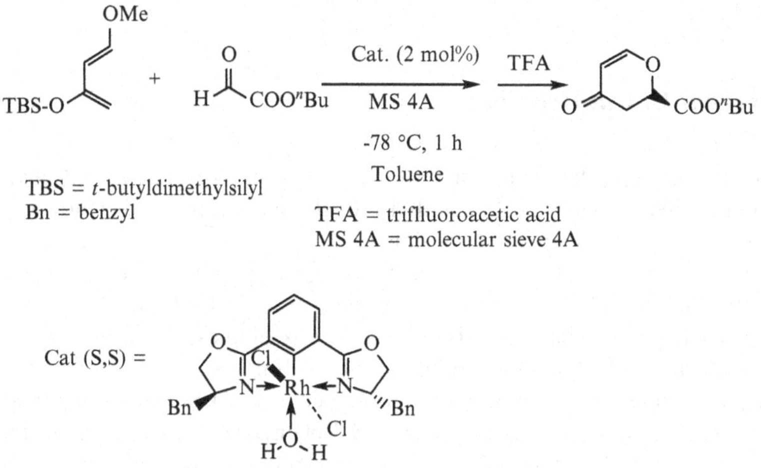

$$\tag{8.8}$$

8.43

Other metal-catalyzed reactions with five-membered pincers include asymmetric aldol-type condensations [26, 28–30], cyclopropanations [31], enantioselective allylations [32], reductive eliminations [33], transfer hydrogenations [28–30, 34], hydroaminations [28], and polymerizations [26, 28, 35].

8.2.6 Michael Addition Reactions

Michael addition reactions generally involve the addition of an active methylene, such as a malonate or nitroalkane, to activated olefins, such as α,β-unsaturated carbonyl compounds, in the presence of a base.

Recently, pincer metal compounds, such as those with rhodium [36], palladium [37, 38], and platinum [39], are used as catalysts for Michael addition reactions. For example, the addition of an α-cyanopropionate to acrolein under mild, neutral conditions in the presence of a bis(oxazolinyl)phenylstannane-derived rhodium complex **8.44** proceeds enantioselectively with a high yield and high TON, as shown in Eq. (8.9) [36].

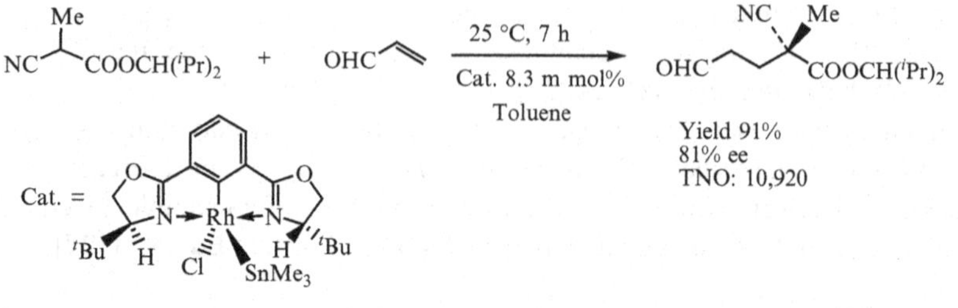

$$ (8.9) $$

8.44

8.2.7 Recent Chiral Reactions

In the first chapter, the synthesis of a representative ligand, BINAP (2,2′-bis(diphenylphosphino)-1,1′-binaphthyl), for chiral reactions with cyclometalation reaction products has already been shown.

In the previous section, three chiral reactions of rearrangements, Diels–Alder reactions, and Michael additions are shown in the presence of Pd, Pt, Rh, and Ru metal catalysts having arylamines, arylimines, aryloxazolines, ferrocenyloxazolines, ferrocenylimines, or ferrocenylamines as their substrates.

Recently, the following two chiral reactions of phenyl alkyl ketone were reported on hydrogenations in the presence of bis(oxazolinyl)phenyl metal compounds (**8.45**, **8.46**), as shown in Eqs. (8.10) [40, 41] and (8.11) [42]. For the latter bis(oxazolinyl) phenyl ruthenium catalyst (**8.46**), Nishiyama reported, furthermore, on the significant enhancement of enantioselectivity with a zinc chloride-bridged ruthenium compound [42].

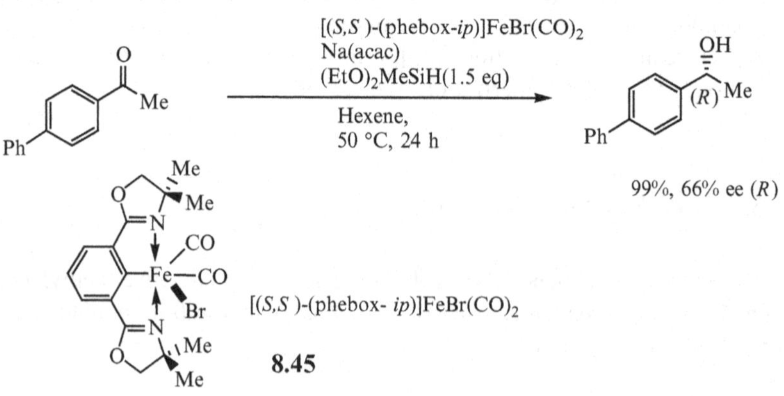

$$ (8.10) $$

99%, 66% ee (R)

8.45

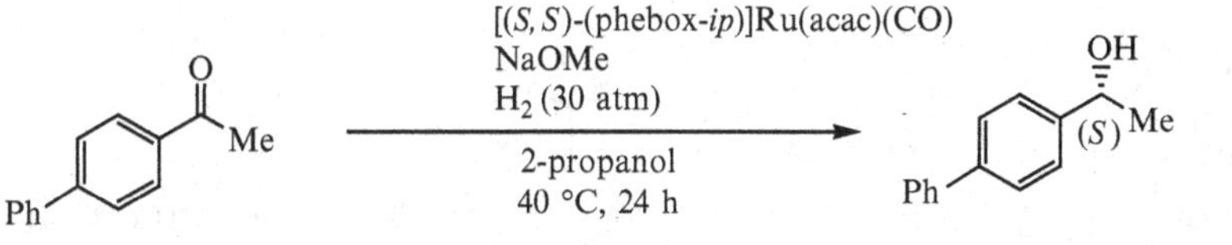

$$99\%, 77\% \text{ ee } (S) \tag{8.11}$$

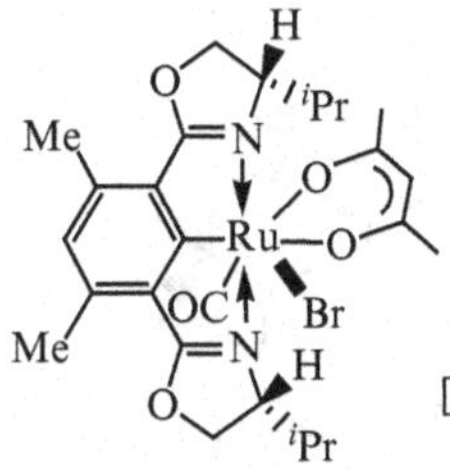

8.46

[(*S,S*)-(phebox-*ip*)]Ru(acac)(CO)

Ohkuma et al. [43] also reported on the hydrogenation of ketones. A benzyl-amine ruthenium compound having bicyclic rings of a five-membered ring and a six-membered ring with the BINAP (2,2′-bis(diphenylphosphino)-1,1′-binaphthyl, see Chap. 1), ligand (**8.47**) shows excellent catalytic activity in the asymmetric hydrogenation of acetophenone, as shown in Eq. (8.12). The turnover frequency of hydrogenation of acetophenone reaches about 35,000 min^{-1} in the best case, affording 1-phenylethanol in >99 % ee.

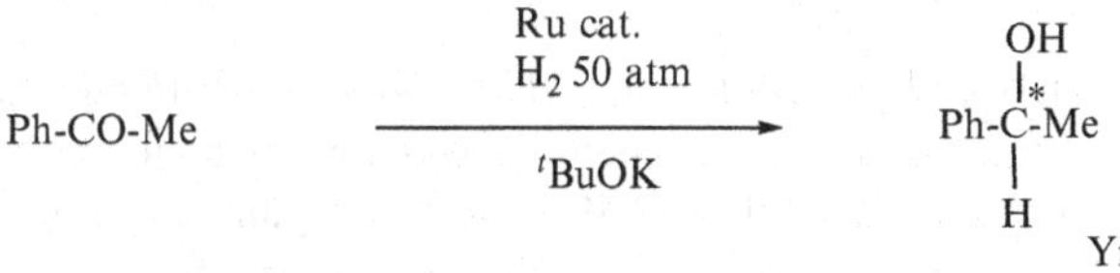

$$\tag{8.12}$$

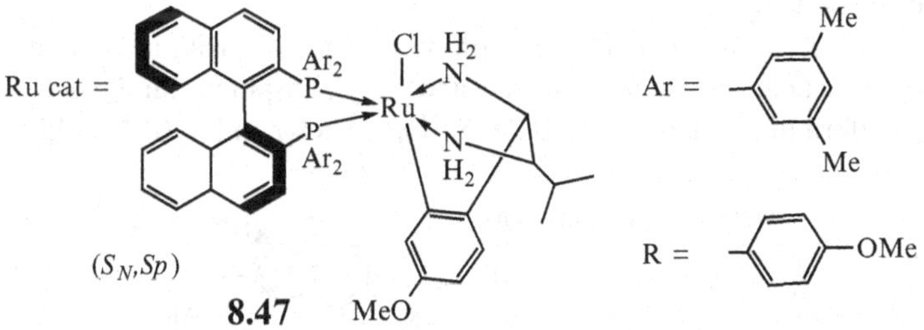

An asymmetric hydroarsination reaction between diphenylarsine and 3-diphenylphosphanyl-but-3-en-1-ol has been achieved using *ortho*-metalated (*R*)-[1-dimethylamino)ethenylnaphthalene as the chiral reaction template (**8.48**) in high stereoselectivities under mild reaction conditions, as shown in Scheme 8.5 [44].

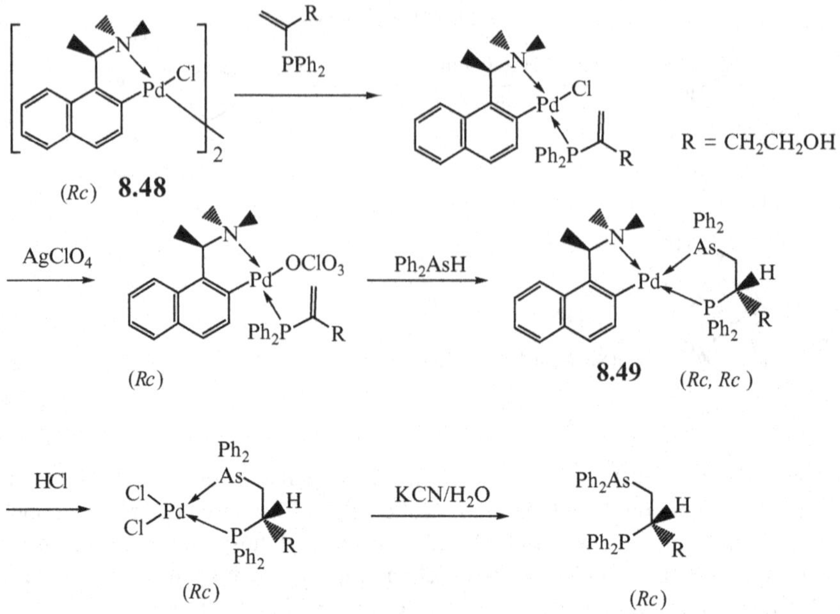

Scheme 8.5 A asymmetric hydroarsenation reaction between diphenylarsine and 3-diphenylphos-phanyl-but-3-en-1-ol with *ortho*-metalated (*R*)-[1-dimethylamino)ethenylnaphthalene as the chiral reaction template [44]

The hydroarsination of 3-diphenylphosphanyl-but-3-en-1-ol with diphenylarsine generated only one stereoisomer as a five-membered As–P bidentate chelate on the chiral naphthylamine palladium template (**8.49**). The naphthylamine auxiliary could be removed chemoselectively by treatment with concentrated hydrochloric acid. The absolute configuration of the final hydroarsination product has been established by single crystal X-ray analysis.

A highly enantioselective alkynylation reaction was carried out with an aromatic ketimine as the cyclometalation reaction substrate and diarylphosphine chiral reaction template (**8.50**) to afford the cyclization product (**8.51**), as shown in Eq. (8.13) [45].

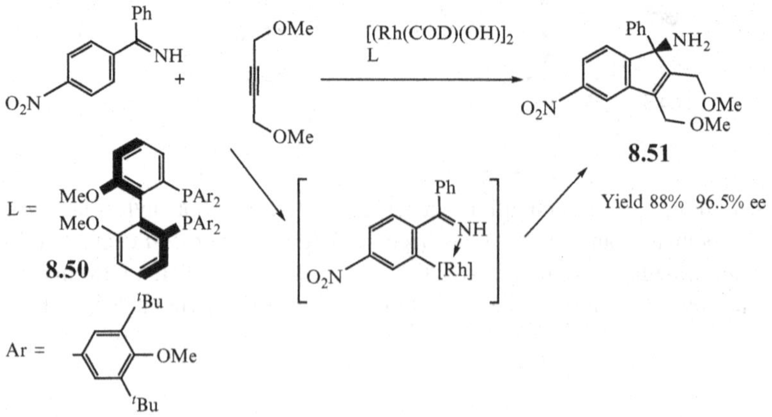

(8.13)

Ligand exchange reactions are usually slower for a Pt center than for a Pd center, because Pt catalysts often show a reduced catalytic activity as compared to their Pd counterparts. Peters et al. [46] investigated the asymmetric Aza–Claisen rearrangement of Z-configuration trifluoroacetimidates with bis-metallic ferrocenyl compounds. A Pt–Pd bis-imidazoleimine cyclometalated ferrocenyl compound is an excellent catalyst for this reaction type, in general allowing for very high enantioselectivities, as shown in Eq. (8.14).

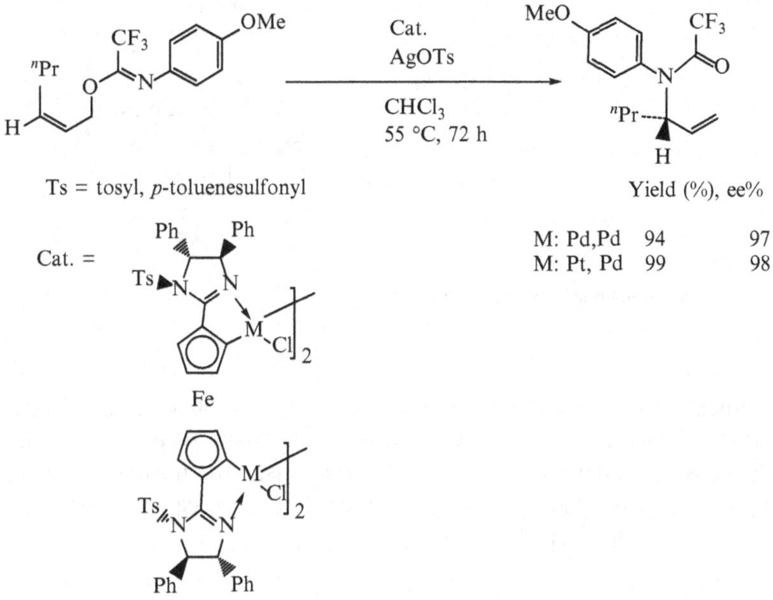

Ts = tosyl, p-toluenesulfonyl

Cat. =

| M: Pd,Pd | 94 | 97 | |
| M: Pt, Pd | 99 | 98 | (8.14) |

Yield (%), ee%

8.3 Metathesis Reactions

Metathesis means "changing places." A metathesis reaction is an exchange reaction occurring mainly between two olefins in which alkylidene groups are interchanged. Metathesis reactions are activated by catalyst systems based on molybdenum, tungsten, or ruthenium. Molybdenum and ruthenium–carbene complexes have been especially useful tools in synthetic organic chemistry since 1990. The 2005 Nobel Prize for chemistry went to Y. Chauvin, who elucidated the reaction mechanism of metathesis, and to R.H. Grubbs and R.R. Schrock, who found the Grubbs' catalyst and Schrock catalyst, respectively [4, 47–50].

The ruthenium–carbene complex **8.52** is an excellent commercially available example of the Grubbs' catalyst. However, intramolecular five-membered ring compounds **8.53–8.55**, which are activated by coordination with an ether oxygen atom, are much more active with respect to electron-deficient olefins and are also stable with respect to air as shown in Fig. 8.2.

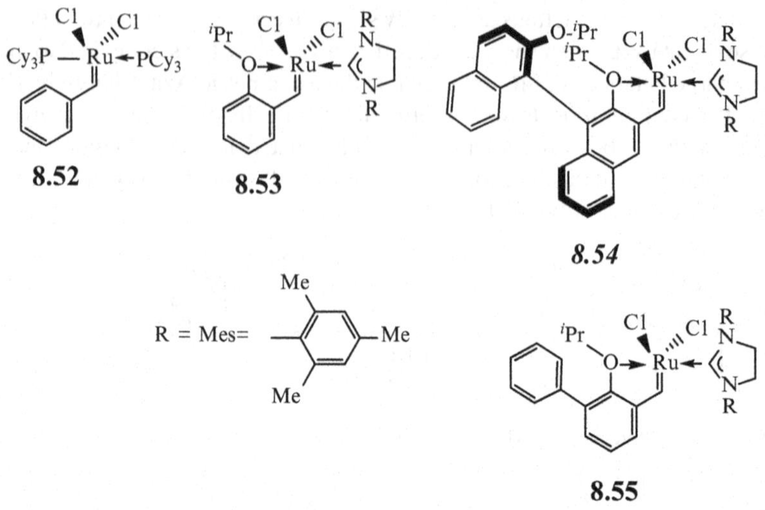

Fig. 8.2 Representative metathesis reaction catalysts

Many ruthenium–carbene five-membered ring compounds have recently been reported to show good activity for metathesis [51–71]. Many types of reactions, such as ring-closing metathesis, ring-opening metathesis, cross metathesis, enyne metathesis, and diyne metathesis, proceed with the help of these catalysts, as shown in Eqs. (8.15), (8.16), (8.17), (8.18), and (8.19) [58].

1. Ring-closing metathesis (RCM)

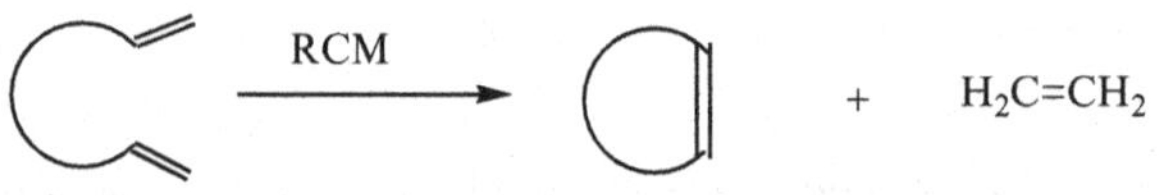

$$ (8.15) $$

2. Ring-opening metathesis (ROM)

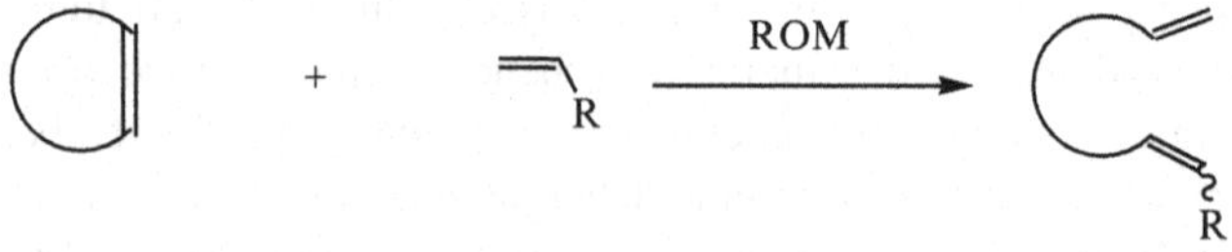

$$ (8.16) $$

3. Cross metathesis (CM)

$$ (8.17) $$

4 .Enyne metathesis

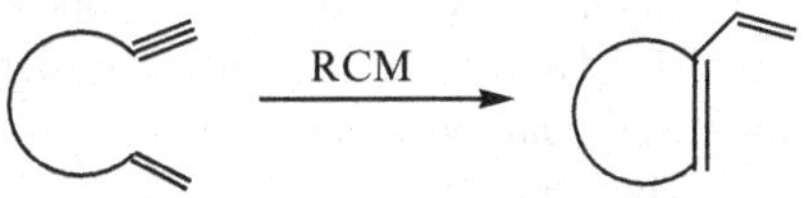

$$(8.18)$$

5. Diyne Metathesis

$$(8.19)$$

The ring-closing metathesis (Eq. (8.15)) of an acyclic diene, for example, proceeds easily at room temperature with a high yield in the presence of the 1,3-dimesityl-4,5-dihydroimidazole-2-ylidene ruthenium catalyst **8.53**, as shown in Eq. (8.20) [51–54].

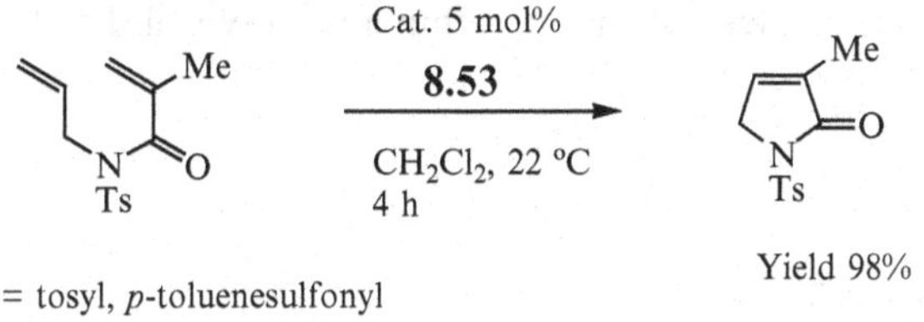

$$(8.20)$$

Ts = tosyl, *p*-toluenesulfonyl

A tandem (domino) cross-metathesis reaction between an enyne and 3 equivalents of a conjugated alkene proceeds in dichloromethane at 40 °C for 12 h with a high yield in the presence of the same ruthenium–carbene five-membered ring compound **8.53**, as shown in Eq. (8.21) [54, 55].

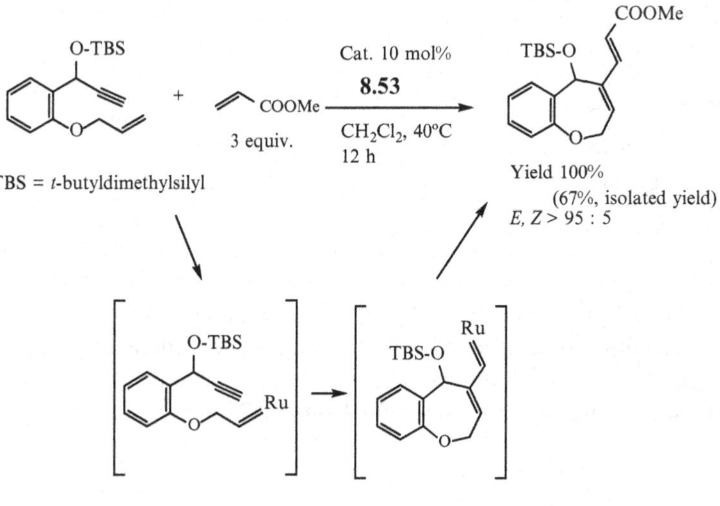

TBS = *t*-butyldimethylsilyl

$$(8.21)$$

Metathesis reactions are reported in some reviews [72–75], a handbook [76], etc. [77–82]. The first well-defined metathesis-active ruthenium–alkylidene complex is a compound, (Ph$_3$P)$_2$Cl$_2$ Ru=CH–CH=CPh$_2$), but Hoveyda–Grubbs' first- and second-generation ruthenium-based catalysts are ruthenium–alkylidene five-membered ring compounds (see Chap. 1).

Recently, these catalysts containing a ruthenium–alkylidene, N-heterocyclic carbene ligand, mainly an alkoxyoxygen σ-donor ligand, or a nitrogen σ-donor ligand as an organometallic intramolecular-coordination five-membered ring compound have also been reported [83–99].

Recent cyclometalation products employed for metathesis reactions are shown in Fig. 8.3, and two interesting reports are also shown below:

Barbasiewicz et al. [99] reported on the Hoveyda–Grubbs' metathesis catalyst bearing a chelating benzylidene ligand assembled on *peri*-substituted naphthalene **8.68**, as shown in Eq. (8.22). In contrast to usual naphthalene-based compounds **8.67**, it exhibits a very fast initiation behavior for a ring-closing metathesis reaction (Eq. (8.22)), which is attributed to a distorted molecular structure and reduced π-electron delocalization within a weakly stabilized six-membered chelate ring.

Ring-closing metathesis

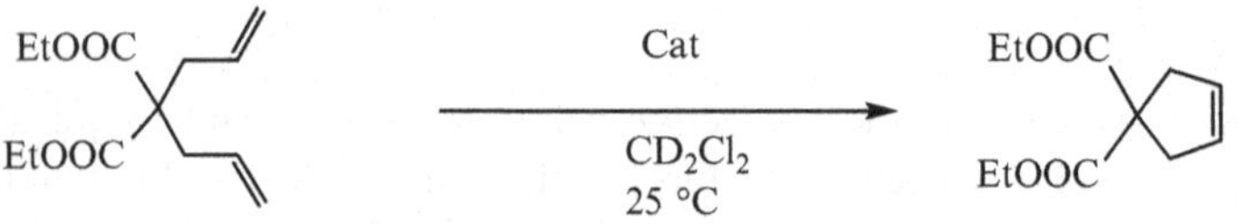

$$(8.22)$$

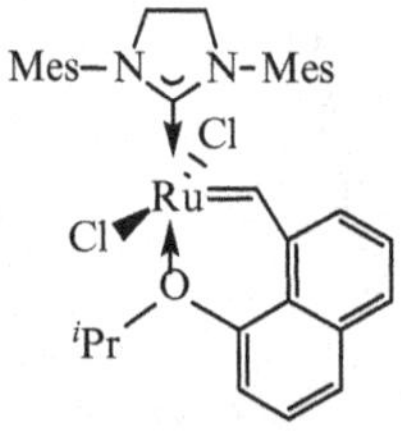

8.68

The reaction of the Grubbs' third-generation complex (1,3-bis(2,6-diisopropylimidazolin)-2-ylidene complex) with 2-ethenyl-N-methylaniline results in the formation of complexes **8.66**. As compared to the respective conventional, O-Hoveyda–Grubbs' complex (such as **8.56–8.64**, **8.67** in Fig. 8.3), the new

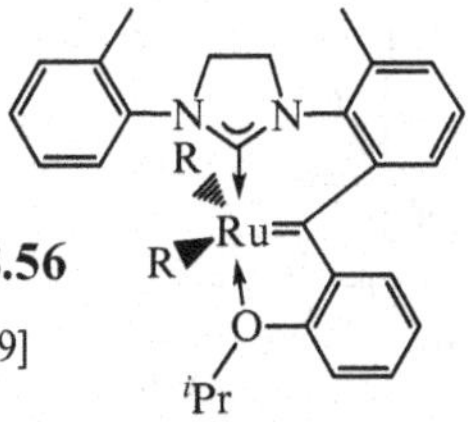

8.56

[89]

R = ^{t}BuCOO, Cl

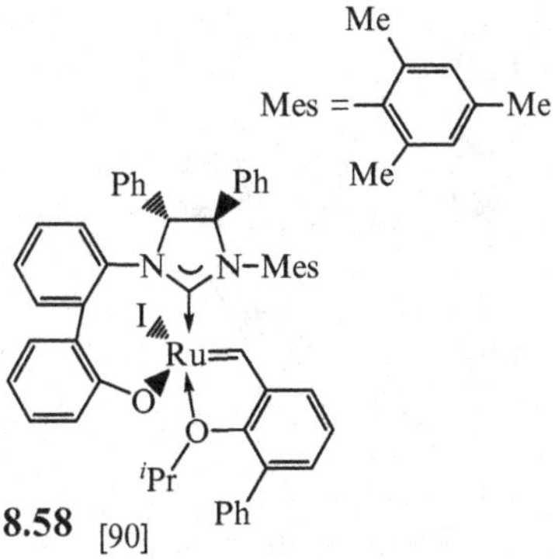

8.58 [90]

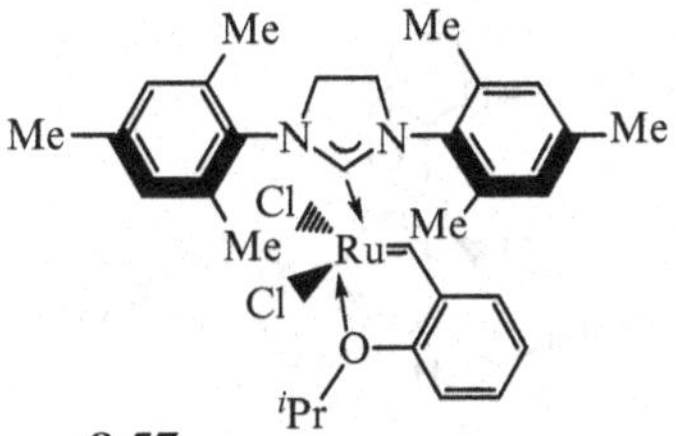

8.57

[82,87,88]

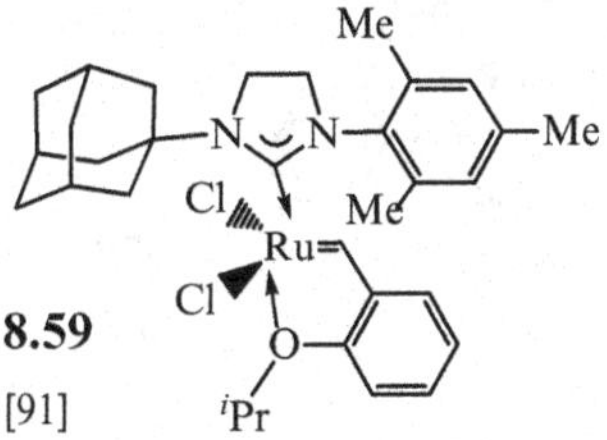

8.59

[91]

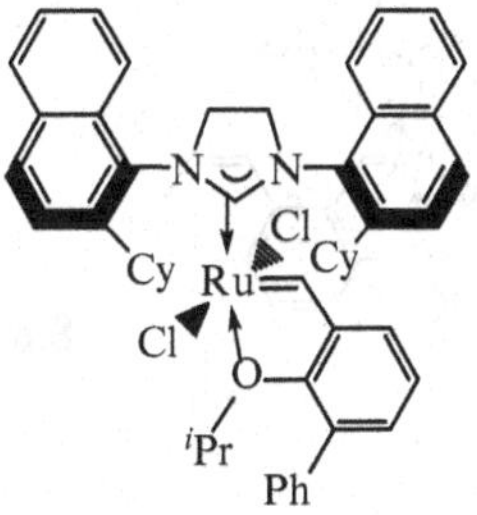

8.60 [92]

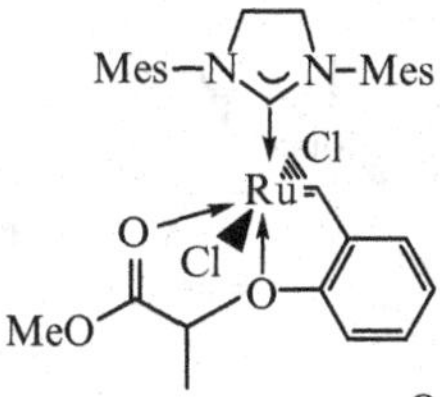

8.61

[93]

Fig. 8.3 Recent cyclometalation products employed for metathesis reactions

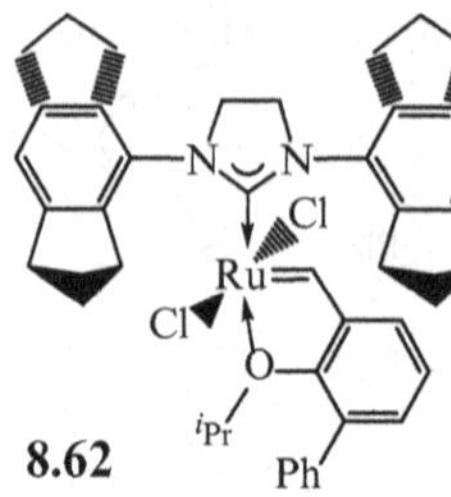

8.62

[94]

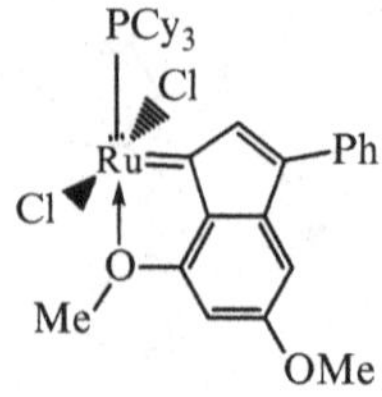

8.63

[95]

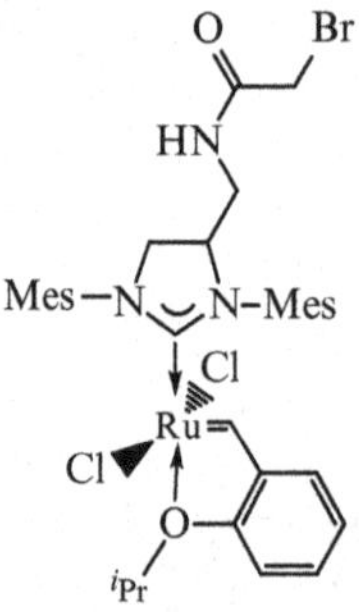

8.64

[96]

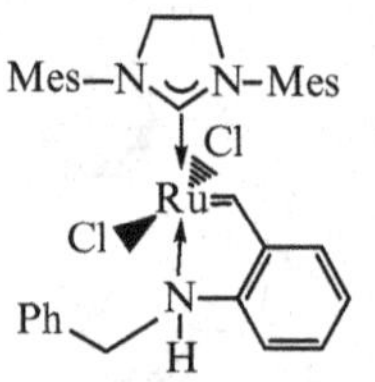

8.65

[97]

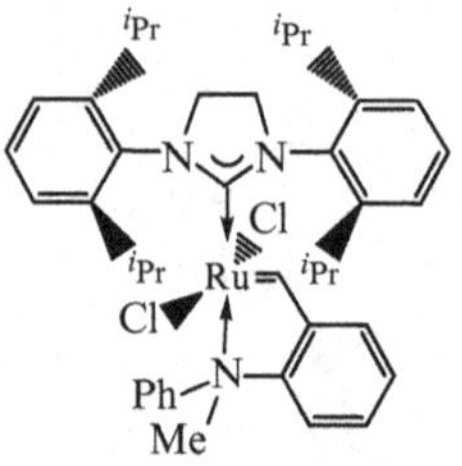

8.66

[98]

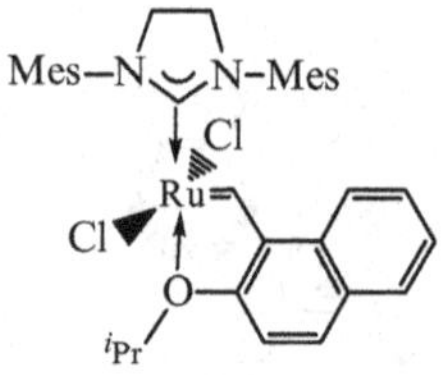

8.67

[99]

Fig. 8.3 (continued)

complexes are characterized by fast catalyst activation that translates into fast and efficient ring-closing metathesis reactivity. Catalyst loadings of 15–150 ppm (0.0015–0.015 mol%) are sufficient for the conversion of a wide range of diolefinic substrates into the respective ring-closing metathesis products after 15 min at 50 °C in toluene as shown in Eq. (8.23). The use of the complex **8.66** in ring-closing metathesis reactions enables the formation of N-protected 2,5-hydropyrroles with turnover numbers (TONs) of up to 58,000 and turnover frequencies (TOFs) of up to 232,000 h^{-1} [98].

Ring-closing metathesis

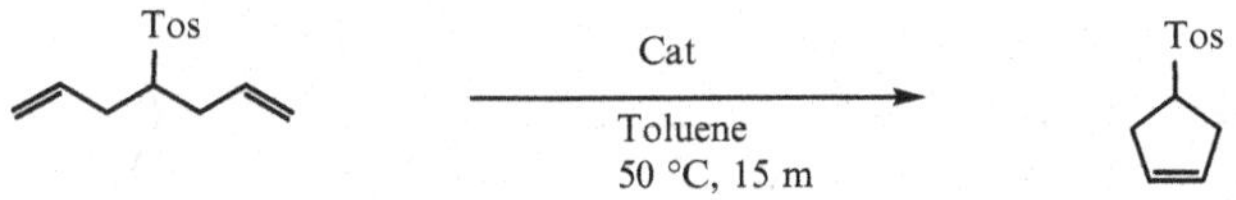

$$TON \quad 58,000 \text{ (yield 80\%)} \qquad (8.23)$$
$$TOF \quad 232,000 \text{ h}^{-1}$$

Cat =

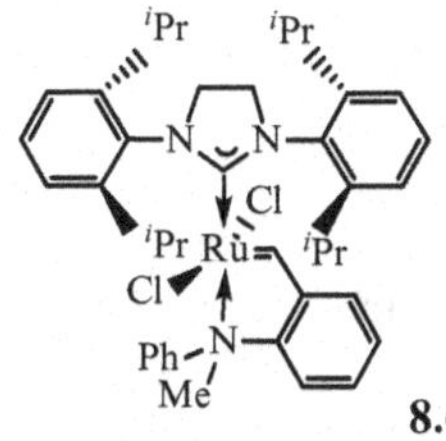

8.66

8.4 Cross-Coupling Reactions

Many cross-coupling reactions have been reported since 1967, as shown in Fig. 8.4 [100, 101].

These reactions are Moritani–Fujiwara reactions, Heck reactions, Mizoroki–Heck reactions, Kumada–Tamao–Corriu reactions, Murahashi reactions, Eto–Hagihara reactions, Negishi reactions, Migita–Kosugi–Stille reactions, Suzuki–Miyaura reactions, and Hiyama reactions.

Among these reactions, three were recognized with a Nobel Prize in Chemistry in 2010. The Nobel Prize was awarded on the basis of practical use in the industrial sector. Heck reactions were used for the synthesis of more than 100 kinds of natural products and physiologically active substances, Negishi reactions were used for the synthesis of pumiliotoxins A (PTXs) and hennoxazole A, and Suzuki–Miyaura reactions were used for the synthesis of dynemicin A and dragmacidin F [100, 101].

An enormous number of articles and many reports [30, 102–108] have also been published on cross-coupling reactions using organometallic intramolecular-coordination five-membered ring compounds as catalysts:

In 1999, "Application of Palladacycles in Heck-Type Reactions," by Herrmann et al. [108]
In 2002, "Palladium-Catalyzed Coupling Reactions of Aryl Chlorides," by Littke and Fu [107]
In 2003 "The Uses of Pincer Complexes in Organic Synthesis," by Singleton [30]
In 2004, "Palladacycles in Catalysis—a Critical Survey," by Beletskaya and Cheprakov [105]

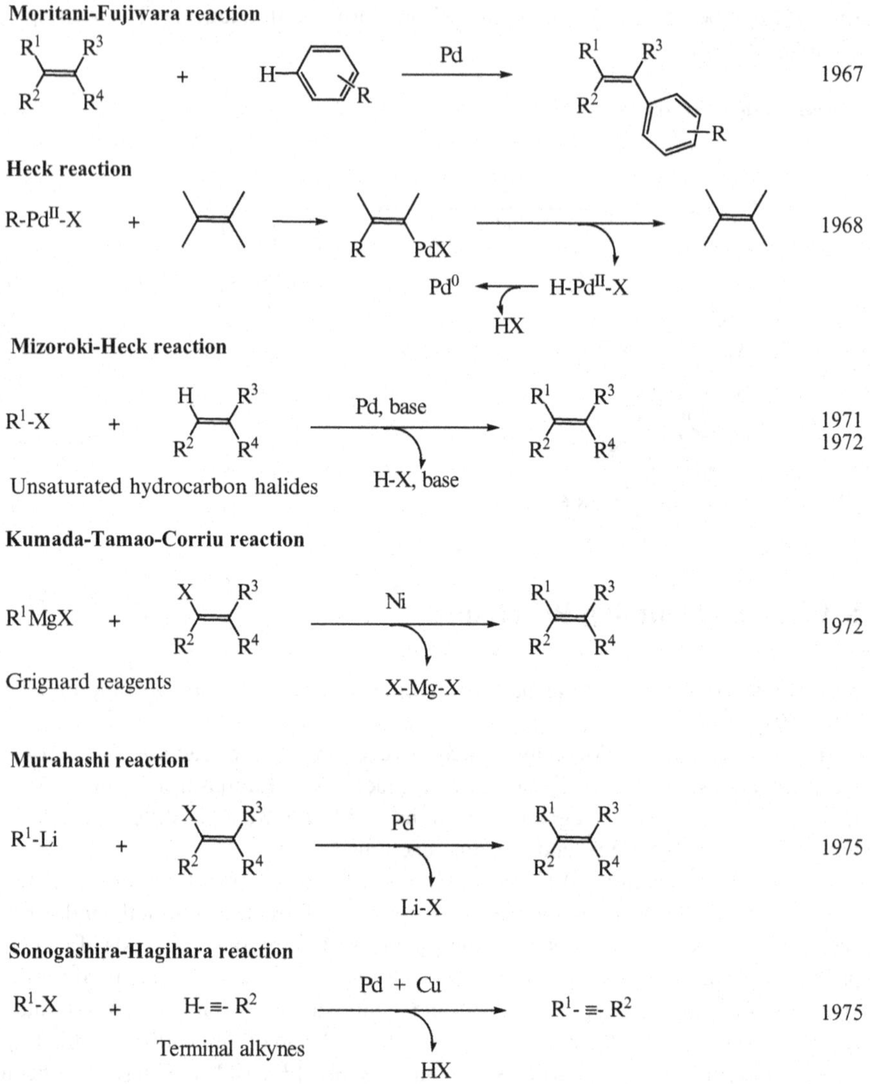

Moritani-Fujiwara reaction

$$R^1 \quad R^3 \qquad + \qquad H- \langle\rangle_R \qquad \xrightarrow{Pd} \qquad R^1 \quad R^3 \qquad\qquad 1967$$

$$R^2 \quad R^4 \qquad\qquad\qquad\qquad\qquad\qquad R^2 \qquad \langle\rangle_R$$

Heck reaction

$$\text{R-Pd}^{II}\text{-X} \quad + \qquad \rangle\!=\!\langle \quad \longrightarrow \quad \underset{R \quad PdX}{\rangle\!=\!\langle} \quad \longrightarrow \quad \rangle\!=\!\langle \qquad\qquad 1968$$

$$Pd^0 \longleftarrow \text{H-Pd}^{II}\text{-X}$$

$$HX$$

Mizoroki-Heck reaction

$$R^1\text{-X} \quad + \quad \underset{R^2 \quad R^4}{\overset{H \quad R^3}{\rangle\!=\!\langle}} \quad \xrightarrow{Pd,\ base} \quad \underset{R^2 \quad R^4}{\overset{R^1 \quad R^3}{\rangle\!=\!\langle}} \qquad \begin{array}{l}1971\\1972\end{array}$$

Unsaturated hydrocarbon halides $\qquad$ H-X, base

Kumada-Tamao-Corriu reaction

$$R^1 MgX \quad + \quad \underset{R^2 \quad R^4}{\overset{X \quad R^3}{\rangle\!=\!\langle}} \quad \xrightarrow{Ni} \quad \underset{R^2 \quad R^4}{\overset{R^1 \quad R^3}{\rangle\!=\!\langle}} \qquad\qquad 1972$$

Grignard reagents $\qquad\qquad$ X-Mg-X

Murahashi reaction

$$R^1\text{-Li} \quad + \quad \underset{R^2 \quad R^4}{\overset{X \quad R^3}{\rangle\!=\!\langle}} \quad \xrightarrow{Pd} \quad \underset{R^2 \quad R^4}{\overset{R^1 \quad R^3}{\rangle\!=\!\langle}} \qquad\qquad 1975$$

Li-X

Sonogashira-Hagihara reaction

$$R^1\text{-X} \quad + \quad H\text{-}\equiv\text{-}R^2 \quad \xrightarrow{Pd\ +\ Cu} \quad R^1\text{-}\equiv\text{-}R^2 \qquad\qquad 1975$$

Terminal alkynes $\qquad\qquad$ HX

Fig. 8.4 Cross-coupling reactions [100, 101]

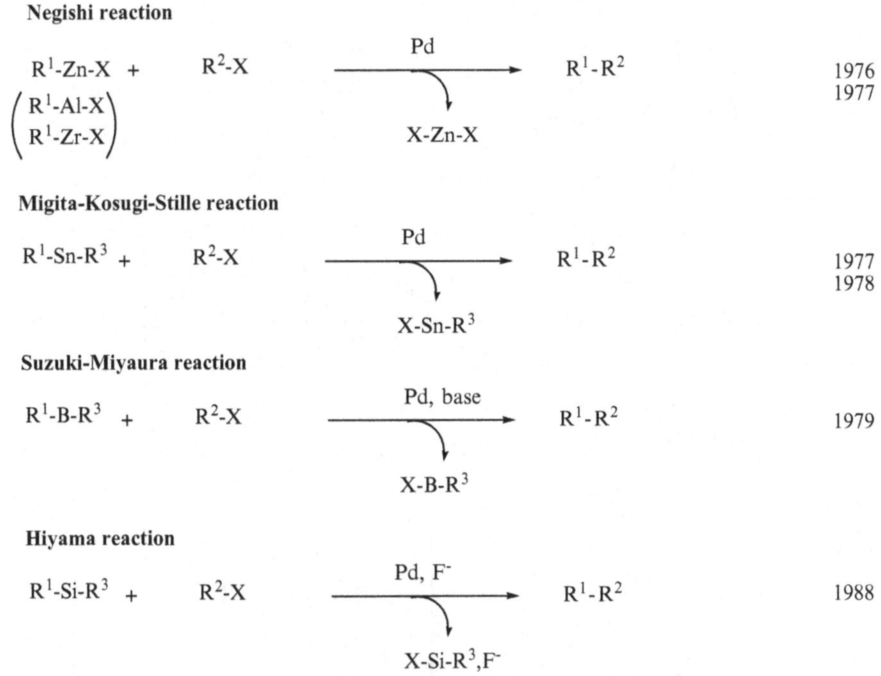

Negishi reaction

$$R^1\text{-Zn-X} + \qquad R^2\text{-X} \xrightarrow{\quad Pd \quad} R^1\text{-}R^2$$

$$\begin{pmatrix} R^1\text{-Al-X} \\ R^1\text{-Zr-X} \end{pmatrix}$$

$$X\text{-Zn-X}$$

1976
1977

Migita-Kosugi-Stille reaction

$$R^1\text{-Sn-}R^3 + \qquad R^2\text{-X} \xrightarrow{\quad Pd \quad} R^1\text{-}R^2$$

$$X\text{-Sn-}R^3$$

1977
1978

Suzuki-Miyaura reaction

$$R^1\text{-B-}R^3 + \qquad R^2\text{-X} \xrightarrow{\quad Pd,\ base \quad} R^1\text{-}R^2$$

$$X\text{-B-}R^3$$

1979

Hiyama reaction

$$R^1\text{-Si-}R^3 + \qquad R^2\text{-X} \xrightarrow{\quad Pd,\ F^- \quad} R^1\text{-}R^2$$

$$X\text{-Si-}R^3, F^-$$

1988

Fig. 8.4 (continued)

"Practical Coupling Reactions of Alkynes," by Mori [109]
In 2005, "The Development of Efficient Catalysts for Palladium-Catalyzed Coupling Reactions of Aryl Halides," by Zapf and Beller [104].
In 2008, "Application of Cyclopalladated Compounds as Catalysts for Heck and Sonogashira Reactions," by Nájera and Alonso [110]. "Palladacyclic Precatalysts for Suzuki Coupling, Buchwald-Hartwig Amination and Related Reactions" Bedford [111]
In 2010, "Oxime-Derived Palladacycles as Sources of Palladium Nanoparticles," by Alonso and Náuera [103]
In 2011, "Palladium-Catalyzed Oxidation Heck Reactions," by Su and Jiao [102]. "Recent Advances in Sonogashira Reactions," by Chinchilla and Nájera [112]. "Nanocatalysts for Suzuki Cross-Coupling reactions," by Fihri, Bouhrara, Nekoueishahraki, Basset, and Polshettiwar [113]

Palladium compounds are mainly used as catalysts, but Ni, Ir, Ru, and Pt compounds are also used as catalysts. Some of these catalysts are shown in Fig. 8.5. These catalysts are mainly palladium compounds containing a phosphorus- or nitrogen-coordinating atom.

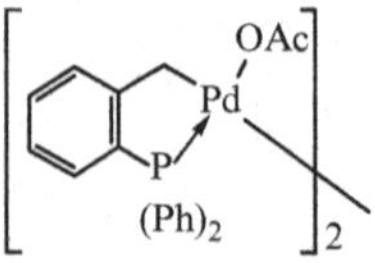

8.69 [108] 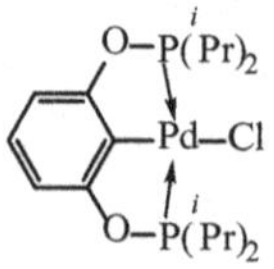   **8.70** [105]

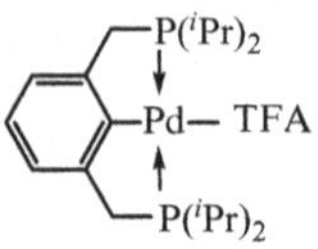

TFA = Trifluoroacetate, (CF$_3$COO$^-$)

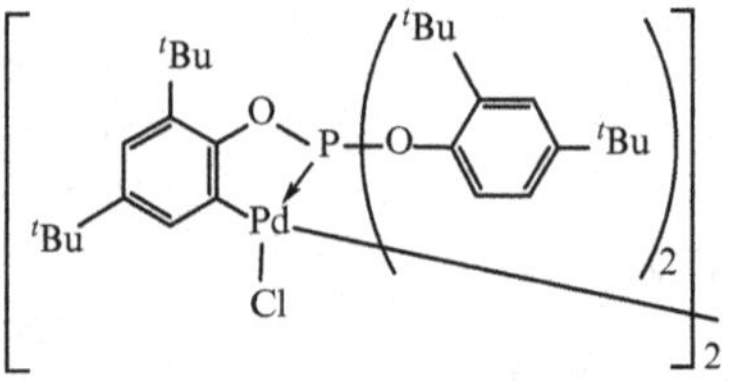

8.72 [105]

8.71 [30]

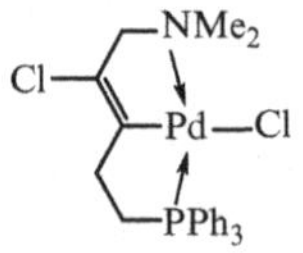

8.73 [104]

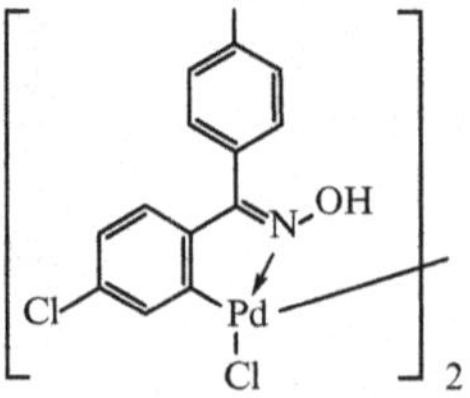

8.74 [114]

Fig. 8.5 Catalysts for cross-coupling reactions

In 1999, cyclopalladated tolylphosphine compounds were reported to show highly catalytic activities, not only for Heck reactions of which the turnover numbers (TONs) are up to 1,000,000, for example, but also for all other metal-catalyzed cross-coupling reactions [115, 116]. In 2005, Heck reactions with naphthyl phosphines [117] or *N*-heterocyclic carbene phosphapalladacycles [118] were also reported to show highly catalytic activities, as evidenced by their TONs of up to 300,000 and 10,800, respectively.

Among Suzuki reactions, palladacycle catalysts exhibit very high TONs, as shown in Eq. (8.24) and Table 8.2 [105, 119–121]. Most notably, phosphite palladacycles **8.76** show the highest TON, 10^8 [105, 120]. *N*-(2-(Diphenylphosphino) phenyl)-2,6-diisopropylaniline palladacycles **8.77** do not have a metal–carbon bond, but they also show high catalytic activities [121].

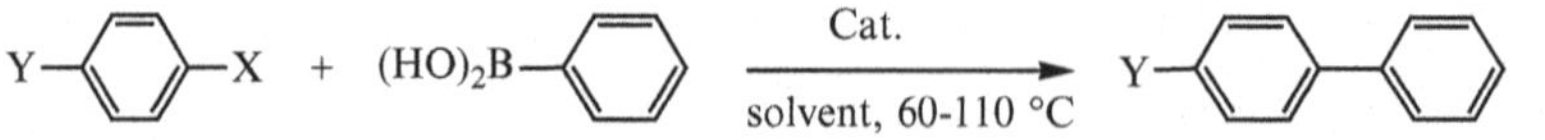

$$Y-\!\!\!\bigcirc\!\!\!-X \ + \ (HO)_2B-\!\!\!\bigcirc \ \xrightarrow[\text{solvent, 60-110 °C}]{\text{Cat.}} \ Y-\!\!\!\bigcirc\!\!\!-\!\!\!\bigcirc \qquad (8.24)$$

Table 8.2 Reaction conditions and catalyst studies on the Suzuki coupling of p-halobenzene derivatives and phenylboronic acid [105, 119–121]

Catalyst (mol%)	Y(X)	Solvent	$T(°C)$	Time (h)	% yield	TON	TOF (h^{-1})
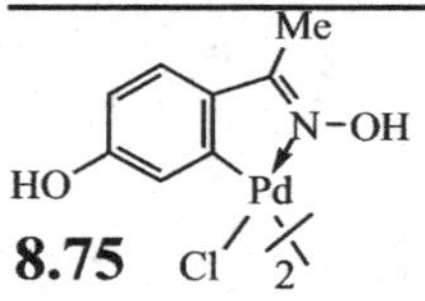 **8.75** (10^{-3})	CH₃CO (Br)	MeOH–H₂O(3/1)	60	2	100	100,000	50,000
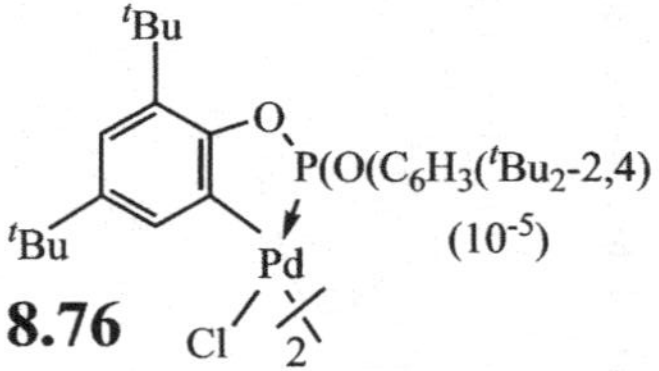 **8.76** (10^{-5})	CH₃CO (Br)	Toluene	110	18	q.y.	100,000,000	
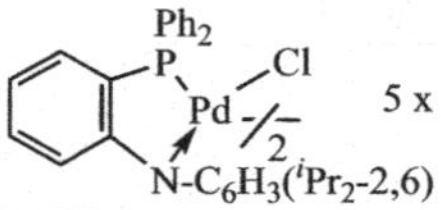 **8.77** 5 x 10^{-6}	MeO (Br)	Dioxane	110	12	100	10,000,000	
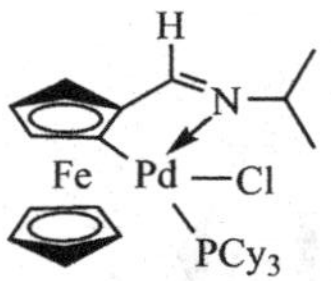 **8.78** 10^{-2}	CH₃CO (Cl)	Dioxane	100	10	98.7		

Recently, many cross-coupling reactions using cyclometalation reaction five-membered ring products as catalysts have been reported, as shown in Fig. 8.6, and two cross-coupling reactions with these catalysts are also shown below:

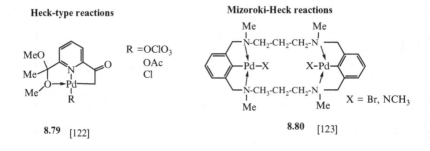

Heck-type reactions

8.79 [122]

Mizoroki-Heck reactions

X = Br, NCH₃

8.80 [123]

Suzuki-Miyaura cross-coupling reactions

8.81 [124]

8.82 [124]

8.83 [124]

Suzuki-Miyaura cross-coupling reactions

8.84 [125]

Suzuki-Miyaura cross-coupling reactions

R¹ = Me,   R² = OH
R¹ = C₆H₄-Cl-p, R² = Cl   8.85 [126]

Suzuki-Miyaura cross-coupling reactions
Sonogashira cross-coupling reactions

8.88
R¹ = Me, R² = H
R¹ = Me, R² = Me
R¹ = C₄H₉, R² = H

Suzuki-Miyaura cross-coupling reactions

8.86 [127]

8.87 [128]

Fig. 8.6 Recent cross-coupling reactions with cyclometalated five-membered products as catalysts

Sonogashira cross-coupling reactions

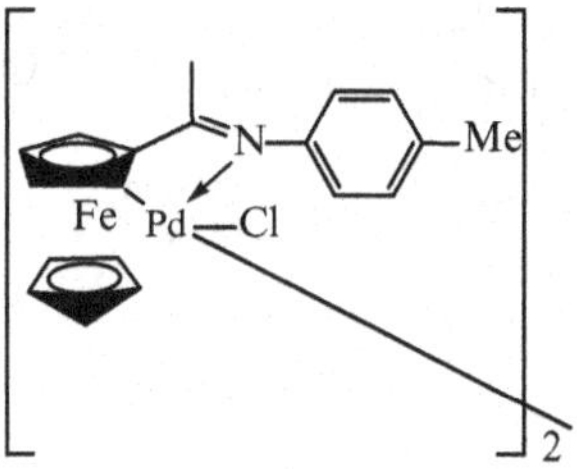

8.89 [129]

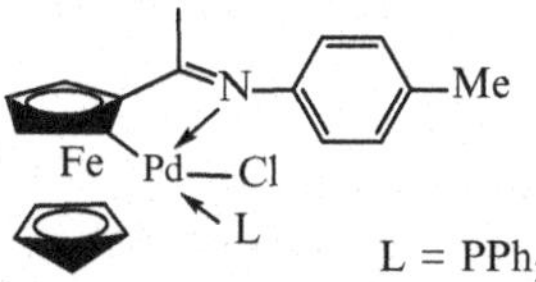

L = PPh$_3$, PCy$_3$

8.90 [129]

C-S cross-coupling reactions

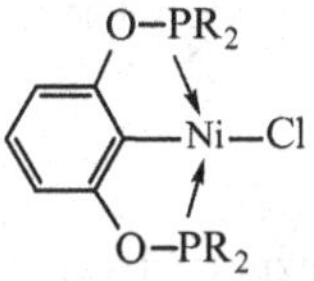

R = Me, *i*-Pr, *t*-Bu, Ph

8.91 [130,131]

C-C cross-coupling reactions

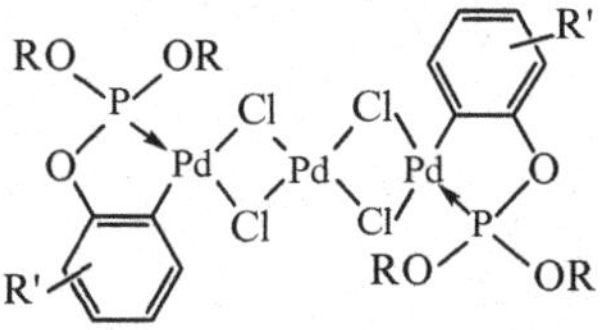

R	R'	
Ph	H	**8.92**
m-MeC$_6$H$_4$	*m*-Me	
o-MeC$_6$H$_4$	*o*-Me	[132]
2,4-*t*-Bu$_2$C$_6$H$_3$	2,4-*t*-Bu$_2$	

Stille cross-coupling reactions
Hiyama cross-coupling reactions

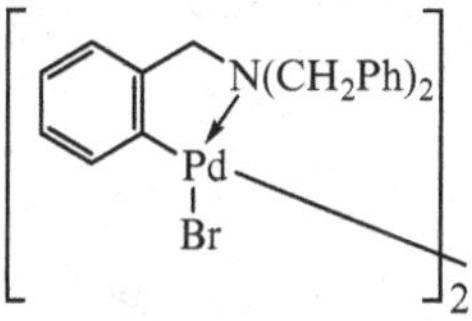

8.93 [133,134]

Fig. 8.6 (continued)

Ohmura et al. [135, 136] reported that enantiospecific Suzuki–Miyaura coupling reaction of enantioenriched α-(acetylamino)benzylboronic esters with aryl bromides can be switched by the choice of acidic additives in the presence of a Pd/XPhos catalyst system. Highly enantiospecific, invertive C–C bond formation takes place with the use of phenol as an additive as shown in Eq. (8.25). In contrast, high enantiospecificity for retention of configuration is attained in the presence of Zr(O^{i}Pr)$_4$,^{i}PrOH as an additive as shown in Eq. (8.26). The reaction proceeds via a cyclization five-membered ring intermediate of the boron atom.

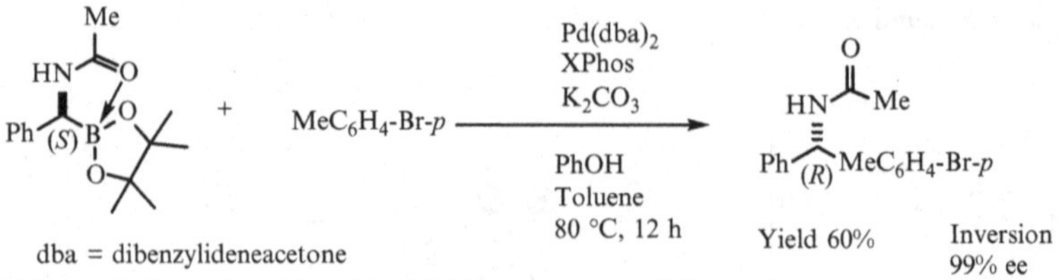

$$
\text{(structure)} \quad + \quad MeC_6H_4\text{-}Br\text{-}p \quad \xrightarrow[\substack{PhOH \\ Toluene \\ 80\ ^\circ C,\ 12\ h}]{\substack{Pd(dba)_2 \\ XPhos \\ K_2CO_3}} \quad \text{(structure)} \tag{8.25}
$$

Yield 60%   Inversion 99% ee

dba = dibenzylideneacetone

XPhos = 2-(Dicyclohexylphosphino)-2',4',6'-triisopropyl-1,1'-biphenyl

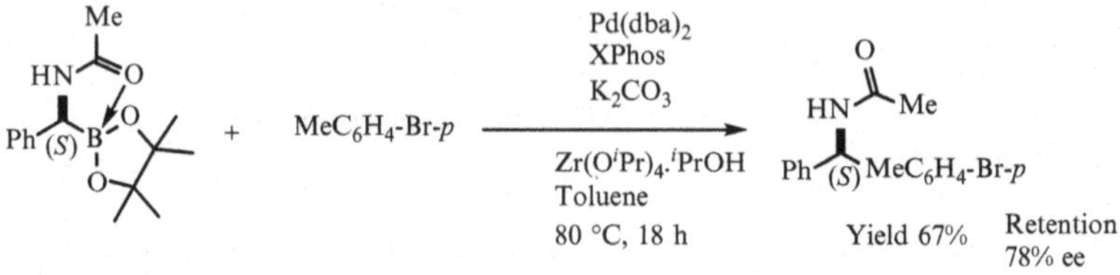

$$
\text{(structure)} \quad + \quad MeC_6H_4\text{-}Br\text{-}p \quad \xrightarrow[\substack{Zr(O^iPr)_4 \cdot {}^iPrOH \\ Toluene \\ 80\ ^\circ C,\ 18\ h}]{\substack{Pd(dba)_2 \\ XPhos \\ K_2CO_3}} \quad \text{(structure)} \tag{8.26}
$$

Yield 67%   Retention 78% ee

Chatani et al. [137] have reported on nickel-catalyzed Suzuki–Miyaura reactions with aryl fluorides as the cyclometalation substrates, as shown in Eqs. (8.27) and (8.28).

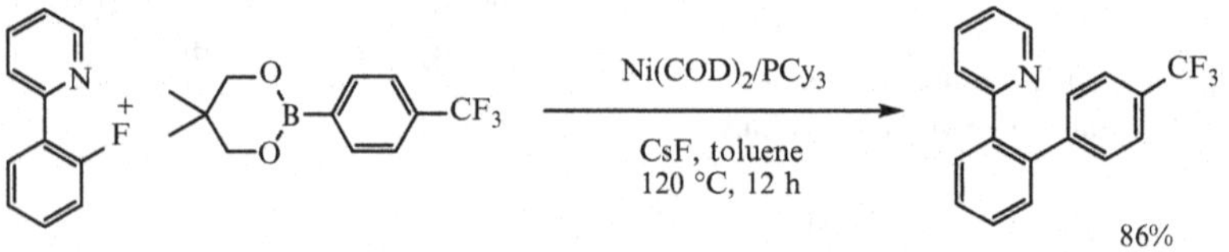

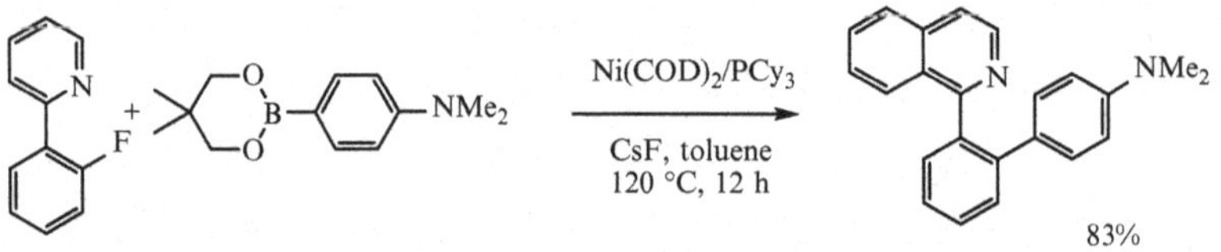

$$
\text{(structure)} \quad + \quad \text{(structure)} \quad \xrightarrow[\substack{CsF,\ toluene \\ 120\ ^\circ C,\ 12\ h}]{Ni(COD)_2/PCy_3} \quad \text{(structure)} \tag{8.27}
$$

86%

$$
\text{(structure)} \quad + \quad \text{(structure)} \quad \xrightarrow[\substack{CsF,\ toluene \\ 120\ ^\circ C,\ 12\ h}]{Ni(COD)_2/PCy_3} \quad \text{(structure)} \tag{8.28}
$$

83%

These cross-coupling reactions are coupling reactions of aryl compounds with aryl compounds, aryl compounds with olefins, olefins with olefins, etc., as shown in Fig. 8.4. The cyclometalation reactions of aryl substrates with aryl compounds give aryl–aryl coupling reaction products, as shown in Eq. (8.29). Stanford et al. [138]

have reported one of the coupling reactions among these reactions. Therefore, these cyclometalation reactions are also coupling reactions.

(8.29)

8.5 Polymerizations

Among polymerization catalysts, Ziegler catalysts (Et_3Al and $TiCl_4$), Natta catalysts (Et_3Al and $TiCl_3$), and metallocene catalysts (Cp_2M(Ti, Zr, Hf, Fe) and $-(O-Al(R))_n-$(R=Me, Et)) are used as vinyl polymerization catalysts. Ca, Hg, Zn, and Cd compounds and Ti, Ge, Sn, Pb, and Sb compounds are used for ester exchange and their polyester polymerization catalysts, respectively.

Recently, however, organometallic intramolecular-coordination five-membered ring compounds are also used as polymerization catalysts. It is considered that the central atoms of organometallic intramolecular-coordination five-membered ring compounds may act as active centers of polymerization catalysts. This is because the metal atoms are activated for the reasons discussed in Chap. 6, that is, for the following three reasons: first is the coordination of lone electron pair of the hetero atom to the metal atom; second, the chelate effect caused by the formation of a five-membered ring, and third, ligands bonding with the central atom, such as a halogen atom, hetero atom group, carbonyl, cyclopentadienyl, aryl, and π-allyl group.

Polymerization reactions of compounds such as ethylene, ethylene/propylene, isoprene, pyridylamino, butadiene, phenylacetylene, lactide, and ε-caprolactone and their catalysts of cyclometalated products are shown in Fig. 8.7.

Cyclometalation products having transition metal compounds of Y, Sc, Nd, Gd, Sc, and Er (Group 3), Zr and Hf (Group 4), Cr (Group 6), Fe and Ru (Groups 8), Co (Group 9), and Pd (Group 10) and main group metals of Al (Group 13) have been used as polymerization catalysts for monoolefins, dienes, conjugated dienes, aryl acetylenes, amides, etc., as shown in Fig. 8.7.

Ethylene polymeriation

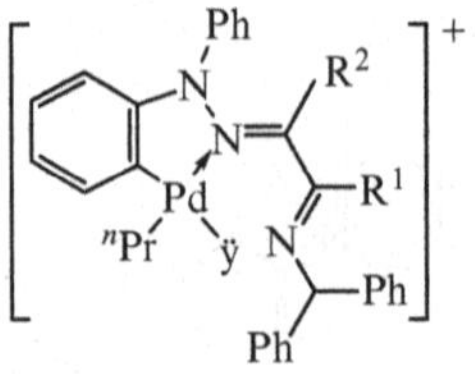

8.94

R¹,R²

[139] a H, H

b Me, H

c Me, Me

Ethylene polymeriation

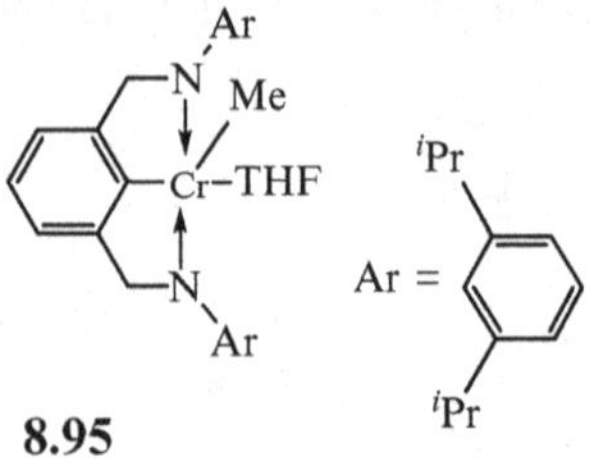

8.95

[140]

Ethylene polymeriation

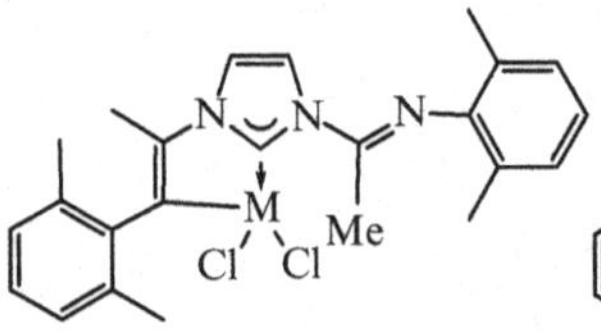

8.96 M = Fe, Co

[141]

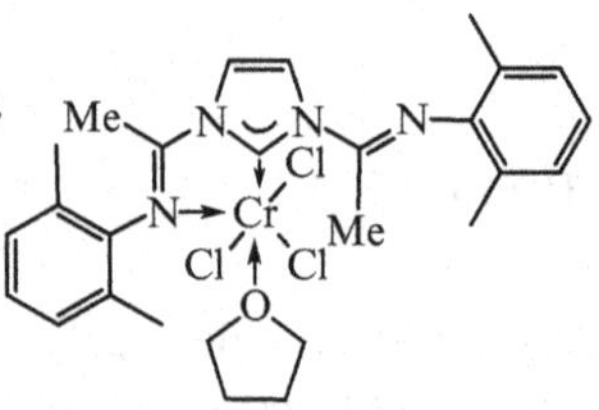

8.97

[141]

Etylene/1-octene and propylene polymerization

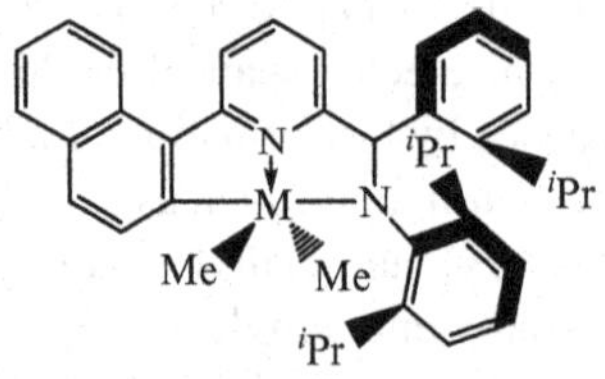

M = Hf, Zr

8.98 [142]

Isoprene polymerization

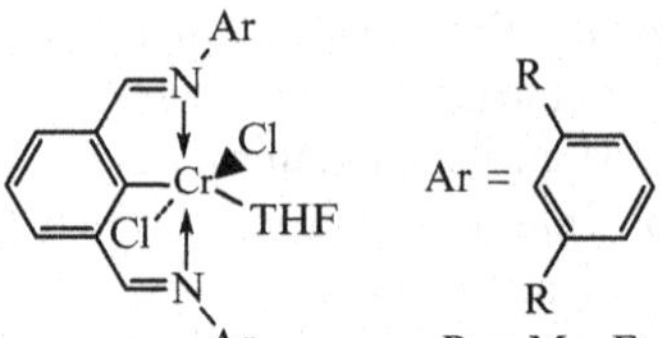

R = Me, Et, ⁱPr

8.99

[143]

Fig. 8.7 Polymerization reactions and their catalysts with cyclometalated products

Olefin polymeriation
(1-Hexene, propylene, ethylene/1-octene)

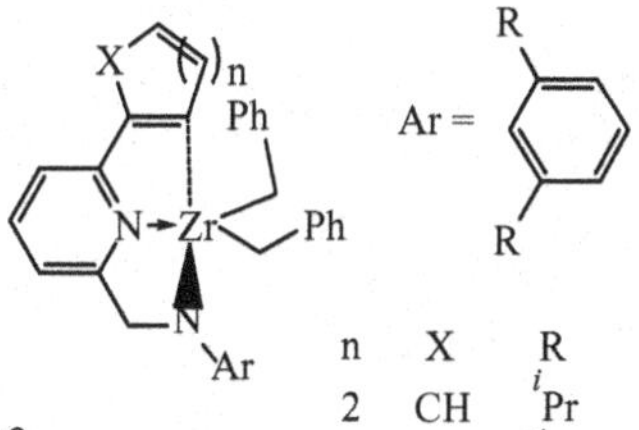

1,3-Butadiene polymerization

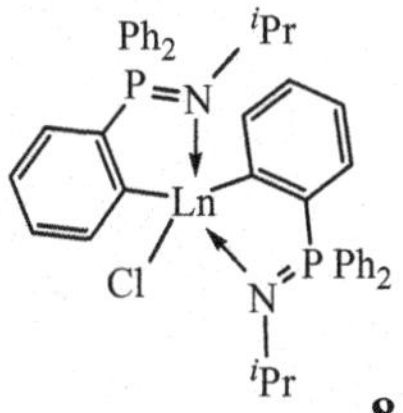

8.100

[144]

n	X	R
2	CH	^{i}Pr
1	O	^{i}Pr
1	S	^{i}Pr
2	CH	Me

Ln = Y, Nd, Gd   **8.101**

[145]

Phenyl acetylene polymeriation

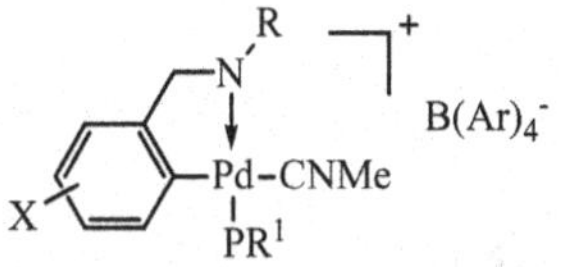

	R	X	R^{1}
8.102	2,6-^{i}Pr-Ph	H	Ph
	2,6-^{i}Pr-Ph	2-Cl	Ph
[146]	2,6-^{i}Pr-Ph	2-Br	Ph
	^{n}Pr	H	Ph
	2,6-^{i}Pr-Ph	2-Cl	Me

L-Lactide polymerization

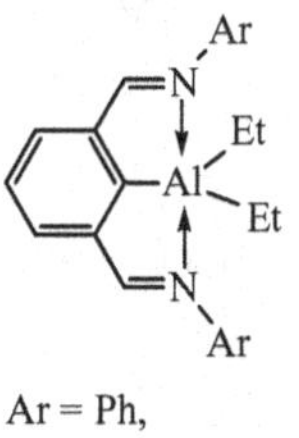

Ar = Ph,
  2,6-Me$_2$-C$_6$H$_3$
  2,6-Et$_2$-C$_6$H$_3$
  2,6-^{i}Pr$_2$-C$_6$H$_3$

8.103

[147]

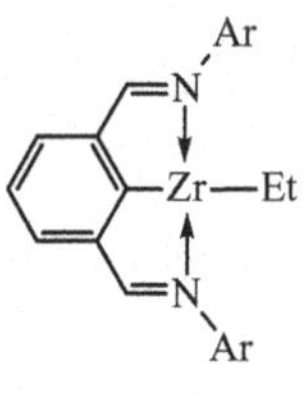

Ar = Ph,
  2,6-Me$_2$-C$_6$H$_3$
  2,6-Et$_2$-C$_6$H$_3$

8.104

[147]

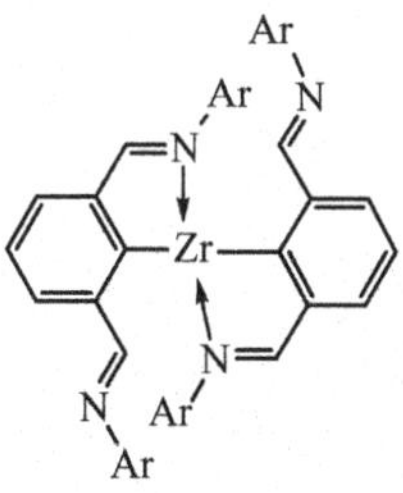

8.105

[147]

Fig. 8.7 (continued)

**Living polymerization of 1,3-conjugated dienes
and copolymerization with ε-caprolactone**

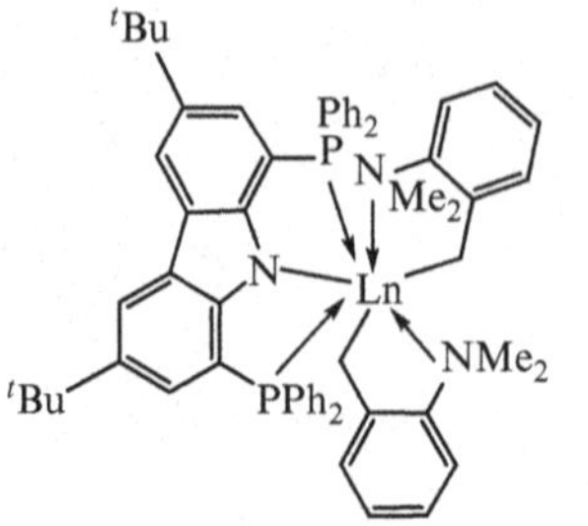

Ln = Y, Sc, Er

8.106

[148]

Polyamidation via dehydrogenation

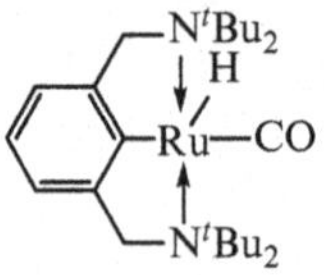

8.107

[149]

Alkene di- and oligomeization

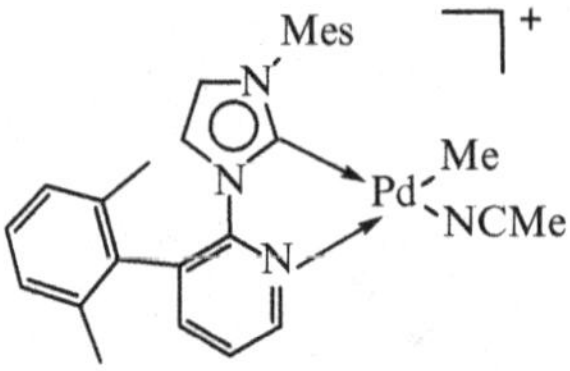

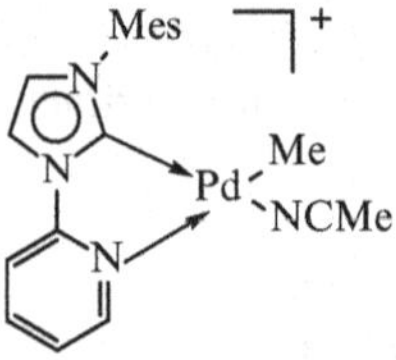

8.108   [150]

8.109   [150]

Fig. 8.7 (continued)

Imine-type cyclopalladated products **8.94** have good activity for ethylene polymerization [139]. The molecular weight and molecular weight distribution of the obtained polymers correspond to single-site catalysts, and the polymers have narrow molecular weight distribution, as shown in Table 8.3.

Al (**8.103**) and Zn (**8.104** and **8.105**) complexes are efficient initiators for L-LACTIDE ring-opening polymerization in the presence of benzyl alcohol, and polymerization reactions take place in an immortal manner, that is, their polydispersity index (PDI) is almost 1 (PDI = 1.02, MW = 13,000–14,000) [147]. Bis(phosphino) carbazole (PNP-carbazolide) rare-earth metal bis(alkyl) complexes (**8.106**) initiated

Table 8.3 Molecular weight, molecular weight distribution, and melting temperature of polyethylene obtained by using the imine cyclometalated palladium catalysts **8.94b** and **8.94c** [139]

Catalyst	Reaction temperature (°C)	T_m (°C)	M_w	M_w/M_n
8.94b	80	136	74,000	1.7
8.94c	40	137	245,000	2.1

cis-1,4-polymerizations of 1,3-conjugated dienes with high activities; the system **8.106** (M=Y)/[Ph$_3$C][(BC$_6$F$_5$)$_4$] displayed especially excellent *cis*-1,4-selectivity (>99 %) and living mode at a broad range of polymerization temperatures (0–80 °C). Remarkably, the living yttrium–polydiene active species could further initiate the ring-opening polymerization of ε-caprolactone to give selectively poly(*cis-1,4-*diene)-*b*-polycaprolactone block copolymer with a controllable molecular weight ($Mn = (10–70) \times 10^4$) and narrow polydispersity (PDI = 1.15–1.47) [148].

8.6 Others

Other reactions employing organometallic intramolecular-coordination five-membered ring compounds as catalysts include reductions and dehydrogenations.

8.6.1 Reductions

Pincer organometallic compounds are reported mainly with regard to two types of compounds, PCP and NCN transition metal complexes [28, 34]. However, ruthenium pincer CNN compounds have also been applied to hydrogen-transfer reductions of ketones.

For example, 6-(4′-methylphenyl)-2-pyridylmethylamine ruthenium pincer compounds **8.110** are highly efficient catalysts in transfer hydrogenation involving 2-propanol to perform the reduction of ketone quantitatively with very low loading and in a short time, as shown in Eq. (8.30) [151].

$$(8.30)$$

8.110

8.6.2　Dehydrogenations and Others

A pincer complex, $(PCPIrH_2$ $(PCP = 2,6-C_6H_3(CH_2P(^tBu_2))_2$ **8.111**, is widely applied for the selective dehydrogenations of alkanes and alkyl groups [34, 151–153]. In dehydrogenations, t-butylethylene acts as a hydrogen acceptor for a substrate in the reaction system with the iridium pincer catalyst **8.111**. Two moles of t-butylethylene convert the iridium monohydride to an active iridium catalyst **8.112**, and t-butylethane and the active iridium catalyst **8.112** react with alkanes or alkyl groups to give the dehydrogenation products, t-butylethylene and the pincer catalyst **8.111**, as shown in Eqs. (8.31) and (8.32), respectively.

Heating the solution of N,N-di(isopropyl)ethylamine, the iridium catalyst **8.111** and t-butylethylene (hydrogen acceptor) at 90 °C in p-xylene for 5 h, for example, yields vinyl N,N-dipropylamine with a 98 % yield, as shown in Eq. (8.33) [153].

$$\text{8.111}\quad PCPIrH_2 \qquad\qquad\qquad \text{8.112}\quad PCH_2PH_2Ir(H)(CH_2=CH\ (t\text{-Bu}))$$

$$(8.31)$$

$$CH_3\text{-}CH_2\text{-}R + (PCP)Ir(H)(CH_2=CH(^tBu)) \longrightarrow CH_2=CH\text{-}R + (PCP)IrH_2 + CH_2=CH\text{-}(^tBu)$$

$$\text{8.112} \qquad\qquad\qquad\qquad \text{8.111}\quad \text{Hydrogen acceptor}$$

$$(8.32)$$

$$(^iPr)_2N\text{-}CH_2CH_3 \quad\xrightarrow[\substack{CH_2=CH\text{-}(t\text{-Bu})\\(200\ m\ mol)}]{\substack{\textbf{8.111}\quad PCPIrH_2\\(10\ m\ mol)\quad 90\ °C,\ 5\ h,\ p\text{-xylene-}d_{10}}}\quad (^iPr)_2N\text{-}CH=CH_2$$

$$(100\ m\ mol) \qquad\qquad\qquad\qquad\qquad\qquad \text{Yield 98\%}$$

$$(8.33)$$

Recently, many articles have also reported on reactions with cyclometalation compounds as the catalysts, including transfer hydrogenations (Eq. (8.34)) [154, 155], N-alkylations of amines with alcohols (Eq. (8.35)) [155], β-alkylation of alcohols (Eq. (8.36)) [155], dehydrogenation of secondary amines (Eq. (8.37))[154, 156], dehydrogenation of primary amines to nitriles (Eq. (8.38)) [154, 157], transfer hydrogenation of ketones, and Oppenauer-type oxidation of ketones (Eqs. (8.39) and (8.40)) [158, 159].

Transfrer hydrogenation [154,155]

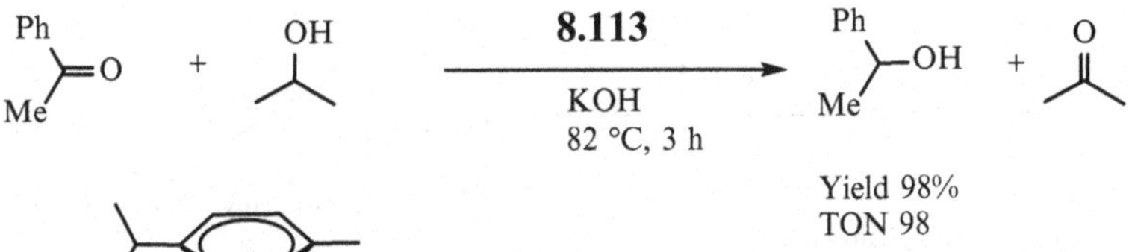

$$\text{(8.34)}$$

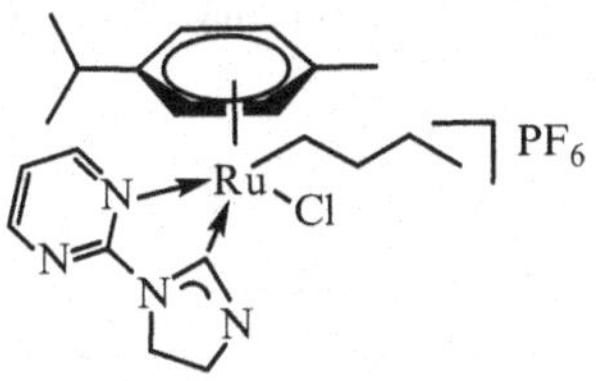

Yield 98%
TON 98

8.113

N-Alkylation of amines with alcohols [155]

$$PhCH_2NH_2 + PhCH_2OH \xrightarrow[\substack{\text{Molecular sieve 4A} \\ \text{Toluene} \\ \text{110 °C, 45 h}}]{\textbf{8.113}} PhCH_2NHCH_2Ph + PhCH=N-CH_2Ph \qquad (8.35)$$

$$86\% \qquad 2\%$$

β-Alkylation of secondary alcohols with primary alcohols

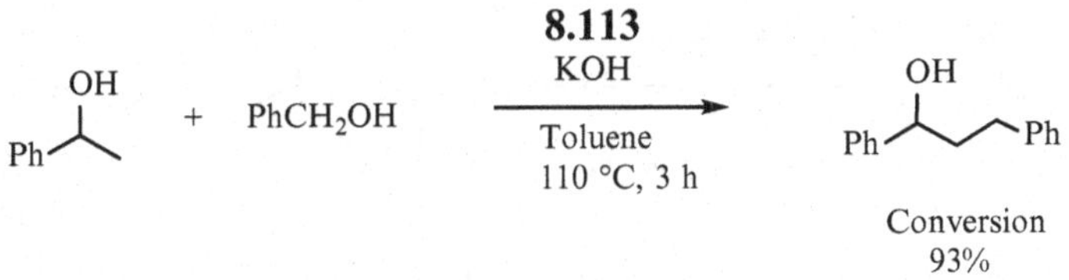

$$\text{(8.36)}$$

Conversion
93%

Dehydrogenation of secondary amines [154,156]

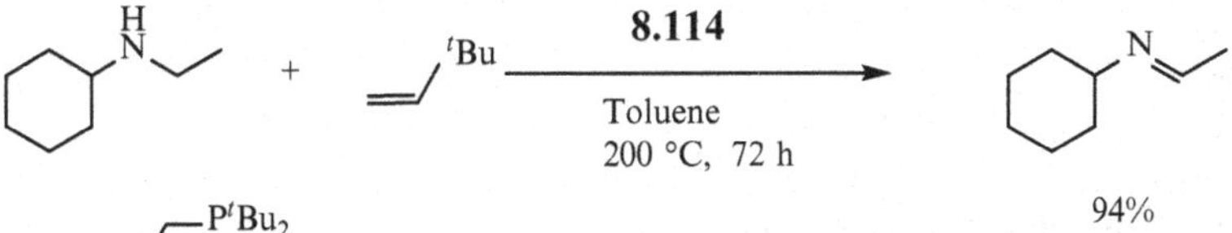

94%　(8.37)

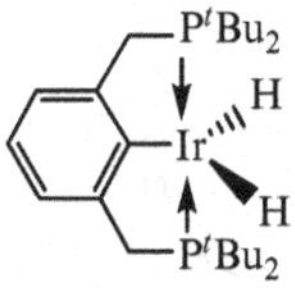

8.114

Dehydrogenation of primary amines to nitriles [154,157]

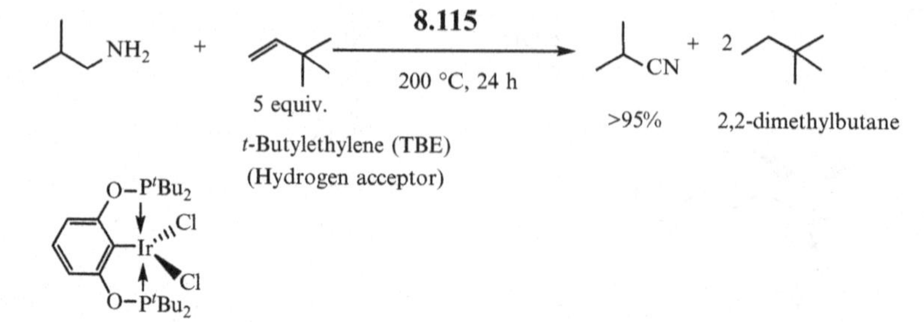
NH2
8.115
200 °C, 24 h
5 equiv.
t-Butylethylene (TBE)
(Hydrogen acceptor)
CN
2
>95%
2,2-dimethylbutane
O−P'Bu2
Ir
Cl
Cl
O−P'Bu2

(8.38)

8.115

In 2012, Z.K. Yu et al. [158] reported on the transfer hydrogenation of ketones and Oppenauer-type oxidation of ketones. Ruthenium compounds containing an unsymmetrical NNC ligand, two phenylphosphines, and a chlorine atom **8.116** have exhibited highly catalytic activity for both reactions, as shown in Eqs. (8.39) and (8.40).

Transfrer hydrogenation of ketones [158]

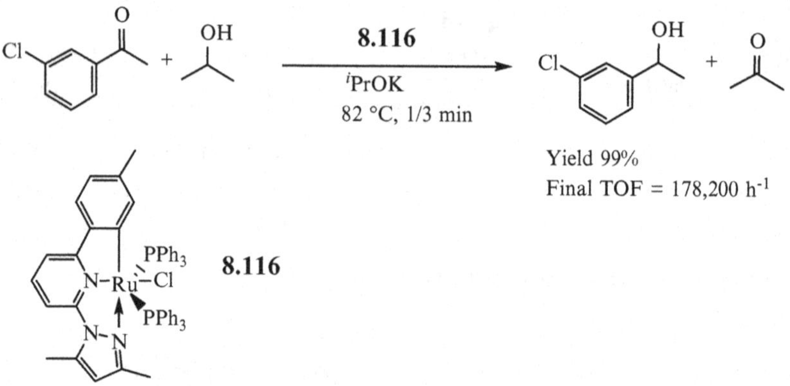
O
Cl
OH
8.116
iPrOK
82 °C, 1/3 min
OH
Cl
O
Yield 99%
Final TOF = 178,200 h-1
PPh3
N−Ru−Cl
PPh3
N−N
8.116

(8.39)

Oppenauer oxidation of alcohols [158]

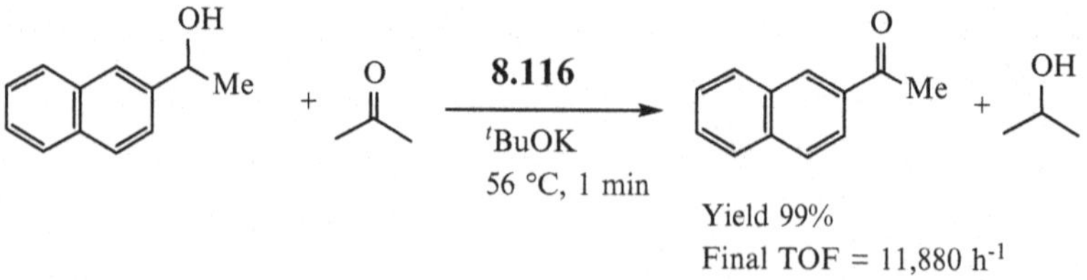
OH
Me
O
8.116
tBuOK
56 °C, 1 min
O
Me
OH
Yield 99%
Final TOF = 11,880 h-1

(8.40)

Other reactions such as alkane dehydrogenation [159, 160], decarbonylation reactions [161], cyclization of alkynoic acids [162, 163], three-component coupling reactions of boronic acids, allenes and imines [164], fluorenone synthesis by sequential reactions of 2-bromobenzaldehydes with arylboronic acids [165], and hydrosilylation reactions [166, 167] using cyclometalation compounds as their catalysts have also been reported.

References

1. Omae I (2004) Coord Chem Rev 248:995
2. Omae I (2007) J Organomet Chem 692:2608
3. Wild SB (1997) Coord Chem Rev 166:291
4. Albert J, Granell J, Muller G (2006) J Organomet Chem 691:2101
5. Djukic J-P (2008) Cyclopalladated compounds as resolving agents for racemic mixtures. In: Dupont J, Pfeffer M (eds) Palladacycles. Synthesis, characterization and applications. Wiley-VCH, Weinheim, p 123
6. Spencer J (2008) Other uses of palladacycles in synthesis. In: Dupont J, Pfeffer M (eds) Palladacycles. Synthesis, characterization and applications. Wiley-VCH, Weinheim, p 227
7. Fischer DF, Barakat A, Xin Z, Weiss MW, Peters R (2009) Chem Eur J 15:8722
8. Dunina VV (2011) Cur Org Chem 15:3415
9. Dunina VV, Gorunova ON, Zykov PA, Kochetkov KA (2011) Russ Chem Rev 80:51
10. Nishiyama H, Ito J (2010) Chem Commun 46:203
11. Djukic J-P, Hijazi A, Flack HD, Bernardinelli G (2008) Chem Soc Rev 37:406
12. Nishiyama H (2007) Chem Soc Rev 36:1133
13. Yu J-Q, Giri R, Chen X (2006) Org Biomol Chem 4:4041
14. Severin K, Bergs R, Beck W (1998) Angew Chem Int Ed 37:1634
15. Ohshima T, Kawabata T, Takeuchi Y, Kakinuma T, Iwasaki T, Yonezawa T, Murakimi H, Nishiyama H, Mashima K (2011) Angew Chem Int Ed 50:6296
16. Hollis TK, Overman LE (1999) J Organomet Chem 576:290
17. Overman LE, Owen CE, Pavan MM, Richards CJ (2003) Org Lett 5:1809
18. Anderson CE, Overman LE (2003) J Am Chem Soc 125:12412
19. Prasad RS, Anderson CE, Richards CJ, Overman LE (2005) Organometallics 24:77
20. Wu Y, Huo S, Gong J, Cui X, Ding L, Ding K, Du C, Liu Y, Song M (2001) J Organomet Chem 637–639:27
21. Wu YJ, Cui XL, Du CX, Wang WL, Guo RY, Chen RF (1998) J Chem Soc Dalton Trans 3727
22. Cui XL, Wu YJ, Yang LR (1999) Chin Chem Lett 10:127
23. Leung P-H, He G, Lang H, Liu A, Loh S-K, Selvaratnam S, Mok KF, White AJP, Williams DJ (2000) Tetrahedron 56:7
24. Yeo W-C, Vittal JJ, White AJP, Williams DJ, Leung P-H (2001) Organometallics 20:2167
25. Qin Y, White AJP, Williams DJ, Leung P-H (2002) Organometallics 21:171
26. Motoyama Y, Nishiyama H (2003) Yuki Gosei Kagaku Kyokaishi 61:343
27. Motoyama Y, Koga Y, Nishiyama H (2001) Tetrahedron 57:853
28. van der Boom ME, Milstein D (2003) Chem Rev 103:1759
29. Motoyama Y, Shimozono K, Aoki K, Nishiyama H (2002) Organometallics 21:1684
30. Singleton JT (2003) Tetrahedron 59:1837
31. Denmark SE, Stavenger RA, Faucher A-M, Edwards JP (1997) J Org Chem 62:3375

32. Motoyama Y, Okano M, Narusawa H, Makihara N, Aoki K, Nishiyama H (2001) Organometallics 20:1580
33. Cohen R, Milstein D, Martin JML (2004) Organometallics 23:2336
34. Albrecht M, van Koten G (2001) Angew Chem Int Ed 40:3750
35. Canty AJ, Denney MC, van Koten G, Skelton BW, White AH (2004) Organometallics 23:5432
36. Motoyama Y, Koga Y, Kobayashi K, Aoki K, Nishiyama H (2002) Chem Eur J 8:2968
37. Stark MA, Richards CJ (1997) Tetrahedron Lett 38:5881
38. Stark MA, Jones G, Richards CJ (2000) Organometallics 19:1282
39. Fossey JS, Richards CJ (2004) Organometallics 23:367
40. Hosokawa S, Ito J, Nishiyama H (2010) Organometallics 29:5773
41. Ito J, Ujiie S, Nishiyama H (2009) Organometallics 28:630
42. Ito J, Teshima T, Nishiyama H (2012) Chem Commun 48:1105
43. Matsumura K, Arai N, Hori K, Saito T, Sayo N, Ohkuma T (2011) J Am Chem Soc 133:10696
44. Cheow YL, Pullarkat SA, Li Y, Leung P-H (2012) J Organomet Chem 696:4215
45. Tran DN, Cramer N (2011) Angew Chem Chem Ed 50:11098
46. Weiss M, Frey W, Peters R (2012) Organometallics 31:6365
47. Samoiłowicz C, Bieniek M, Grela K (2009) Chem Rev 109:3708
48. Chauvin Y (2006) Angew Chem Chem Ed 45:3740
49. Grubbs RH (2006) Angew Chem Chem Ed 45:3760
50. Schrock RR (2006) Angew Chem Chem Ed 45:3748
51. Garber SB, Kingsbury JS, Gray BL, Hoveyda AH (2000) J Am Chem Soc 122:8168
52. Yao Q, Zhang Y (2003) Angew Chem Int Ed 42:3395
53. Yao Q, Sheets M (2005) J Organomet Chem 690:3577
54. Clavier H, Audic N, Guillemin J-C, Mauduit M (2005) J Organomet Chem 690:3585
55. Roye F, Vilain C, Elkaïm L, Grimaud L (2003) Org Lett 5:2007
56. Harrity JPA, La DS, Cefalo DR, Visser MS, Hoveyda AH (1998) J Am Chem Soc 120:2343
57. Kingsbury JS, Harrity JPA, Bonitatebus PJ Jr, Hoveyda AH (1999) J Am Chem Soc 121:791
58. Mori M (2005) Yuki Gosei Kagaku Kyokaishi 63:423
59. Fukuyama T, Ryu T (2005) Yuki Gosei Kagaku Kyokaishi 63:503
60. Pietraszuk C, Fischer H, Rogalski S, Marciniec B (2005) J Organomet Chem 690:5912
61. Michaelis S, Blechert S (2005) Org Lett 7:5513
62. Mix S, Blechert S (2005) Org Lett 7:2015
63. Vehlow K, Maechling S, Blechert S (2006) Organometallics 25:25
64. Gillingham DG, Kataoka O, Garber SB, Hoveyda AH (2004) J Am Chem Soc 126:12288
65. BouzBouz S, Simmons R, Cossy J (2004) Org Lett 6:3465
66. Advanced Information on the Nobel Prize in Chemistry 2005, 30 November 2005
67. Tabari DS, Tolentino DR, Schrodi Y (2013) Organometallics 32:5
68. Trillo B, Gulías M, López F, Castedo L, Mascareñas JL (2005) J Organomet Chem 690:5609
69. Arisawa M, Terada Y, Theeraladanon C, Takahashi K, Nakagawa M, Nishida A (2005) J Organomet Chem 690:5398
70. Grela K, Harutyunyan S, Michrowska A (2002) Angew Chem Int Ed 41:4038
71. Rouhi AM (2003) Chem Eng News 23:34
72. Volugioukalakis GC, Grubbs RH (2010) Chem Rev 110:1746
73. Miki K, Inoue T, Ohe K (2013) Yuki Gosei Kagaku Kyokaishi 71:601
74. Schrodi Y, Pederson RL (2007) Adv Synth Catal 249:1
75. Grubbs RH (2004) Tetrahedron 60:7117
76. Grubbs RH (ed) (2003) Handbook of metathesis, vol 1–3. Wiley-VCH, Weinheim
77. Hryniewicka A, Kozłowska A, Witkowski S (2012) J Organomet Chem 701:87
78. Hérisson J-I, Chauvin Y (1971) Macromol Chem 141:161

79. Stewart IC, Ung T, Pletnev AA, Berlin JM, Gruffs RH, Schrodi Y (2007) Org Lett 9:1589
80. Murelli RP, Snapper ML (2007) Org Lett 9:1749
81. Ritter T, Hejl A, Wenzel AG, Funk TW, Grubbs RH (2006) Organometallics 25:5740
82. Benitez D, Goddard WA III (2005) J Am Chem Soc 127:12218
83. Haibach MC, Kundu S, Brookhart M, Goldman AS (2012) Acc Chem Res 45:947
84. Čusak A (2012) Chem Eur J 18:5800
85. Vougioukalakis GC (2012) Chem Eur J 18:8868
86. Kress S, Blechert S (2012) Chem Soc Rev 41:4389
87. Peeck LH, Leuthäusser S, Plenio H (2010) Organometallics 29:4339
88. Fustero S, Bello P, Miró J, Simón A, del Pozo C (2012) Chem Eur J 18:10991
89. Herbert MB, Lan Y, Keitz BK, Liu P, Endo K, Day MW, Houk KN, Grubbs RH (2012) J Am Chem Soc 134:7861
90. Khan RKM, O'Brien RV, Torker S, Li B, Hoveyda AH (2012) J Am Chem Soc 134:12774
91. Endo K, Grubbs RH (2011) J Am Chem Soc 133:8525
92. Vieille-Petit L, Clavier H, Linden A, Blumentritt S, Nolan SP, Dorta R (2010) Organometallics 29:775
93. Bieniek M, Samojłowicz C, Sashuk V, Bujok R, Śledź P, Lugan N, Lavigne G, Arlt D, Grela K (2011) Organometallics 30:4144
94. Savka RD, Plenio H (2012) J Organomet Chem 710:68
95. Jimenez LR, Gallon BJ, Schrodi Y (2010) Organometallics 29:3471
96. Mayer C, Gillingham DG, Ward TR, Hilvert D (2011) Chem Commun 47:12068
97. Żukowska K, Szadkowska A, Pazio AE, Woźniak K, Grela K (2012) Organometallics 31:462
98. Barbasiewicz M, Grudzień K, Malinska M (2012) Organometallics 31:3171
99. Peeck LH, Savka RD, Plenio H (2012) Chem Eur J 18:12845
100. Ohshima K (2005) Kagaku 65:12
101. Kagaku's editors (2005) Kagaku 65:18
102. Su Y, Jiao N (2011) Curr Org Chem 15:3362
103. Alonso DA, Nájera C (2010) Chem Soc Rev 39:2891
104. Zapf A, Beller M (2005) Chem Commun 431
105. Beletskaya IP, Cheprakov AV (2004) J Organomet Chem 689:4055
106. Bedford RB (2003) Chem Commun 1787
107. Littke AF, Fu GC (2002) Angew Chem Int Ed 41:4176
108. Herrman WA, Böhm VPW, Reisenger C-P (1999) J Organomet Chem 576:23
109. Mori A (2004) Yuki Gosei Kagaku Kyokaishi 62:355
110. Nájera C, Alonso DA (2008) Application of cyclopalladated compounds as catalysts for Heck and Sonogashira reactions. In: Dupont J, Pfeffer M (eds) Palladacycles. Synthesis, characterization and applications. Wiley-VCH, Weinheim, p 155
111. Bedford RB (2008) Palladacyclic Pre-catalysts for Suzuki coupling, Buchwald-Hartwig amination and related reactions. In: Dupont J, Pfeffer M (eds) Palladacycles. Synthesis, characterization and applications. Wiley-VCH, Weinheim, p 209
112. Chinchilla R, Nájera C (2011) Chem Soc Rev 40:5084
113. Fihri A, Bouhrara M, Nekoueishahraki B, Basset J-M, Polshettiwar V (2011) Chem Soc Rev 40:5181
114. Sun C-L, Li B-J, Shi Z-J (2010) Chem Commun 46:677
115. Herrmann WA, Böhm VPW, Reisinger C-P (1999) J Organomet Chem 576:23
116. Herrmann WA, Brossmer C, Reisinger C-P, Riermeier TH, Öfele K, Beller M (1997) Chem Eur J 3:1357
117. Frey GD, Reisinger C-P, Herdtweck E, Hermann WA (2005) J Organomet Chem 690:3193
118. Frey GD, Schütz J, Herdweck E, Herrmann WA (2005) Organometallics 24:4416
119. Botella L, Nájera C (2002) J Organomet Chem 663:46

120. Albisson DA, Bedford RB, Lawrence SE, Scully PN (1998) Chem Commun 2095
121. Liang L-C, Chien P-S, Huanf M-H (2005) Organometallics 24:353
122. Juliá-Hernández F, Arcas A, Vincente J (2012) Chem Eur J 18:7780
123. Tsubomura T, Chiba M, Nagai S, Ishihira M, Matsumoto K, Tsukuda T (2011) J Organomet Chem 696:3657
124. Kozlov VA, Aleksanvan DV, Nelyubina YV, Lyssenko KA, Petrovskii PV, Vasil'ev AA, Odinets IL (2011) Organometallics 30:2920
125. Serrano JL, García L, Pérez J, Pérez E, García J, Sánchez G, Sehnal P, Ornellas SD, Williams TJ, Fairlamb IJS (2011) Organometallics 30:5095
126. Cívicos JF, Gholinejad M, Alonso DA, Nájera C (2011) Chem Lett 40:907
127. Yu A, Li X, Peng D, Wu Y, Chang J (2012) Appl Organomet Chem 26:301
128. Błaszczyk I, Gniewek A, Trzeciak AM (2011) J Organomet Chem 696:3601
129. Huang M, Feng Y, Wu Y (2012) Tetrahedron 68:376
130. Zhang J, Medley CM, Krause JA, Guan H (2010) Organometallics 29:6393
131. Xu G, Li X, Sun H (2011) J Organomet Chem 696:3011
132. Błaszczyk I, Gniewek A, Trzeciak AM (2012) J Organomet Chem 710:44
133. Hajipour AR, Karami K, Rafiee F (2012) Appl Organomet Chem 26:27
134. Hajipour AR, Rafiee F (2012) Appl Organomet Chem 26:51
135. Awano T, Ohmura T, Suginome M (2011) J Am Chem Soc 133:20738
136. Huang X, Anderson KW, Zim D, Jiang L, Klapars A, Buchwald SL (2003) J Am Chem Soc 125:6653
137. Tobisu M, Xu T, Shimasaki T, Chatani N (2011) J Am Chem Soc 133:19505
138. Lyons TW, Hull KL, Sanford MS (2011) J Am Chem Soc 133:4455
139. Pérez MA, Quijada R, Ortega-Jiménez F, Alvarez-Toledano C (2005) J Mol Catal A Chem 226:291
140. McGowan KP, Veige AS (2012) J Organomet Chem 711:10
141. Thagfi JA, Lavoie GG (2012) Organometallics 31:2463
142. Frazier KA, Froese RD, He Y, Klosin J, Theriault CN, Vosejpka PC, Zhou Z, Abboud KA (2011) Organometallics 30:3318
143. Liu Z, Gao W, Liu X, Luo X, Cui D, Mu Y (2011) Organometallics 30:752
144. Luconi L, Rossin A, Tuci G, Tritto I, Boggioni L, Klosin JJ, Theriault CN, Giambastiani G (2012) Chem Eur J 18:671
145. Martinez-Arripe E, Jean-Baptiste-dit-Dominique F, Auffrant A, Goff X-FL, Thuilliez J, Nief F (2012) Organometallics 31:4854
146. Mungwe N, Swarts AJ, Mapolie SF, Westman G (2011) J Organomet Chem 696:3527
147. Liu Z, Gao W, Zhang J, Cui D, Wu Q, Mu Y (2010) Organometallics 29:5783
148. Wang L, Cui D, Hou Z, Li W, Li Y (2011) Organometallics 30:760
149. Zeng H, Guan Z (2011) J Am Chem Soc 133:1159
150. Khlebnikov V, Meduri A, Mueller-Bunz H, Montini T, Fornasiero P, Zangrando E, Milani B, Albrecht M (2012) Organometallics 31:976
151. Jensen CM (1999) Chem Commun 2443
152. Renkema KB, Kissin YV, Goldman AS (2003) J Am Chem Soc 125:7770
153. Zhang X, Fried A, Knapp S, Goldman AS (2003) Chem Commun 2060
154. Dobereiner GE, Crabtree RH (2010) Chem Rev 110:681
155. Gnanamgari D, Sauer ELO, Schley ND, Butler C, Incarvito CD, Crabtree RH (2009) Organometallics 28:321
156. Gu X-Q, Chen W, Morales-Morales D, Jensen CM (2002) J Mol Catal A Chem 189:119
157. Bernskoetter WH, Brookhart M (2008) Organometallics 27:2036
158. Du W, Wang L, Wu P, Yu Z (2012) Chem Eur J 18:11550
159. Gruver BC, Adams JJ, Warner SJ, Arulsamy N, Roddick DM (2011) Organometallics 30:5133
160. Punji B, Emge TJ, Goldman AS (2010) Organometallics 29:2702

161. Adams JJ, Arulsamy N, Roddick DM (2012) Organometallics 31:1439
162. Ogata K, Sasano D, Yokoi T, Isozaki K, Seike H, Yasuda N, Ogawa T, Kurata H, Takaya H, Nakamura M (2012) Chem Lett 41:194
163. Ogata K, Sasano D, Yokoi T, Isozaki K, Seike H, Takaya H, Nakamura M (2012) Chem Lett 41:498
164. Shiota A, Malinakova HC (2012) J Organomet Chem 704:9
165. Liu T-P, Liao Y-X, Xing C-H, Hu Q-S (2011) Org Lett 13:2452
166. Bhattacharya P, Krause JA, Guan H (2011) Organometallics 30:4720
167. Chakraborty S, Krause JA, Guan H (2009) Organometallics 28:582

Chapter 9
Applications of Five-Membered Ring Products in Cyclometalation Reactions for Other Purposes

Abstract Applications of five-membered ring products in cyclometalation reactions for other purposes include organic electronic devices, pharmaceuticals, dye-sensitized solar cells, carbon dioxide utilizations, sensors, dendrimers, liquid crystals, resolving agents, and photosensitizers for hydrogen production.

Keywords Carbon dioxide • Dye-sensitized solar cells • Five-membered ring • OLED • Organic electric devices • Pharmaceuticals • Photosensitizers • Sensors

9.1 Introduction

This monograph has already been reported in Chaps. 7 and 8 on the synthetic and catalytic applications of cyclometalation reactions and their five-membered ring products, respectively.

This chapter reports on other applications, such as organic electronic devices, especially organic light-emitting devices (OLEDs), pharmaceuticals, dye-sensitized solar cells, carbon dioxide utilizations, sensors, and photosensitizers for hydrogen production.

9.2 Organic Electric Devices

Many compounds with luminescent properties have been found among organometallic intramolecular-coordination five-membered ring compounds. An enormous number of articles have been published on organic light-emitting devices (OLEDs), in particular. Many reviews [1–11] and a few books [12–14] have also been

I. Omae, *Cyclometalation Reactions: Five-Membered Ring Products as Universal Reagents*, DOI 10.1007/978-4-431-54604-7_9, © Springer Japan 2014

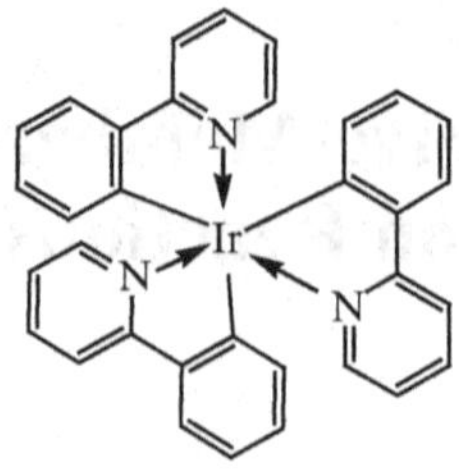

Ir(ppy)$_3$, green

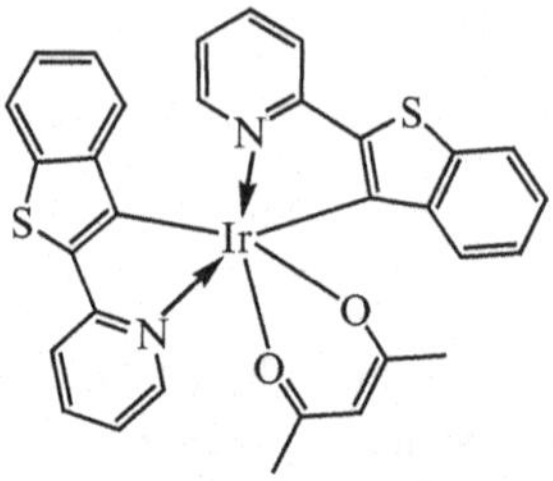

(btp)$_2$Ir(acac), red

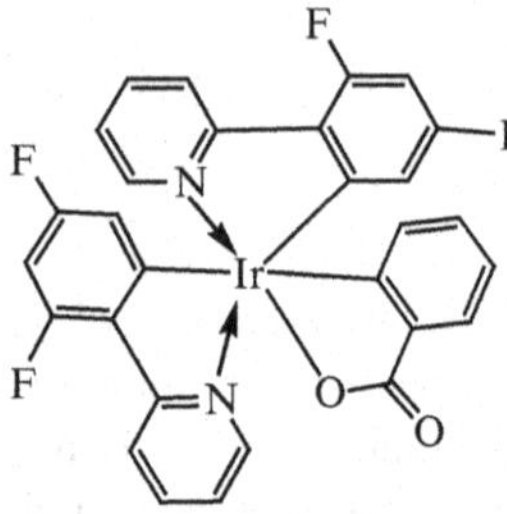

Flrpic, blue

Fig. 9.1 The representative iridium OLED materials for the three primary colors [12]

published. Reports regarding the luminescence of organometallic intramolecular-coordination five-membered ring compounds have been published for many kinds of metal compounds, such as Ir [3, 5, 6, 9, 15–49], Pt [3, 5, 8, 50–56], Pd [52, 54, 55, 57, 58], Ru [3, 5, 7, 59, 60], Re [52, 61], Os [3, 5, 39, 62], Pt/Ir [48], Au [40, 63–67], Hg [68], Al [14], B [24, 53], Eu [14], Tb [14], and Dy [14]. Iridium compounds, especially, are already used as representative iridium OLED materials for the three primary colors, green, Ir(ppy)$_3$, red, (btp)$_2$Ir(acac), and blue, Flrpic, for use in TV screens and mobile phone displays, as shown in Fig. 9.1 [12].

These compounds have the representative substrates shown in Fig. 9.2 and the ancillary ligands shown in Fig. 9.3 [1].

Some other representative compounds exhibiting luminescent properties are shown in Fig. 9.4.

These compounds exhibiting luminescent properties for organic light-emitting devices (OLEDs) not only have 2-phenylpyridine and 2-thiophenylpyridine as substrates but also benzoates, acetylacetonate, and bis-pyridine-, phenanthroline-, imidazole-, and triazole-based ancillary ligands as shown in Figs. 9.2 and 9.3, respectively [1].

Sony exhibited an experimental super high-resolution television (4 K TV) with this type of organic electroluminescent display at an international consumer electronic product trade show (CES) held in the United States from January 8 to 10, 2013.

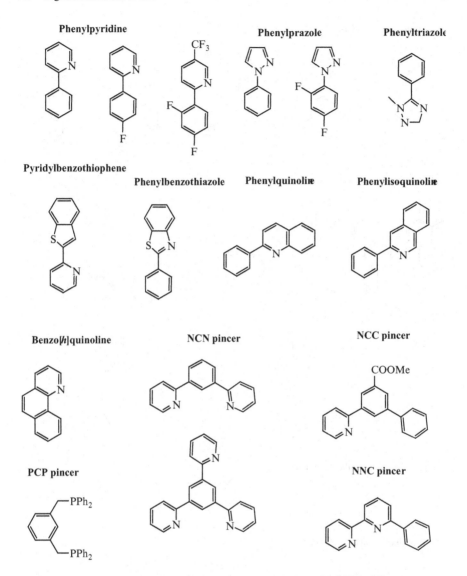

Fig. 9.2 Some of the most widely utilized representative substrates for OLEDs [1]

Iridium metal compounds have shown some remarkable effects recently for luminescent properties such as very high photoluminescent quantum yields [27, 29, 44] and high stability [46]. Two reviews [2, 4] have recently reported on phosphorescent heavy-metal complexes (Re, Ru, Os, Ir, and Rh) as bioimaging probes, including their photophysical properties, cytotoxicity, and cellular uptake mechanisms [2], and on transition metals (Ir, Re, Ru) in fluorescent cell imaging applications such as uptake and toxicity [4], respectively.

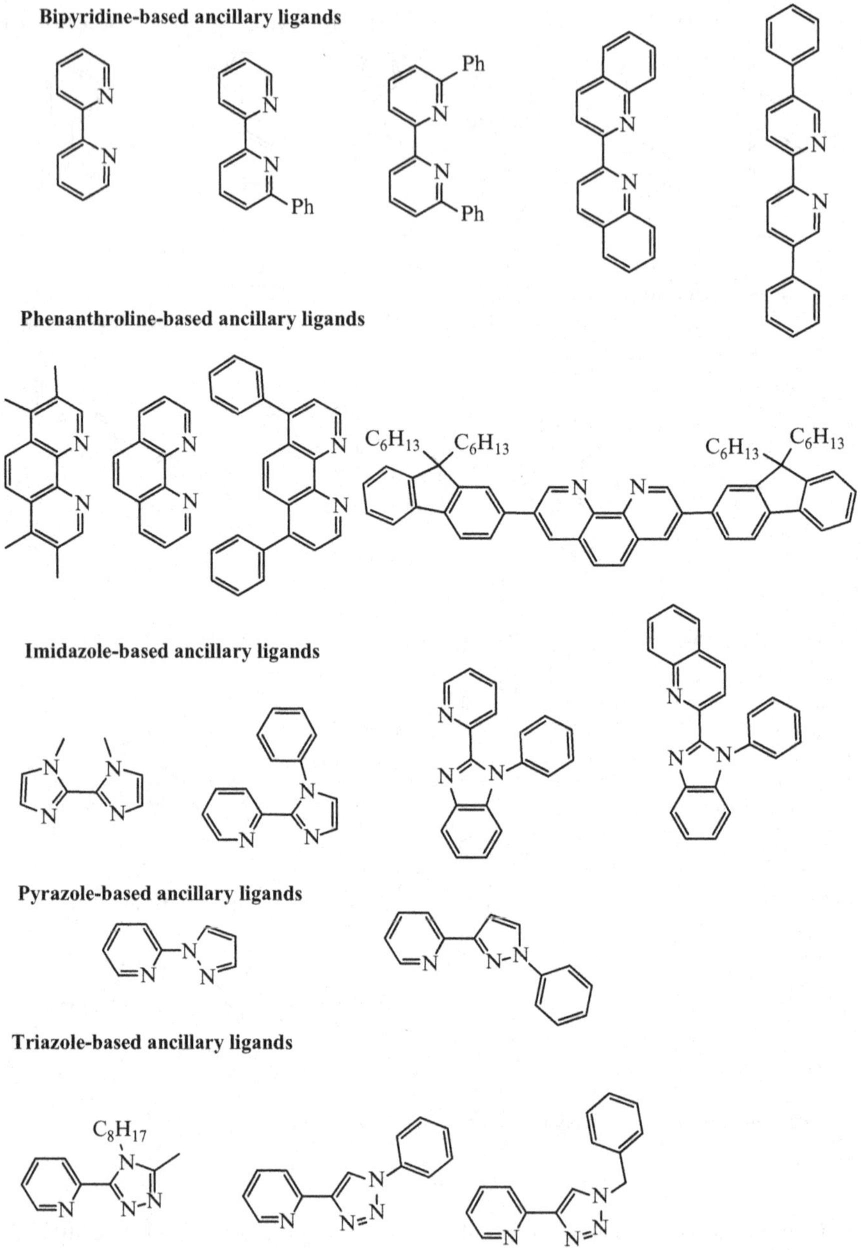

Bipyridine-based ancillary ligands

Phenanthroline-based ancillary ligands

Imidazole-based ancillary ligands

Pyrazole-based ancillary ligands

Triazole-based ancillary ligands

Fig. 9.3 Some of the most widely utilized ancillary ligands for OLEDs [1]

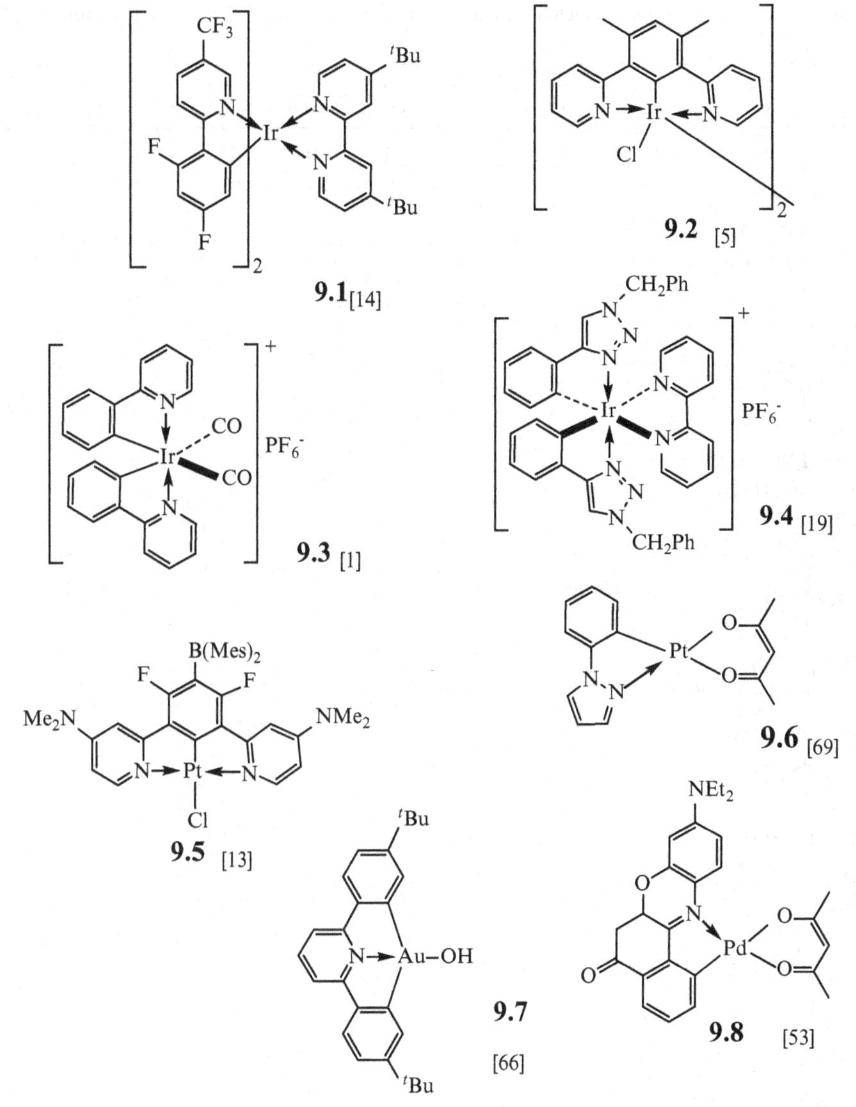

Fig. 9.4 Some other representative compounds exhibiting luminescent properties

9.3 Pharmaceuticals

With respect to medical applications of cyclopalladation compounds, Ryabov [70] reported in 2008 that the first systematic study on the cytotoxicity of cyclometalated palladium compounds was undertaken by Higgins III et al. [71]. The compounds were screened for cytotoxicity against a panel of seven human tumor cell lines, as shown in Table 9.1.

Table 9.1 Cytotoxicity of selected palladacycles against a panel of seven human tumor cell lines [70, 71]

Compound		Cell line/IC$_{50}$ (mg mL^{-1})						
		SW6020	SW1176	SW403	ZR75-1	HT29/219	HT1376	SK-OV-3
(quinoline N→Pd–OAc)	L1	7	8	6	10	6	6	12
	L2	6	6	6	10	6	5	6
L1 = py L2 = H$_2$N(CH$_2$)$_3$OH								
(phenylpyridine N→Pd–Cl/L)	L1	7	8	7	16	5	5	11
	L2	12	12		7	8	7	8
	L3	7	7		7	6	5	6
L1 = H$_2$N(CH$_2$)$_3$OH L2 = Me$_2$CHNH$_2$ L3 = py								
(NMe$_2$ Pd L Cl)	L1	44	44		50	33	8	20
	L2	30	26		9	18	9	10
L1 = py (**9.9**) L 2 = Me$_2$CHNH$_2$ (**9.10**)								
(NMe$_2$ Pd$^+$ L L Cl$^-$)		51	59	61	49	36	35	46
L = Me$_2$CHNH$_2$								
(pyrrole–pyridine N N→Pd py Cl)		10	7		7	6	4	5
(OMe phenanthroline N N→Pd Cl)		5	7	6	7		8	7

Most of the compounds are quite cytotoxic toward the tumor panel, having IC$_{50}$ values in the 10 mg mL^{-1} range. The *N,N*-dimethylbenzylamine derivatives **9.9** and **9.10** were observed to show a 3–5-fold differential response, for example, between the HT1376 and SW6020 cell lines. These palladacycles **9.9** and **9.10** with a tertiary amine aliphatic chelate arm are less cytotoxic than pyridine and quinoline compounds. The nature of auxiliary non-metalated ligands has a minor effect on cytotoxicity. The differential cytotoxicity observed for *N,N*-dimethylbenzylamine derivatives **9.9** and **9.10** suggests that this structural type might be the most promising.

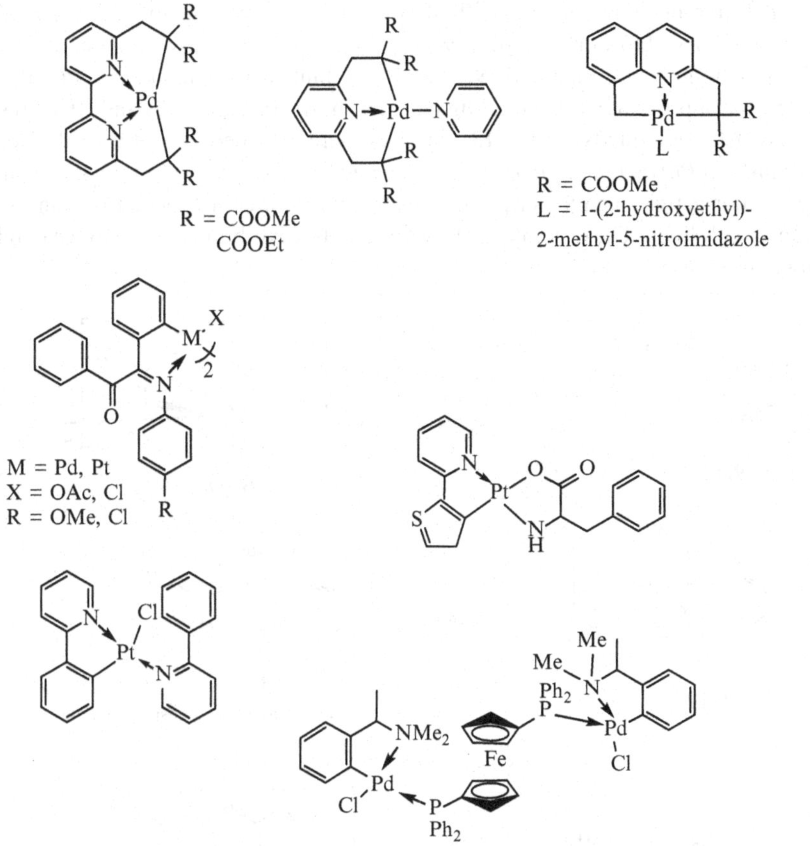

Fig. 9.5 Some cyclometalated compounds for medical uses containing antitumor reagents [70]

Ryabov [70] also reported on other cyclometalated compounds for medical uses containing antitumor reagents, as shown in Fig. 9.5.

Recently, Esmaeilbeig et al. [72] reported on an antitumor activity study of cycloplatinum compounds containing 2-phenylpyridine derivatives, as shown in Eq. (9.1). Antitumor effects of 2-phenylpyridine platinum 1-inidazole, 4-methylpyridine, pyridine, and 4,4'-bipyridine derivatives were determined for Jukat, K562, and Raji cell lines, and the results showed reasonable cytotoxicities.

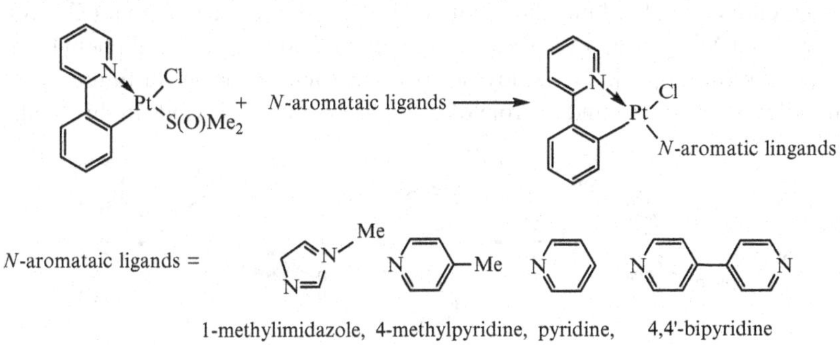

$$(9.1)$$

1-methylimidazole, 4-methylpyridine, pyridine,   4,4'-bipyridine

The tetranuclear compound **9.12** was prepared by reaction of 3,4-dichloro-acetophenone thiosemicarbazone **9.11** with K_2PtCl_4, as shown in Eq. (9.2) [73]. Two mononuclear compounds **9.13** and two dinuclear compounds **9.14** were isolated for appropriate phosphine ligands, as shown in Eqs. (9.3) and (9.4), respectively. All compounds **9.12**, **9.13**, and **9.14** were screened for antiparasitic activity against the *Plasmodium falciparum* strain and *Trichomonas vaginalis* strain and for antitumor activity against cisplatin-sensitive and cisplatin-resistant human ovarian cancer cell lines. These compounds were found to exhibit moderate to weak inhibitory activities.

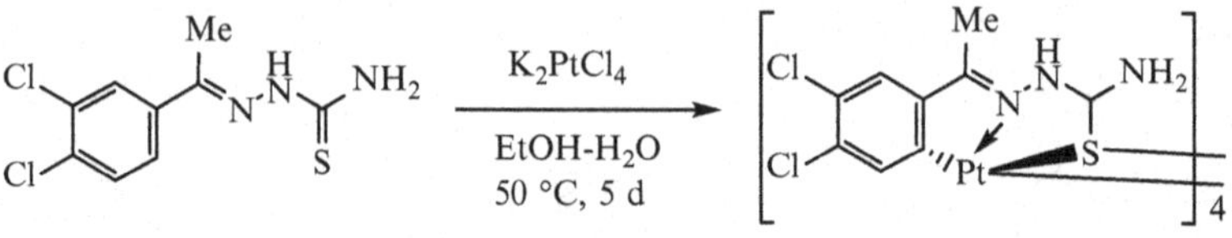

$$(9.2)$$

9.11

9.12

$$(9.3)$$

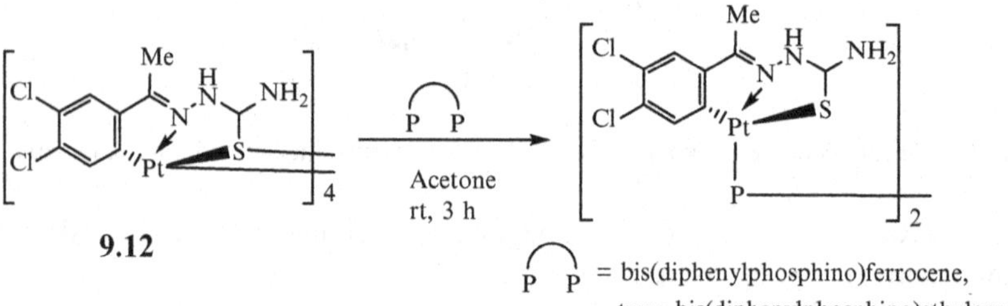

9.12

9.13 L

L = PPh₃, 1,3,5-triaza-7-phosphaadmantane

$$(9.4)$$

9.12

P P = bis(diphenylphosphino)ferrocene, *trans*-bis(diphenylphosphino)ethylene

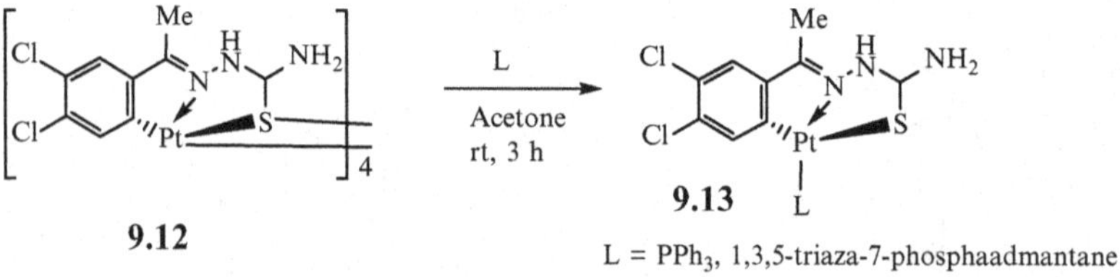

9.14

In the derivatives of 2-phenylpyridine platinum compounds **9.15**, as shown in Eq. (9.5), cytotoxicity was also studied in three human cancer cell lines derived from ovarian carcinoma (CHI), lung carcinoma (A549), and colon carcinoma (SW480) by means of an MTT assay (MTT = 3-(4,5-dimethyl-2-thiazolyl)-2,5-diphenyl-2H-tetrazolium bromide) [74]. This study needs to be pursued further in future, however, in order to be able to find the role of different phosphine ligands in this field.

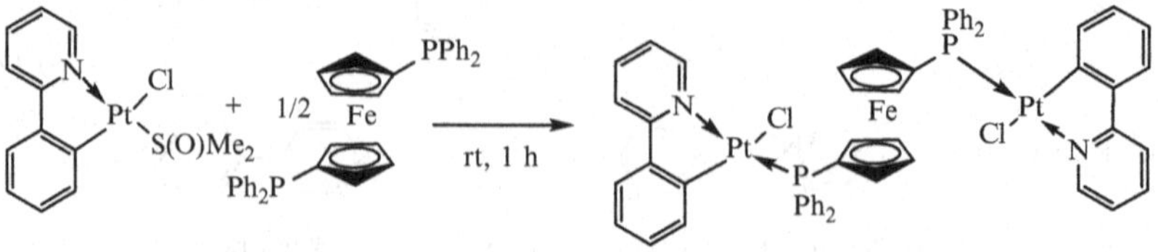

$$(9.5)$$

Almost quantitative yield   **9.15**

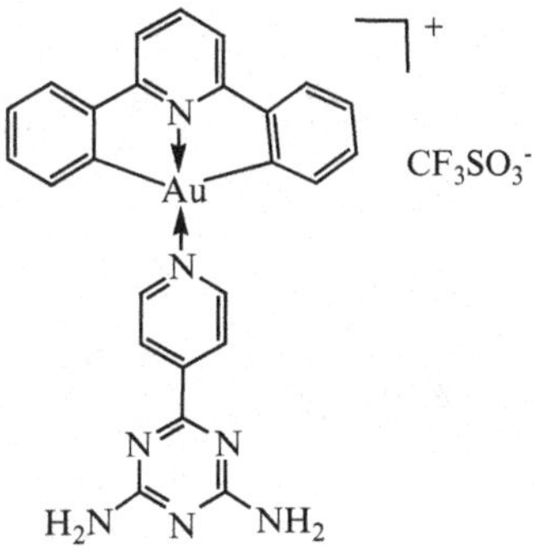

Fig. 9.6 2,4-Diamino-6-
(4-pyridyl)-1,3,5-triarizine
gold compound **9.16** [75]

9.16

At the same time, a gold compound, that is, 2,6-diphenylpyridine, 2,4-diamino-6-(4-pyridyl)-1,3,5-triazine gold compound **9.16**, was found to self-assemble into a supramolecular polymer (showing formation of a viscous fluid) in CH_3CN. This polymer **9.16** displayed sustained cytotoxicity and selective cytotoxicity toward cancerous cells as shown in Fig. 9.6 [75].

Park et al. [76] reported on other medicinal uses employing antimicrobial activity. The five pincer palladium compounds **9.17–9.21** are prepared by substitution reactions, as shown in Eq. (9.6). These compounds were tested for antimicrobial activity on a wide range of bacterial strains. Compound **9.17** was found to be potential in this respect [76].

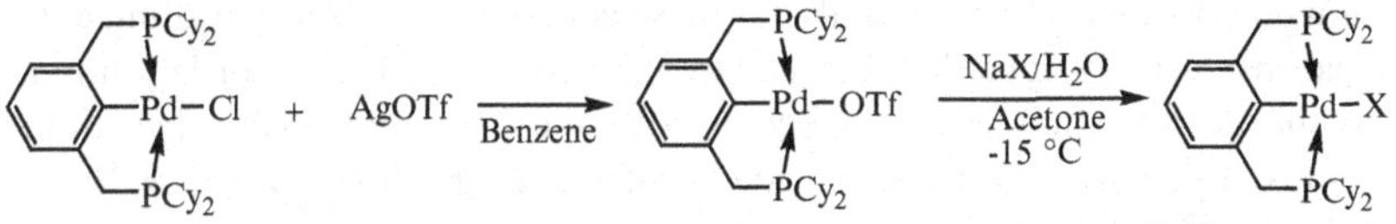

$$X = N_3 \quad \textbf{9.18}$$
$$X = NCO \quad \textbf{9.19}$$
$$X = NCS \quad \textbf{9.20}$$
$$X = SeCN \quad \textbf{9.21}$$

$$(9.6)$$

Ozerianskyi et al. [77] reported on antibacterial and antifungal activities in medical studies on main group metal cyclometalation compounds, e.g., organotin compounds, **9.22**, **9.23**, and **9.24**, as shown in Eqs. (9.7, 9.8, and 9.9). They found that five-membered ring compounds **9.23** of the di-*n*-butyl-substituted compound are the most efficient in inhibiting growth of yeasts, molds, and G^+ bacteria strains.

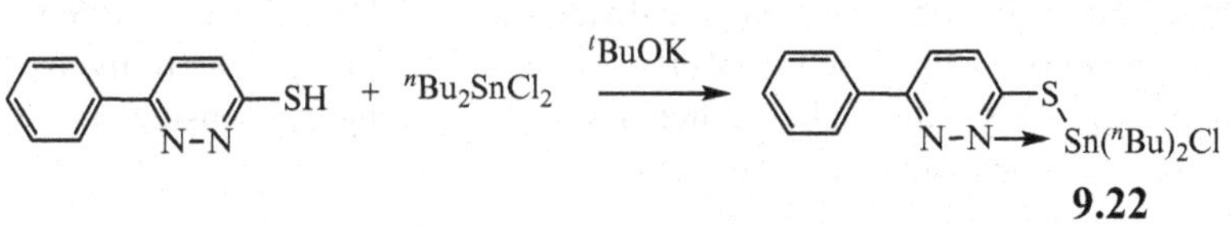

$$(9.7)$$

9.22

$$(9.8)$$

$$(9.9)$$

9.4   Dye-Sensitized Solar Cells

As the fact that cyclometalated products contain such substrates as phenylpyridines, phenylpyrazoles, and phenylquinolines shown in Sect. 9.2 suggests, these compounds show strong absorption in the visible light range. These compounds are therefore able to absorb energy from the visible light range. However, these cyclometalated compounds are able not only to convert electricity to light but also to convert light to electricity for energy-generating molecular photovoltaic cells, that is, dye-sensitized solar cells.

The OLED applications in Sect. 9.2 are used mainly for cyclometalated iridium compounds. However, in the case of dye-sensitized solar cells, the cyclometalated ruthenium compounds are used mainly for these cells. This is because iridium compounds are not usually considered to be strong absorbers, which is of key importance to the device efficiency provided by dye-sensitized solar cells [6]. Some representative cyclometalated dye-sensitized solar cells are shown in Fig. 9.7.

In 2011, Kim et al. [81] reported on a cyclometalated ruthenium sensitizer incorporating a CNN ligand and conjugated 2,2'-bipyridyl,4,4'-thiopenyl ancillary ligands 9.31, as shown in Fig. 9.7. The photovoltaic performance of compound 9.31 gave a short-circuit photocurrent density of 19.63 mA cm^{-2}, an open-circuit voltage of 0.74 V, and a fill factor of cells based 0.72 m, affording an overall conversion efficiency of 10.39 %. This efficiency is the highest reported for dye-sensitized solar cells based on the type-CNN cyclometalated ruthenium sensitizer. Moreover, the same device using a polymer gel electrolyte exhibited a remarkable stability when soaked with light for a 1,000 h at 60 °C, retaining 91 % of its initial efficiency of 7.14 %.

Furthermore, in 2012, Funaki et al. [82] reported on a cyclometalated ruthenium compound of 2-phenylpyrimidine containing terpyridine and NCS ligands as ancillary ligands 9.32 as dye-sensitized solar cells. The dye-sensitized solar cells sensitized with the 2-phenylpyrimidine compound 9.32 show 10.7 % efficiency, which is higher than the N749 (9.40) benchmark (10.1 %) as shown in Fig. 9.8. To the best of their knowledge, this is the highest efficiency to date among dye-sensitized solar cells based on cyclometalated Ru sensitizers.

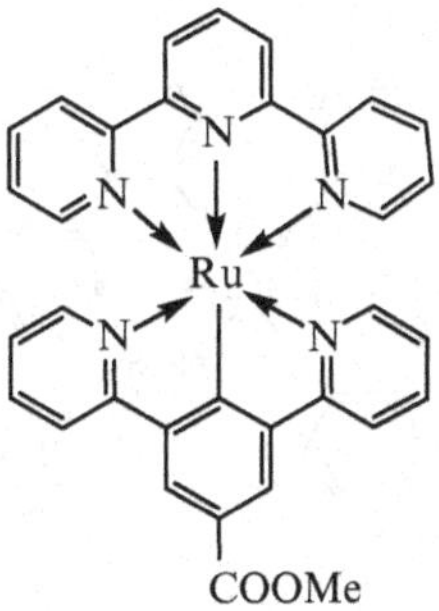

9.25 [78]

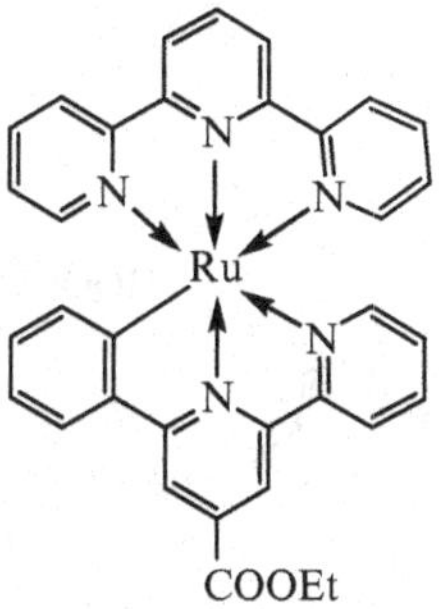

9.26 [78]

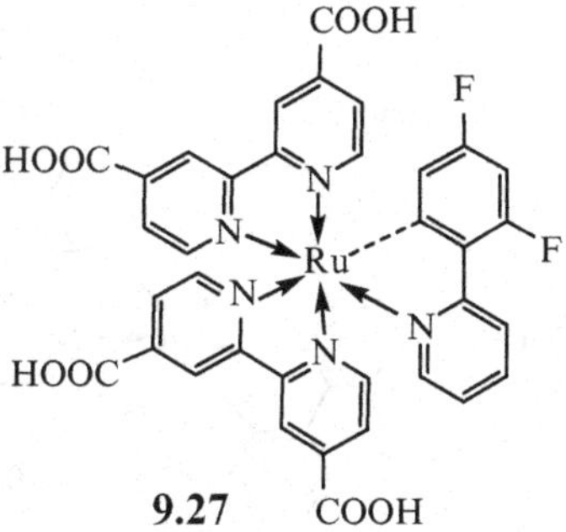

9.27 [78]

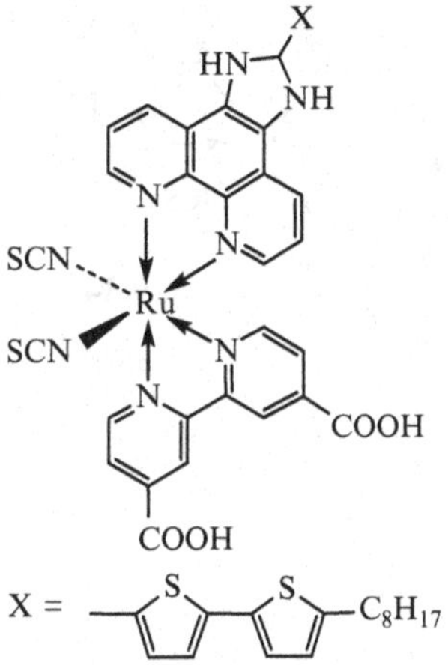

9.28 [78]

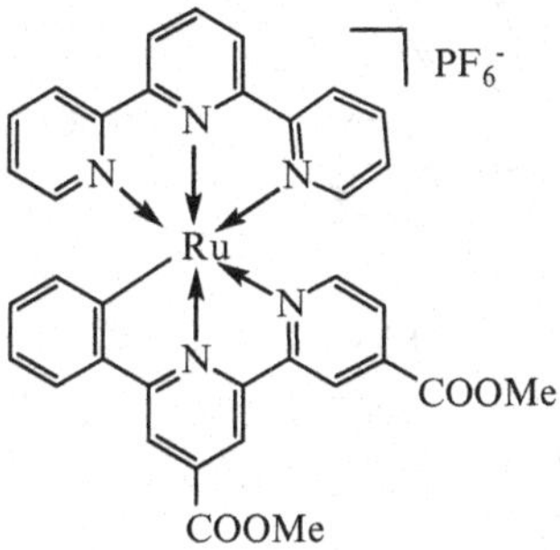

9.29 [79]

9.30 [80]

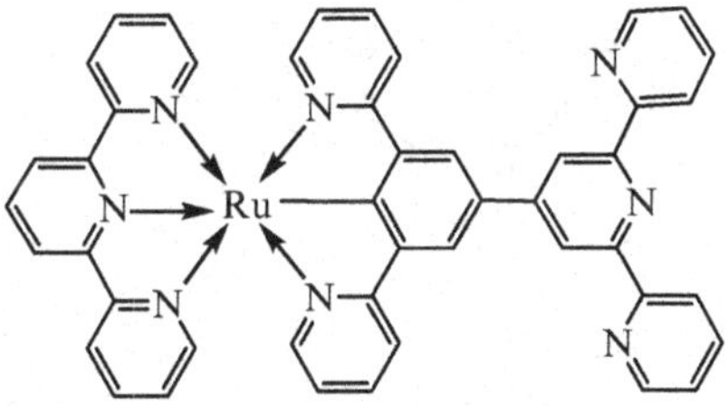

Fig. 9.7 Some representative cyclometalated dye-sensitized solar cells

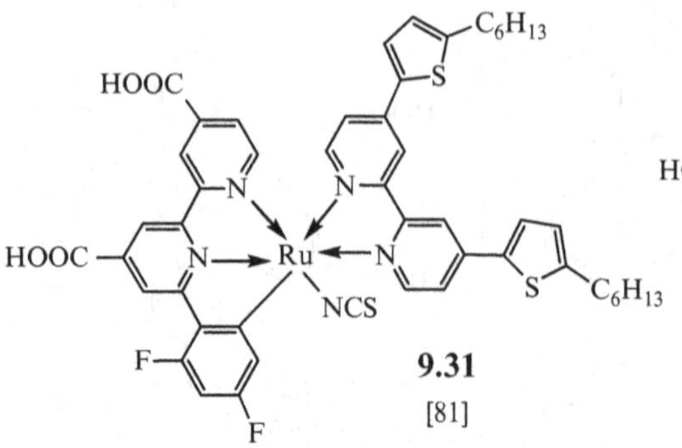

9.31

[81]

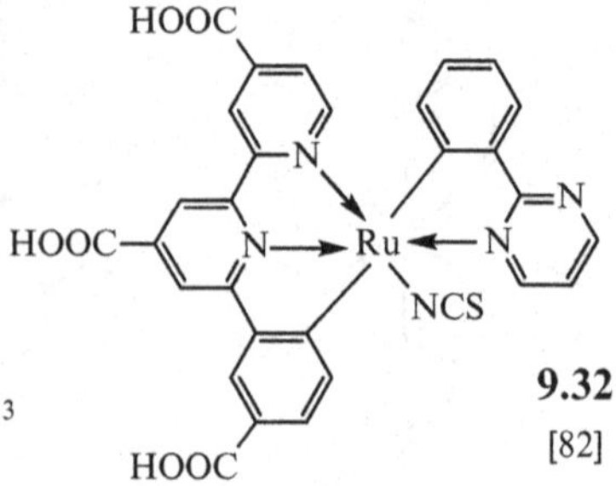

9.32

[82]

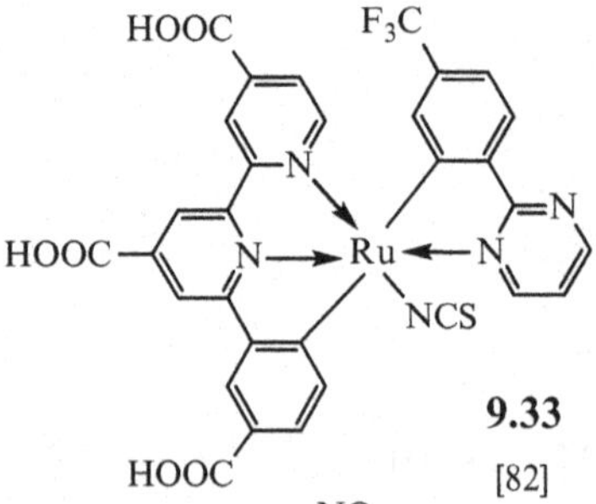

9.33

[82]

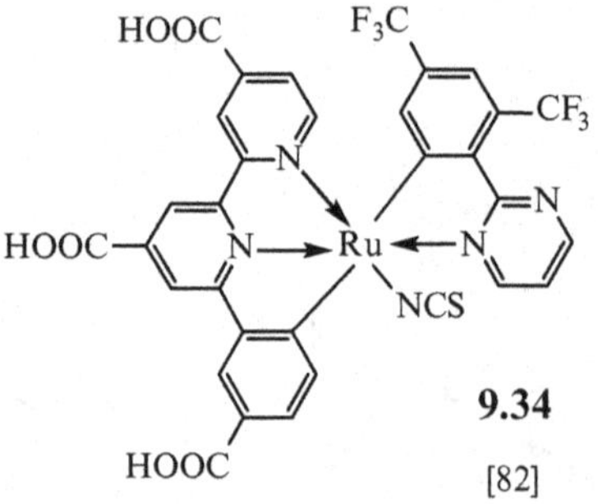

9.34

[82]

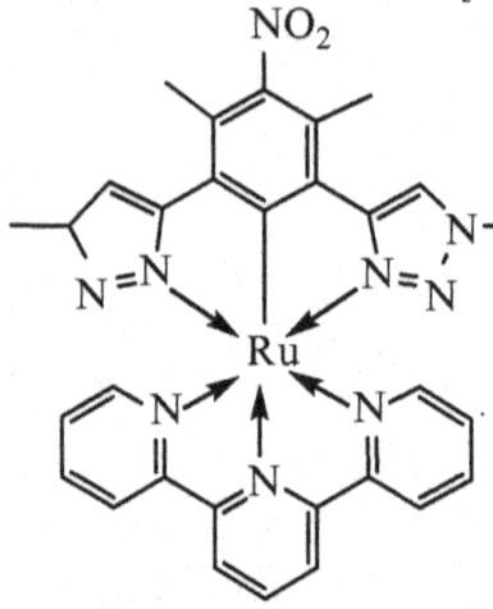

9.35

[83]

9.36

[83]

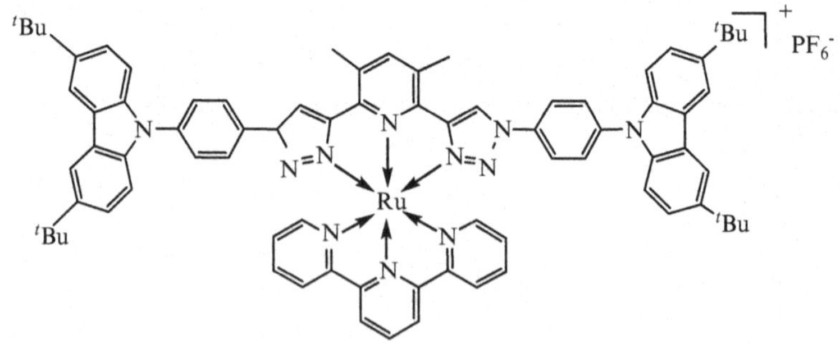

Fig. 9.7 (continued)

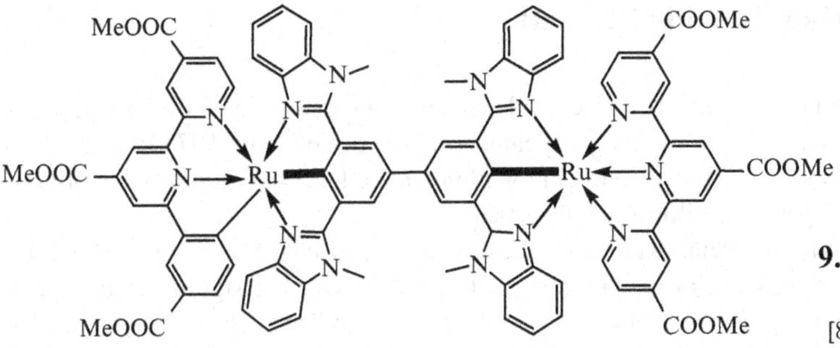

9.37

[84]

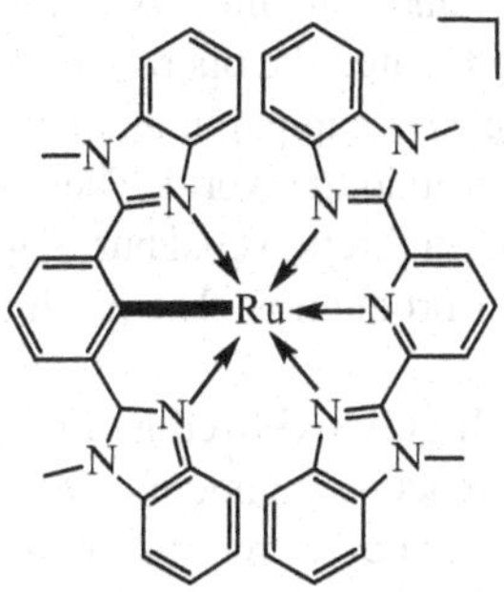

9.38

[85]

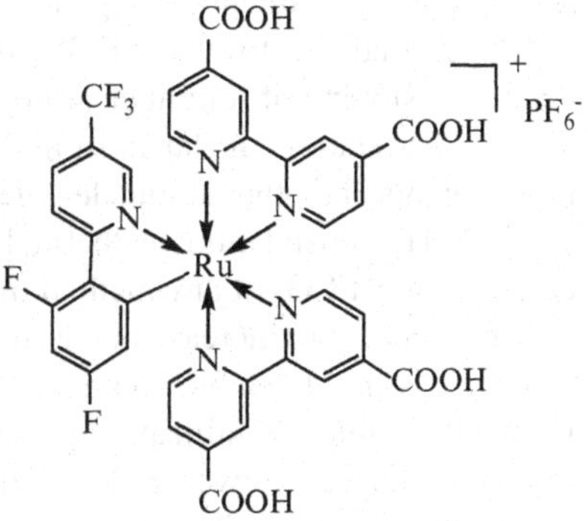

9.39

[86]

Fig. 9.7 (continued)

Fig. 9.8 N749 (**9.40**) is a benchmark compound for dye-sensitized solar cells, exhibiting an efficiency of 10.1 % [82]

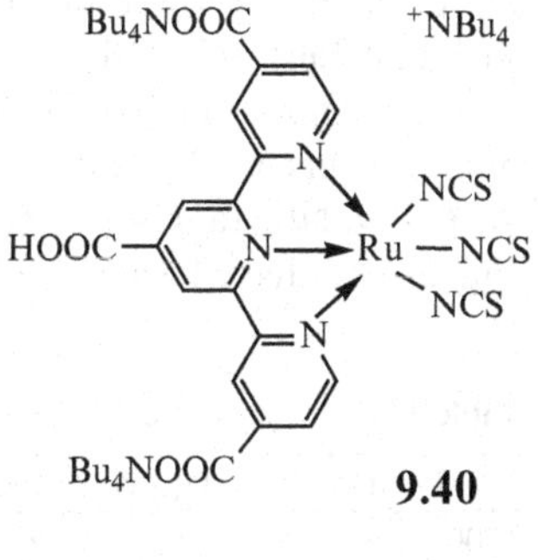

9.40

9.5   Carbon Dioxide Utilization

As shown in Table 9.2, carbon dioxide in the atmosphere increased by approximately 1.5 ppm (8 billion tons) per year during the period from 1975 to 2002. Very recently, however, carbon dioxide in the atmosphere has increased much more rapidly, by 2.25 ppm (12 billion tons) per year.

Two important events have occurred in the area of climate change since 2005. First, the most advanced countries agreed in 1997, with the signing of the Kyoto Protocol, to reduce their collective greenhouse gas emissions by 5.2 % from 1990 levels. However, the overall emission levels actually increased rapidly after 1990 because greenhouse gas emissions by developing countries increased rapidly, and the two countries with the largest emissions, the United States and China, did not accept any CO_2 reduction obligations.

The second event was the Fukushima earthquake and tsunami, which forced the decommissioning of four nuclear power stations and halted the operation of almost all 50 other stations in Japan in mid-2012 because of a campaign against nuclear power plants and other antinuclear demonstrations around the world. Therefore, we are forced to depend more on thermal power generation through coal burning, which emits the most CO_2 of any method of electric power production. The production of CO_2 is expected to increase in the future.

CO_2 is a highly stable compound, because it is highly oxidized and thermodynamically stable. Its utilization generally requires reactive compounds and highly reactive catalysts. However, due to the electron deficiency of the carbonyl carbon, CO_2 has a strong affinity toward nucleophiles and electron-donating reagents.

The chemical produced in the greatest amount through CO_2 utilization is urea. According to the international fertilization association, 157 million tons of urea were produced in 2010. Other chemicals produced through CO_2 utilization are cyclic carbonates, acyclic carbonates, polyalkylene carbonates, Asahi Kasei polycarbonates, carbamic acid esters, oxazolidinones, polyurethanes, carboxylic acids and esters, lactones, formic acid, and methanol. The amounts of various organic chemicals produced through carbon dioxide utilization throughout the world are shown in Table 9.3 [88].

Investigations of reactions between CO_2 and metal compounds, including transition elements, G10: Ni, Pd, Pt; G9: Co, Rh, Ir; G8: Fe, Ru; G7: Mn, Re; G6: Cr, Mo, W; G5: V, Nb, Ta; G4: Ti, Zr, and U, and main group metal elements, Mg, Zn, Sn, Cu, and Ag, have been carried out [87, 88].

Table 9.2 Increase in rates of CO_2 concentrations in the atmosphere over the past 1,000 years [87, 88]

Year (years)	Period (ppm)	Concentration (ppm)	Increase (ppm/year)	Increase rate
1000–1800	800	270–280	10	0.01
1800–1950	150	280–310	30	0.2
1958–1975	17	315–330	15	0.9
1975–2002	27	330–370	40	1.5 (8 billion tons[a])
2002–2010	8	370–388	18	2.25 (12 billion tons[a])

[a]Increased concentrations in the atmosphere

Table 9.3 Worldwide production of organic chemicals utilizing CO_2 [88]

Chemical	Production (tons)
Cyclic carbonates	80,000
Polypropylene carbonate	70,000
Polycarbonate (Asahi Kasei process)	605,000
Urea	157,000,000
Acetylsalicylic acid	16,000
Salicylic acid	90,000
Methanol	4,000

Cyclometalation reaction five-membered ring products are utilized with two kinds of methods for CO_2 utilization. The first is utilized for an active metal atom coordinated with a hetero atom and other ligands. The second is utilized as a catalyst for CO_2 utilization reactions.

The first type of application is already described in Chap. 6. The metal elements in the active metal center react easily with CO_2 to give carboxylic acid derivatives. For example, the cyclometalation of 2-phenylpyridine as a substrate in the presence of a rhodium compound proceeds easily to give a five-membered ring rhodium intermediate. CO_2 can be inserted into the rhodium–phenyl carbon bond, and a methyl ester is then formed from rhodium and a carboxylate through reaction with $TMSCH_2N_2$, as shown in Eq. (6.5) [89]. The reaction mechanism is proposed as shown in Scheme 6.2 [89].

As an example of the second catalytic utilization for cyclometalation reaction five-membered ring products, pincer iridium compounds are used as catalysts for CO_2 to methane with trialkylsilanes, as shown in Eq. (9.10) [90]. Using less bulky silanes such as Me_2EtSiH or Me_2PhSiH results in rapid formation of CH_4 and siloxane with no detection of bis(silyl)acetal and methyl silyl ether intermediates. The catalyst system is long-lived, and 8,300 turnovers can be achieved using Me_2PhSiH with 0.0077 mol % loading of iridium [90].

$$CO_2 + Me_2PhSiH \xrightarrow[\substack{\text{Solvent : PhCl} \\ 23\ ^\circ\text{C, 72 h}}]{\text{Cat.}} CH_4 + Me_2PhSiOSiPhMe_2$$

(1 atm)

TON 8293
TOF   115 h^{-1} (9.10)
Isolated yield 2.97 g

Cat. = (structure: pincer iridium complex with $O-P^tBu_2$ groups, Ir center with $+H$, $B(C_6F_4)_4^-$, $O=CMe_2$)

9.41

A pincer nickel phosphinite **9.42** similar to the above iridium catalyst **9.41** acts for catalytic hydroboration of CO_2 with the highest TOF (495 h^{-1} based on B–H) reported to date for the reduction of CO_2 to the methoxide level, as shown in Eq. (9.11) [91].

$$\text{CO}_2 \;+\; \text{HBat} \;\xrightarrow[\substack{\text{C}_6\text{D}_6 \\ \text{rt, 1 h}}]{\text{Cat.}}\; \text{CH}_3\text{OBcat} \;+\; \text{catBOBcat}$$

$$\xrightarrow{\text{H}_2\text{O}}\; \text{CH}_3\text{OH} \;+\; \text{catBOBcat}$$

$$\text{CO}_2 \text{ (1 atm)} \qquad \text{HBat (500 equiv)}$$

HBcat = catecholborane

Cat. =

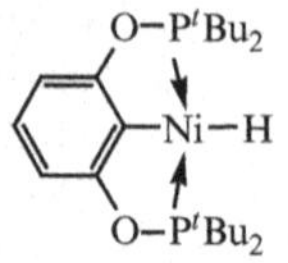

9.42

(9.11)

The synthesis of dimethyl carbonate by reaction of CO_2 with methanol generally results in a low yield, because many catalysts are deactivated by the formation of water. However, Sakakura et al. [92, 93] reported an excellent process using an acetal and a molecular sieve 3A as dehydrating agents in the presence of a $Bu_2Sn(OMe)_2$ catalyst, as described in a previous review [87, 88]. The reactions with the organotin cyclometalated with *N,N*-dimethylbenzylamine compounds **9.43** show promising yields of dimethyl carbonate, as shown in Eq. (9.12) [94].

$$\text{CO}_2 \;+\; 2\,\text{MeOH} \;\xrightarrow[\substack{\text{MeOH} \\ 150\,^\circ\text{C, 15 h}}]{\text{Cat.}}\; (\text{MeO})_2\,\text{C=O}$$

200 bar
32.7 g

Promising yield

(9.12)

Cat. =

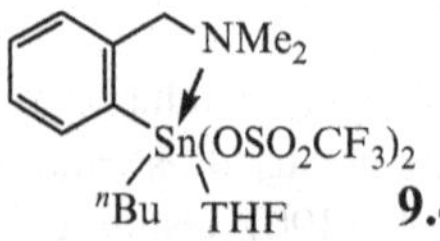

9.43

As other applications for cyclometalation reaction five-membered ring products, they can be utilized for the compounds for CO_2 fixation. For example, the pincer *N,N*-dimethylbenzylamine-type tin compound **9.44** readily absorbs CO_2 at room temperature to yield organotin carbonate **9.45** [95]. Easy desorption and reversible CO_2 fixation is achieved.

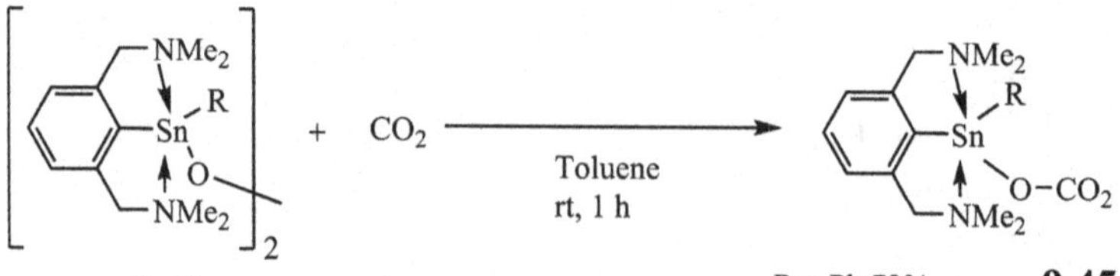

9.44
R = Ph, "Bu

R = Ph 72%   **9.45**
"Bu 61%

(9.13)

Pincer PCP nickel compound **9.46** also react easily with CO_2 at room temperature to form the carboxylate product **9.47** in quantitative yields, as shown in Eq. (9.14) [96].

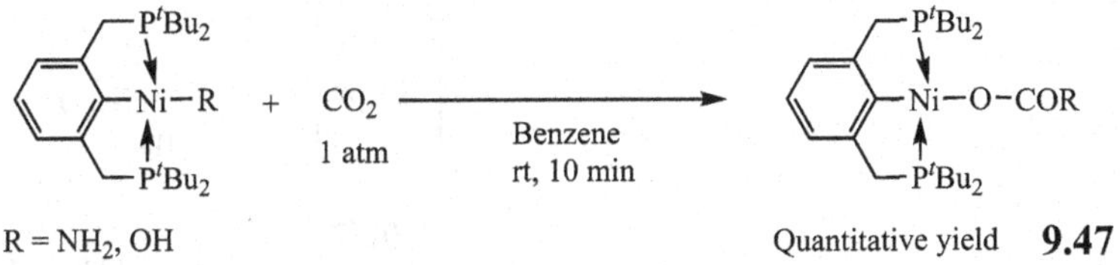

$$(9.14)$$

R = NH$_2$, OH

9.46

The pincer PCP palladium compound **9.48** also reacts with CO_2. CO_2 inserted into a palladium metal and methyl group bond gives the acetate product **9.49**, as shown in Eq. (9.15) [97].

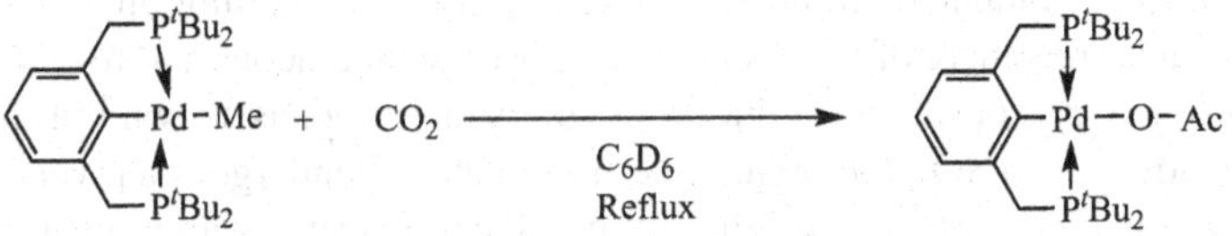

$$(9.15)$$

9.48   **9.49**

Tris(2-pyridylthio)methane zinc cyclometalation product **9.51** also reacts easily with CO_2 to form a carboxylate product, as shown in Eq. (9.16) [98, 99].

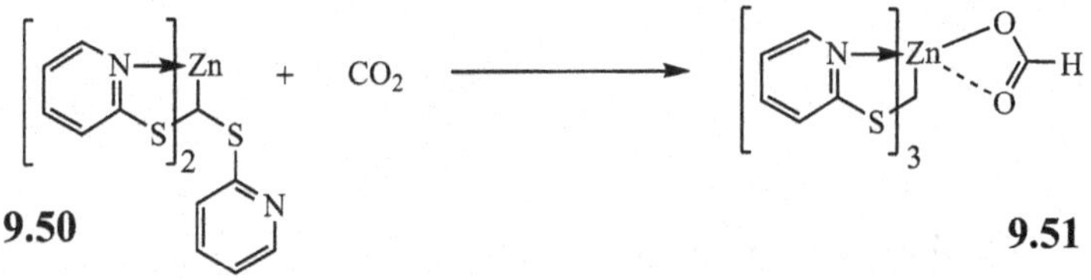

$$(9.16)$$

9.50   **9.51**

9.6 Sensors and Others

As described in Sects. 9.2 and 9.4, the uses for organometallic intramolecular-coordination five-membered ring compounds are related to their ability to exhibit absorption across a wide range of visible light wavelengths. That is to say, these compounds can show a wide range of colors and change the voltage of electric current as solar cells. They can also be used as sensors.

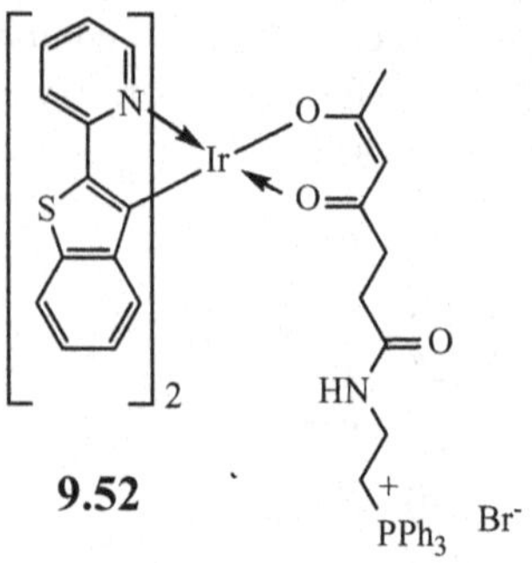

Fig. 9.9 2-Benzothienyl-
pridine iridium cation **9.52** as
a mitochondria-specific
oxygen sensor [102]

9.52

These sensors are mainly utilized to detect specific colors or color changes of cyclometalation reaction five-membered ring products. These sensors can detect specific materials, such as hazardous molecules, ions, radicals, and compounds. Some examples follow:

- Hazardous Materials: SO_2
  On exposing a square-planar platinum or nickel complex containing an NCN pincer ligand to an atmosphere of SO_2, for example, the instantaneous and reversible formation of a pentacoordinate adduct (orange) is observed, as shown in Eq. (9.17). Upon binding with SO_2, the organoplatinum material undergoes a reversible color change from colorless to bright orange, which is used as an indicator of the presence (or absence) of the above gas [100, 101].

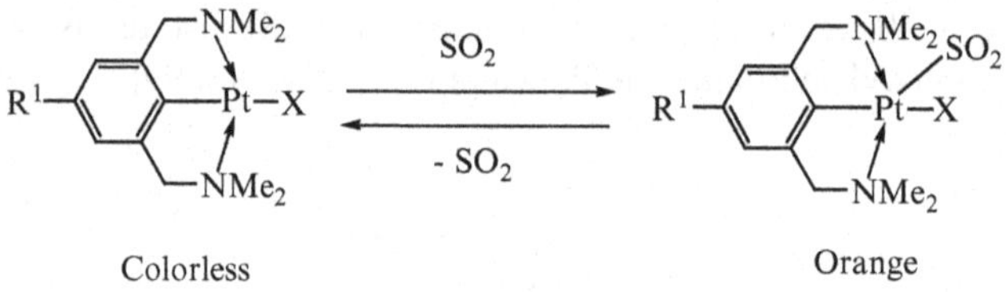

(9.17)

- Specific Materials: Mitochondria-Specific Oxygen
- The phosphorescence of the 2-benzothienylpridine iridium cation **9.52** was significantly quenched by molecular oxygen in living cells, demonstrating that 2-benzothienylpridine iridium cation **9.52** can be used as a mitochondria-specific oxygen sensor as shown in Fig. 9.9 [102].
- Specific Materials: Sugar
  Nitrogen-15 NMR spectroscopy showed strongly upfield values for chemical shifts for one of the azo nitrogen atoms of a boron–nitrogen (B–N) dative bond in the cyclometalated boron compound **9.53**. The B–N dative bond was cleaved by sugar addition, as shown in Eq. (9.18) [103].

N^{15} NMR chemical shift value change

9.53   (The B-N dative bond was cleaved by sugar addition)

$$(9.18)$$

- Specific Ion: CN$^-$

  N-(2-Anthryl)-2-[bis(pentafluorophenyl)boryl]benzylideneamine changed its fluorescence color from yellow to green upon addition of an equimolar amount of cyanide ion, in contrast to the N-phenyl derivatives, which showed quenching of the emission as shown in Eq. (9.19). The benzylamine cycloborated compound shows a cyanide fluorescence sensor [104].

$$(9.19)$$

The other sensors are reported on dual anions of CN$^-$ and CH$_3$COO$^-$ [105], free radical [106], and copper(II) ion [107].

The other applications for cyclometalation reaction five-membered ring products include dendrimers [108–112], liquid crystals (metallomesogens) [113–116], resolving agents (chiral auxiliaries) [117–119], and photosensitizers for hydrogen production [120, 121].

These cyclometalation compounds are van Koten's carbosilane pincer nickel dendrimer **9.54** [109, 111], pincer SCS palladium dendrimer **9.55** [110, 112], azobenzene platinum chloro-bridged liquid crytal **9.56** [113, 114], and N,N-dimethylnaphthylene palladium resolving agent **9.57** [117–119] as shown in Fig. 9.10, and photosensitizers for hydrogen production, bis(2-phenylpyridine-4-methyl,4'-fluoride) **9.59** and other photosensitizers with bis(2-phenylpyridine) derivatives **9.60** as shown in Fig. 9.11 [120, 121].

Among uses for these compounds, especially the last two applications as photosensitizers for hydrogen production, we anticipate bright prospects for the chemical industry in the future.

Second, uses as photosensitizers in catalytic photoinduced hydrogen generation via the reduction of water are also extremely promising.

Dendrimers

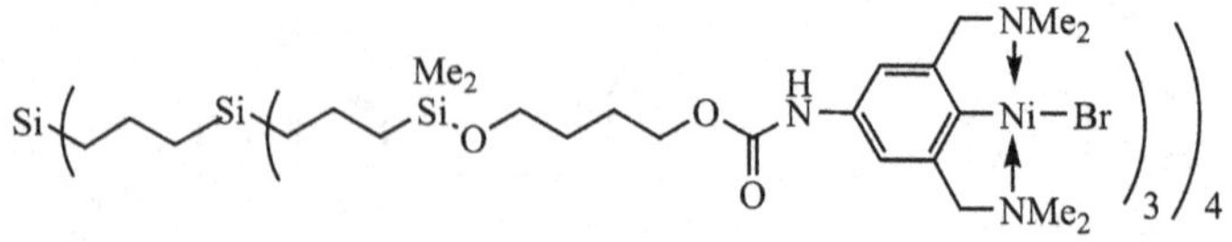

9.54   [109,111]

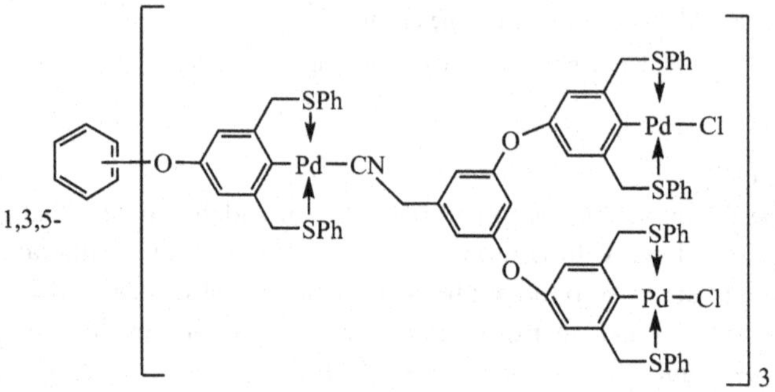

Liquid Crystals

9.55   [110,112]

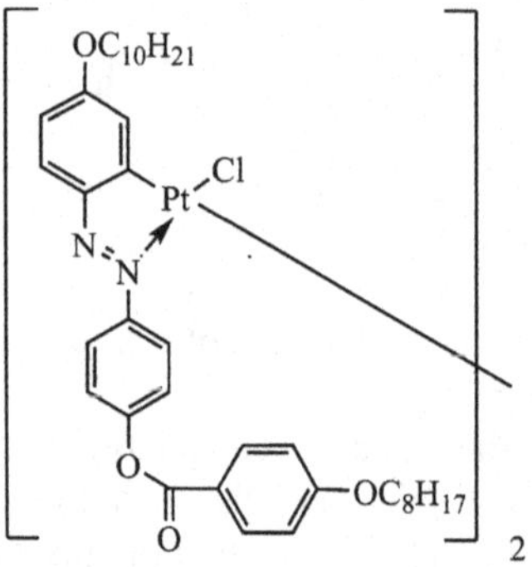

9.56   [113,114]

Resolving Agents

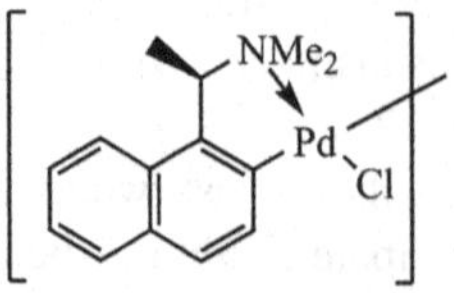

9.57   [117–119]

Fig. 9.10 Some representative cyclometalation reaction five-membered ring products are used for the other applications such as dendrimers, liquid crystals and resolving agnets

Example 1 [120]:

Photosensitizer: Bis(2-phenylpyridine-4-methyl,4'-fluoride) **9.58**, 50 μM

Catalyst: Co(bpy)$_3^{2+}$ 2.5 mM

H$_2$ evolved: 460 μmol

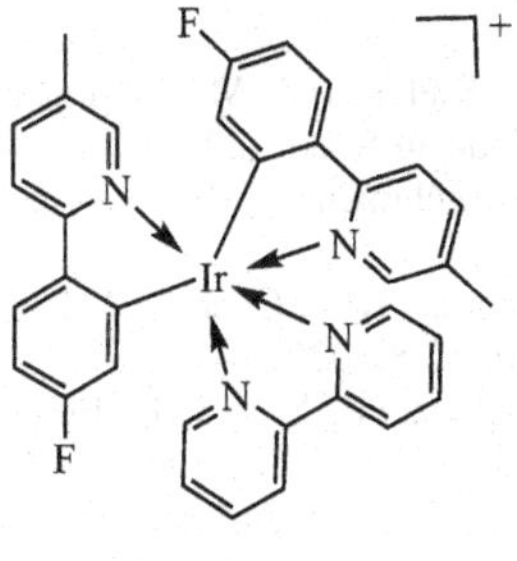

9.58

Example 2 [121]:

Photosensitizer: Bis(2-phenylpyridine) **9.59** 10^{-4} M

Catalyst: Cobalt diimine-dioxime **9.60** 10^{-4} M, PPh$_3$ 2-equiv

Time: 10 h

TON H$_2$: 696

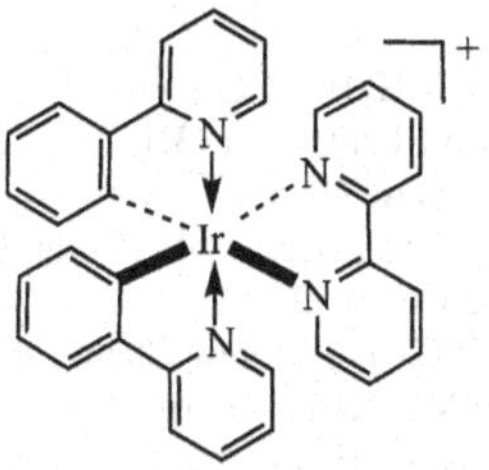

9.59 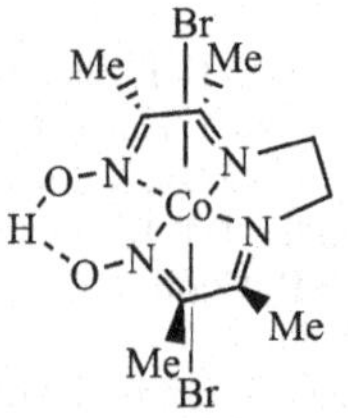   **9.60**

Fig. 9.11 Two examples of 2-phenylpyridine derivatives as photosensitizers **9.58**, and **9.59** for hydrogen production [120, 121]

References

1. Costa RD, Ortí E, Bolink HJ, Monti F, Accorsi G, Armaroli N (2012) Angew Chem Int Ed 51:8178
2. Zhao Q, Huang C, Li F (2011) Chem Soc Rev 40:2508
3. Chi Y, Chou P-T (2010) Chem Soc Rev 39:638
4. Fernández-Moreira V, Thorp-Greenwood FL, Coogan MP (2010) Chem Commun 46:186
5. Williams JAG (2009) Chem Soc Rev 38:1783
6. Baranoff E, Yum J-H, Graetzel M, Nazeeruddin MK (2009) J Organomet Chem 694:2661
7. Djukic J-P, Sortais J-B, Barloy L, Pfeffer M (2009) Eur J Inorg Chem 7:817
8. Williams JAG, Develay S, Rochester DL, Murphy L (2008) Coord Chem Rev 252:2596
9. Lowry MS, Bernhard S (2006) Chem Eur J 12:7970
10. Ghedini M, Aiello I, Crispini A, Golemme A, Deda ML, Pucci D (2006) Coord Chem Rev 250:1373

11. von Zelewsky A, Belser P, Hayoz P, Dux R, Hua X, Suckling A, Stoeckli-Evans H (1994) Coord Chem Rev 132:75
12. Mori T (2008) Organic electroluminescence (Yuki EL no Hon). Nikkan Kogyo Shinbun
13. Tokito S, Adachi T, Murata H (2004) Organic electroluminescence display [Yuki EL Display], Ohmusha
14. Sato Y (2004) Organic electroluminescence technology and material development (Yuki EL Gijyutu to Zairyo Kaihatsu), C M C Shuppann
15. St-Pierre G, Ladouceur S, Fortin D, Zysman-Colman E (2011) Dalton Trans 40:11726
16. Wilkinson AJ, Pushchmann H, Howard AK, Foster CE, Williams JAG (2006) Inorg Chem 45:8685
17. Velusamy M, Thomas KRJ, Chen C-H, Lin JT, Wen YS, Hsieh W-T, Lai C-H, Chou P-T (2007) Dalton Trans 28:3025
18. Fernández-Hernández JM, Yang C-H, Beltrán JI, Lemaur V, Polo F, Fröhlich R, Cornil J, Cola LD (2011) J Am Chem Soc 133:10543
19. Ladouceur S, Fortin D, Zysman-Colman E (2011) Inorg Chem 50:11514
20. Aoki S, Matsuo Y, Ogura S, Ohwada H, Hisamatsu Y, Moromizato S, Shiro M, Kitamura M (2011) Inorg Chem 50:806
21. Liu S, Müller P, Takase MK, Swager TM (2011) Inorg Chem 50:7598
22. Lo KK-W, Zhang KY, Li SP-Y (2011) Pure Appl Chem 83:823
23. Huang W-S, Lin JT, Chien C-H, Tao Y-T, Sun S-S, Wen Y-S (2004) Chem Mater 16:2480
24. Rao Y-L, Wang S (2011) Organometallics 30:4453
25. Wilkinson AJ, Goeta AE, Foster CE, Williams JAG (2004) Inorg Chem 43:6513
26. Beydoun K, Zaarour M, Williams JAG, Doucet H, Guerchais V (2012) Chem Commun 48:1260
27. Shan G-G, Li H-B, Cao H-T, Zhu D-X, Su Z-M, Liao Y (2012) J Organomet Chem 713:20
28. Li C, Yu M, Sun Y, Wu Y, Huang C, Li F (2011) J Am Chem Soc 133:11231
29. Yu L, Huang Z, Liu Y, Zhou M (2012) J Organomet Chem 718:14
30. Fernández-Hernández JM, Beltrán JI, Lemaur V, Gálvez-López M-D, Chien G-H, Polo F, Orselli E, Fröhlich R, Cornil J, Cola LD (2013) Inorg Chem 52:1812
31. Prokhorov AM, Santoro A, Williams JAG, Bruce DW (2012) Angew Chem Int Ed 51:95
32. Zheng Y, Batsanov AS, Edkins RM, Beeby A, Bryce MR (2012) Inorg Chem 51:290
33. Baranoff E, Curchod BFE, Frey J, Scopelliti R, Kessler F, Tavernelli I, Rothlisberger U, Grätzel M, Nazeeruddin MK (2012) Inorg Chem 51:215
34. Smith ARG, Riley JJ, Burn PL, Gentle IR, Lo S-C, Powell BJ (2012) Inorg Chem 51:2821
35. Shavaleev NM, Monti F, Costa RD, Scopelliti R, Bolink HJ, Ortí E, Accorsi G, Armaroli N, Baranoff E, Grätzel M, Nazeeruddin MK (2012) Inorg Chem 51:2263
36. Chen J-L, Wu Y-H, He L-H, Wen H-R, Liao J, Hong R (2010) Organometallics 29:2882
37. Moriuchi T, Katano C, Hirao T (2012) Chem Lett 41:310
38. Brulatti P, Gildea RJ, Howard JAK, Fattori V, Cocchi M, Williams JAG (2012) Inorg Chem 51:3813
39. Lin C-H, Chiu Y-C, Chi Y, Tao Y-T, Liao L-S, Tseng M-R, Lee G-H (2012) Organometallics 31:4349
40. Schwartz KR, Chitta R, Bohnsack JN, Ceckanowicz DJ, Miró P, Cramer CJ, Mann KR (2012) Inorg Chem 51:5082
41. Wiegmann B, Jones PG, Wagenblast G, Lennartz C, Münster I, Metz S, Kowalsky W, Johannes H-H (2012) Organometallics 31:5223
42. Ren X, Giesen DJ, Rajeswaran M, Madaras M (2009) Organometallics 28:6079
43. Shavaleev NM, Monti F, Scopelliti R, Armaroli N, Grätzel M, Nazeeruddin MK (2012) Organometallics 31:6288
44. Marchi E, Sinisi R, Bergamini G, Tragni M, Monari M, Bandini M, Ceroni P (2012) Chem Eur J 18:8765
45. Kessler F, Curchod BFE, Tavernelli I, Rothlisberger U, Scopelliti R, Censo DD, Grätzel M, Nazeeruddin MK, Baranoff E (2012) Angew Chem Int Ed 51:8030

46. Costa RD, Ortí E, Bolink HJ, Graber S, Housecroft CE, Constable EC (2011) Chem Commun 47:3207
47. Bronner C, Veiga M, Guenet A, Cola LD, Hosseini MW, Straqssert CA, Baudron SA (2012) Chem Eur J 18:4041
48. Kozhevnikov VN, Durrant MC, Williams JAG (2011) Inorg Chem 50:6304
49. Lin H-C, Kim H, Barlow S, Hales JM, Perry JW, Marder SR (2011) Chem Commun 47:782
50. Jenkins DM, Bernhard S (2010) Inorg Chem 49:11297
51. Díez A, Forniés J, Larraz C, Lalinde E, López JA, Martin A, Moreno MT, Sicilia V (2010) Inorg Chem 49:3239
52. Ainscough EW, Allcock HR, Brodie AM, Gordon KC, Hindenlang MD, Horvath R, Otter CA (2011) Eur J Inorg Chem 25:3691
53. Pugliese T, Godbert N, Aiello I, Deda ML, Ghedini M, Amati M, Belviso S, Lelj F (2008) Dalton Trans 6563
54. Hudson ZM, Wang S (2011) Organometallics 30:4695
55. Hirani B, Li J, Djurovich PI, Yousufuddin M, Oxgaard J, Persson P, Wilson SR, Bau R, Goddard WA III, Thompson ME (2007) Inorg Chem 46:3865
56. Wadman SH, Lutz M, Tooke DM, Spek AL, Hartl F, Havenith RWA, van Klink RGPM, van Koten G (2009) Inorg Chem 48:1887
57. Ghedini M, Aiello I, Deda ML, Grisolia A (2003) Chem Commun 2198
58. Ogawa Y, Taketoshi A, Kuwabara J, Okamoto K, Fukuda T, Kanbara T (2010) Chem Lett 39:385
59. Fuertes S, Brayshaw SK, Raithbby PR, Schiffers S, Warren MR (2012) Organometallics 31:105
60. Robson KCD, Koivisto BD, Yella A, Sporinova B, Nazeeruddin MK, Baumgartner T, Grätzel M, Berlinguette CP (2011) Inorg Chem 50:5494
61. Li X-W, Li H-Y, Wang G-F, Chen F, Li Y-Z, Chen X-T, Zheng Y-X, Xue ZL (2012) Organometallics 31:3829
62. Crespo O, Eguillor B, Esteruelas MA, Fernández I, García-Raboso J, Gómez-Gallego M, Martín-Ortiz M, Oliván M, Sierra MA (2012) Chem Commun 48:5328
63. Wong KM-C, Hung L-L, Lam WH, Zhu N, Yam VW-W (2007) J Am Chem Soc 129:4350
64. Au VK-M, Wong KM-C, Tsang DP-K, Chan M-Y, Zhu N, Yam VW-W (2010) J Am Chem Soc 132:14273
65. Kriechbaum M, List M, Berger RJF, Patzschke M, Monkowius U (2012) Chem Eur J 18:5506
66. Roşca D-A, Smith DA, Bochmann M (2012) Chem Commun 48:7247
67. Muñoz-Rodríguez R, Buñuel E, Williams JAG, Cárdenas DJ (2012) Chem Commun 48:5980
68. Baligar RS, Sharma S, Singh HS, Butcher RJ (2011) J Organomet Chem 696:3015
69. Wu W, Guo H, Wu W, Ji S, Zhao J (2011) Inorg Chem 50:11446
70. Ryabov AD (2008) Cyclopalladated compounds as enzyme prototypes and anticancer drugs. In: Dupont J, Pfeffer M (eds) Palladacycles. Synthesis, characterization and applications. Wiley-VCH, Weinheim, p 307
71. Higgins JD III, Neely L, Fricker S (1993) J Inorg Biochem 49:149
72. Esmaeilbeig A, Samouei H, Abedanzadeh S, Amirghofran Z (2011) J Organomet Chem 696:3135
73. Chellan P, Land KM, Shokar A, Au A, An SH, Chavel CM, Dyson PJ, de Kock C, Smith PJ, Chibale K, Smith GS (2012) Organometallics 31:5791
74. Samouei H, Rashidi M, Heinemann FW (2011) J Organomet Chem 696:3764
75. Zhang J-J, Lu W, Sun RW-W, Che C-M (2012) Angew Chem Int Ed 51:4882
76. Lee HJ, Lee SH, Kim HC, Lee Y-E, Park S (2012) J Organomet Chem 717:164
77. Ozerianskyi A, Švec P, Vaňkátová H, Vejsová M, Česlová L, Padělková Z, Růžička A, Holeček J (2011) Appl Organomet Chem 25:725
78. Xie P, Guo F (2011) Curr Org Chem 15:3849
79. Wadman SH, Kroom JM, Bakker K, Havenith RWA, van Klink GPM, van Koten G (2010) Organometallics 29:1569

80. Wadman SH, van Leeuwen YM, Havenith RWA, van Klick GPM, van Koten G (2010) Organometallics 29:5635
81. Kim J-J, Choi H, Paek S, Kim C, Lim K, Ju M-J, Kang HS, Kang M-S, Ko J (2011) Inorg Chem 50:11340
82. Funaki T, Funakoshi H, Kitao O, Onozawa-Komatsuzaki N, Kasuga K, Sayama K, Sugihara H (2012) Angew Chem Int Ed 51:7528
83. Schulze B, Escudero D, Friebe C, Siebert R, Görls H, Sinn S, Thomas M, Mai S, Popp J, Dietzek B, González L, Schubert US (2012) Chem Eur J 18:4010
84. Shao J-Y, Yang W-W, Yao J, Zhong Y-W (2012) Inorg Chem 51:4343
85. Yang W-W, Zhong Y-W, Yoshikawa S, Shao J-Y, Masaoka S, Sakai K, Yao J, Haga M (2012) Inorg Chem 51:890
86. Dragonetti C, Valore A, Colombo A, Roberto D, Trifiletti V, Manfredi N, Salomone MM, Ruffo R, Abbotto A (2012) J Organomet Chem 714:88
87. Omae I (2006) Cat Today 115:33
88. Omae I (2012) Coord Chem Rev 256:1384
89. Mizuno H, Takaya J, Iwasawa N (2011) J Am Chem Soc 133:1251
90. Park S, Bézier D, Brookhart M (2012) J Am Chem Soc 134:11404
91. Chakraborty S, Zhang J, Krause JA, Guan H (2010) J Am Chem Soc 132:8872
92. Choi J-C, Sakakura T, Sako T (1999) J Am Chem Soc 121:3793
93. Choi J-C, He L-N, Yasuda H, Sakakura T (2002) Green Chem 4:230
94. Švec P, Olejník R, Padělková Z, Růžjčka A, Plasseraud L (2012) J Organomet Chem 708–709:82
95. Mairychová B, Dostál L, Růžjčka A, Beneš L, Jambor R (2012) J Organomet Chem 699:1
96. Schmeier TJ, Nova A, Hazari N, Maseras F (2012) Chem Eur J 18:6915
97. Johnson MT, Johansson R, Kondrashov MV, Steyl G, Ahlquist MSG, Roodt A, Wendt OF (2010) Organometallics 29:3521
98. Sattler W, Parkin G (2011) J Am Chem Soc 133:9708
99. Sattler W, Parkin G (2012) J Am Chem Soc 134:17462
100. Albrecht M, Gossage RA, Lutz M, Spek AL, van Koten G (2000) Chem Eur J 6:1431
101. Albrecht M, Schlupp M, Bargon J, van Koten G (2001) Chem Commun 1874
102. Murase T, Yoshihara T, Tobita S (2012) Chem Lett 41:262
103. Egawa Y, Tanaka Y, Gotoh R, Niina S, Kojima Y, Shimomura N, Nakagawa H, Seki T, Anzai J (2010) Chem Lett 39:1188
104. Yoshino J, Kano N, Kawashima T (2010) Bull Chem Soc Jpn 83:1185
105. Schmittel M, Qinghai S (2012) Chem Commun 48:2707
106. Yoshikawa H, Kobayashi M, Takahashi T, Awaga K (2010) Bull Chem Soc Jpn 83:762
107. You Y, Han Y, Lee Y-M, Park SY, Nam W, Lippard SJ (2011) J Am Chem Soc 133:11488
108. Pijnenburg NJM, Korstanje TJ, van Koten G, Gebbink RJMK (2008) Palladacycles on dendrimers and starshaped molecules. In: Dupont J, Pfeffer M (eds) Palladacycles. Synthesis, characterization and applications. Wiley-VCH, Weinheim, p 361
109. Chase PA, Gebbink RJMK, van Koten G (2004) J Organomet Chem 689:4016
110. Newkome GR, He E, Moorefield CN (1999) Chem Rev 99:1689
111. Knapen JWJ, van der Made AW, de Wilde JC, van Leeuwen PWNM, Wijkens P, Grove DM, van Koten G (1994) Nature 372:659
112. Huck WTS, van Veggel FCJM, Reinhoudt DN (1996) Angew Chem Int Ed 35:1213
113. Donnio B, Bruce DW (2008) Liquid crystalline ortho-palladated complexes. In: Dupont J, Pfeffer M (eds) Palladacycles. Synthesis, characterization and applications. Wiley-VCH, Weinheim, p 239
114. Ghedini M, Pucci D, Crispini A, Barberio G (1999) Organometallics 18:2116
115. Espinet P, Esteruelas MA, Oro LA, Serrano JE, Sola E (1992) Coord Chem Rev 117:215
116. Date RW, Iglesias EF, Rowe KE, Elliott JM, Bruce DW (2003) Dalton Trans 1914

117. Djukic J-P (2008) Cyclopalladated compounds as resolving agents for racemic mixtures. In: Dupont J, Pfeffer M (eds) Palladacycles. Synthesis, characterization and applications. Wiley-VCH, Weinheim, p 123
118. Alcock NW, Hulmes DI, Brown JM (1995) J Chem Soc D Chem Commun 395
119. Wild SB (1997) Coord Chem Rev 166:291
120. Goldsmith JI, Hudson WR, Lowey MS, Anderson TH, Bernhard S (2005) J Am Chem Soc 127:7502
121. Zhang P, Jacques P-A, Chavarot-Kerlidou M, Wang M, Sun L, Fontecave M, Arteo V (2012) Inorg Chem 51:2115

Chapter 10
Concluding Remarks

1. It is considered that the reason why cyclometalation reactions for producing organometallic intramolecular-coordination five-membered ring compounds proceed extremely easily is that the reactions first proceed through metal activation initiated by the coordination of lone electron pair, such as N, P, O, or S, to a metal atom. This is followed by γ-C–H agostic interactions, C–H activation, and the chelate effect in this order (Eq. (6.7)). A recent report entitled "Chelation-Assisted Reactions (Eqs. (6.42)–(6.51))" provides evidence that the chelate effect is a strong activation source for cyclometalation reactions.
2. It is considered that metals are also activated by bonding with ligands (ancillary ligands) such as hetero atom groups (bipyridines, benzoquinolines, phenanthrolines, benzothazoles, phosphines, phosphinites, carboxylates, etc.), unsaturated groups (aryl, allyl, cyclopentadienyl, etc.), carbonyl groups, halogen atoms (F, Cl, Br, or I), and N-heterocyclic carbenes in addition to metal activation due to the coordination of a hetero atom to a metal in cyclometalation reactions.
3. Cyclometalation reactions can be applied extremely easily to various kinds of synthetic organic reactions, because almost all the transition metal compounds and main group metal compounds (Fig. 5.10), as well as many substrates (Fig. 5.11, Tables 5.5 and 5.6), can be used in these reactions.
4. There are two applications for cyclometalation reactions and five-membered ring products for synthetic purposes. The first is synthesis of five-membered ring products by cyclometalation reactions. The second is synthesis of the derivatives of five-membered ring products or their intermediates during cyclometalation reactions. Pincer products are also used for synthesis of their derivatives.
5. Applications of five-membered ring products as catalysts in cyclometalation reactions include chiral reactions, metathesis reactions, cross-coupling reactions, and polymerization reactions. Additional reactions include reductions, Michael addition reactions, dehydrogenations, Diels–Alder reactions, and hydrogenations.

I. Omae, *Cyclometalation Reactions: Five-Membered Ring Products as Universal Reagents*, DOI 10.1007/978-4-431-54604-7_10, © Springer Japan 2014

6. Applications of five-membered ring products in cyclometalation reactions for other purposes include organic electronic devices, pharmaceuticals, dye-sensitized solar cells, carbon dioxide utilizations, sensors, dendrimers, liquid crystals, resolving agents, and photosensitizers for hydrogen production.

7. The author was the first to discover the intramolecular coordination bond, because he studied a subsidiary research subject, that is, not the main reactions of halo-dicarboxylic acid esters with tinfoil but the main reactions of halo-monocarboxylic acid esters with tinfoil. He was fortunate to be able to recognize a large shift in the IR spectrum, because the products contain two kinds of absorptions in carbonyl groups in a single molecule (Eq. (2.4)).

Printed in the United States
By Bookmasters